K. Ozaki, J. Yukawa, T. Ohgushi, P.W. Price (Eds.)

Galling Arthropods and Their Associates

Ecology and Evolution

K. Ozaki, J. Yukawa,
T. Ohgushi, P.W. Price (Eds.)

Galling Arthropods and Their Associates

Ecology and Evolution

With 68 Figures

Springer

Kenichi Ozaki, Ph.D.
Forestry and Forest Products Research Institute
1 Matsunosato, Tsukuba 305-8687, Japan

Junichi Yukawa, D.Agr.
Former Director
Kyushu University Museum
1-5-12 Matsuzaki, Higashi-ku, Fukuoka 813-0035, Japan

Takayuki Ohgushi, Ph.D.
Professor
Center for Ecological Research, Kyoto University
2 Hirano, Otsu 520-2113, Japan

Peter W. Price, Ph.D.
Regents' Professor Emeritus
Department of Biological Sciences, Northern Arizona University
Flagstaff, Arizona 86011-5640, USA

Library of Congress Control Number: 2006921176

ISBN-10 4-431-32184-5 Springer-Verlag Tokyo Berlin Heidelberg New York
ISBN-13 978-4-431-32184-2 Springer-Verlag Tokyo Berlin Heidelberg New York

Springer is a part of Springer Science+Business Media
springer.com

Printed in Japan

Typesetting: Camera-ready by the editors and authors
Printing and binding: Nikkei Printing Inc., Japan

Printed on acid-free paper

Preface

This book is the result of an international symposium on gall-inducing arthropods, which was held September 5–9, 2005, in Kyoto, Japan. It was organized as the 4th international symposium on gall-inducing insects and as the symposium of the International Union of Forestry Research Organizations (IUFRO) working group, 7.03.02, Gall-Inducing Insects.

The book addresses recent developments in the ecology, evolution, systematics, physiology, and biodiversity of gall-inducing arthropods, with individual contributions ranging in scope from detailed descriptions to profoundly synthetic studies. One underlying theme of the book is the various impacts of gall induction that indirectly affect insect communities on the host plant. The other important contribution is the highly intricate and dynamic interactions between galling arthropods and their uniquely specialized host plants. Included also are chapters that discuss biodiversity and distribution patterns of gall-inducing arthropods, and biological control of invasive gall-inducing arthropods and of invasive trees.

We believe that this book makes an important contribution to the knowledge of galling arthropods and their associates and to the development of robust, general principles of the ecology and evolution of these organisms.

We gratefully acknowledge the Japan Society for the Promotion of Science (JSPS), the International Union of Forestry Research Organizations Japan (IUFRO-J), and the 21st Century COE Programs, Kyoto University (Formation of a Strategic Base for the Multidisciplinary Study of Biodiversity, and Innovative Food and Environmental Studies Pioneered by Entomomimetic Sciences), for generous financial support. We also acknowledge the following colleagues who reviewed chapters of the book: Robert Anderson, Joseph Bailey, Randy Bangert, Daniel Burckhardt, Jonathan Brown, Tim Craig, Luc De Bruyn, Paul Dykstra, Phil Fay, Ray Gagné, Keith Harris, Moshe Inbar, Masato Ito, Kaoru Maeto, Masahiro Nakamura, Dan Quiring, Heikki Roininen, Ei'ichi Shibata, Joseph Shorthouse, Graham Stone, Ken Tabuchi, Koichi Tanaka, and Gina Wimp. Masato Ito and Ken Tabuchi took on the role of editing each chapter for consistency in style. Nami Uechi provided pictures for the cover. The editors also wish to thank all the participants of the symposium, whose penetrating and spirited contributions helped make this scientific exchange highly worthwhile.

Kenichi Ozaki
Junichi Yukawa
Takayuki Ohgushi
Peter W. Price

Sapporo, Japan, January 2006

Contents

2. Biological Control and Galling Arthropods

3. Galling Arthropods - Plant Interactions

4. Indirect Effects of Galling Arthropods

5. Evolution and Taxonomy

Contributors

Yoshihisa Abe, Laboratory of Applied Entomology, Graduate School of Agriculture, Kyoto Prefectural University, Kyoto 606-8522, Japan

Lawrence P. Abrahamson, State University of New York's College of Environmental Science and Forestry, 241 Illick Hall, Syracuse, NY 13210, USA

Robin J. Adair, Department of Primary Industries, Primary Industries Research Victoria, PO Box 48, Frankston 3199, Australia

Alexandre Aebi, Institute of Evolutionary Biology, The Kings Buildings, West Mains Road, Edinburgh, EH9 3JT, UK

Alberto Alma, Department of Exploitation and Protection of the Agricultural and Forestry Resources, Entomology and Zoology Applied to the Environment "Carlo Vidano", Via Leonardo da Vinci 44, Grugliasco 10095, Italy

Hiroshi Amano, Laboratory of Applied Entomology and Zoology, Faculty of Horticulture, Chiba University, 648 Matsudo, Chiba 271-8510, Japan

Tracey Begg, Institute of Evolutionary Biology, The Kings Buildings, West Mains Road, Edinburgh, EH9 3JT, UK

William J. Boecklen, Laboratory of Ecological Chemistry, Department of Biology, New Mexico State University, Las Cruces, New Mexico 88003, USA

Giovanni Bosio, Phytosanitary Service, Regione Piemonte, Via Livorno 60, Torino 10144, Italy

Jonathan M. Brown, Department of Biology, Grinnell College, Grinnell, IA 50112, USA

Wei-Chung Chen, Department of Entomology, National Chung Hsing University, Taichung 40227, Taiwan

Idelle Cooper, Department of Biology, Grinnell College, Grinnell, IA 50112, USA

Nobuyuki Endo, National Agricultural Research Center for Kyushu Okinawa Region, 2421 Suya, Nishigoshi, Kumamoto 861-1192, Japan

Leah Flaherty, Population Ecology Group, Faculty of Forestry and Environmental Management, University of New Brunswick, Fredericton, New Brunswick E3B 6C2, Canada

Tomoko Ganaha, Entomological Laboratory, Graduate School of Bioresource and Bioenvironmental Sciences, Kyushu University, Fukuoka 812-8581, Japan

Shigekazu Haitsuka, Saga Prefectural Agriculture Research Center, Saga 840-2205, Japan

Naoki Hijii, Laboratory of Forest Protection, Graduate School of Bioagricultural Sciences, Nagoya University, Nagoya 464-8601, Japan

Teruaki Hino, Kansai Research Center, Forestry and Forest Products Research Institute, 68 Nagaikyutaro, Fushimi, Kyoto 612-0855, Japan

Shih Shu Huang, Department of Entomology, National Chung Hsing University, Taichung 40227, Taiwan

Noriyuki Ikai, Laboratory of Forest Protection, Graduate School of Bioagricultural Sciences, Nagoya University, Nagoya 464-8601, Japan

Kensuke Imai, Laboratory of Insect Ecology, Graduate School of Agriculture, Kyoto University, Kitashirakawa-Oiwakecho, Sakyo, Kyoto 606-8502, Japan

Moshe Inbar, Department of Evolutionary & Environmental Biology, University of Haifa, Mount Carmel, Haifa 31905, Israel

Masato Ito, JSPS Research Fellow, Hokkaido Research Center, Forestry and Forest Products Research Institute, 7 Hitsujigaoka, Toyohira, Sapporo 062-8516, Japan

Motomi Ito, Ito Laboratory, Department of General Systems Studies, Graduate School of Arts and Sciences, The University of Tokyo, 3-8-1 Komaba, Meguro-ku, Tokyo 153-8902, Japan

Rob Johns, Population Ecology Group, Faculty of Forestry and Environmental Management, University of New Brunswick, Fredericton, New Brunswick E3B 6C2, Canada

Naoto Kamata, Laboratory of Ecology, Graduate School of Natural Science and Technology, Kanazawa University, Ishikawa 920-1192, Japan

Takahiro Kamikado, Kagoshima Prefectural Plant Protection Office, Kagoshima 891-0116, Japan

Toshihide Kato, Ito Laboratory, Department of General Systems Studies, Graduate School of Arts and Sciences, The University of Tokyo, 3-8-1 Komaba, Meguro-ku, Tokyo 153-8902, Japan

Jiro Kodani, Ishikawa Forest Experiment Station, Sannomiya, Hakusan, Ishikawa 920-2114, Japan

Ling-Hsiu Liao, Department of Entomology, National Chung Hsing University, Taichung 40227, Taiwan

Mei-Fiang Lou, Department of Entomology, National Chung Hsing University, Taichung 40227, Taiwan

Masaya Matsumura, National Agricultural Research Center for Kyushu Okinawa Region, 2421 Suya, Nishigoshi, Kumamoto 861-1192, Japan

Kotaro Matsushita, The Institute of Low Temperature Science, Hokkaido University, Sapporo 060-0819, Japan

George Melika, Systematic Parasitoid Laboratory, Vas County Plant Protection and Soil Conservation Service, Kelcz-Adelffy St. 6, Kőszeg 9730, Hungary

Katsuhiko Miyaji, Agricultural Management Division Kagoshima Prefectural Agricultural Experiment Station, Kagoshima 891-0116, Japan

Seiichi Moriya, National Agricultural Research Center, Tsukuba, Ibaraki 305-8666, Japan

Andrew Morrison, Population Ecology Group, Faculty of Forestry and Environmental Management, University of New Brunswick, Fredericton, New Brunswick E3B 6C2, Canada

Masahiro Nakamura, Tomakomai Research Station, Field Science Center for Northern Biosphere, Hokkaido University, Takaoka, Tomakomai 053-0035, Japan

Ottilie C. Neser, Plant Protection Research Institute, PB X134, Pretoria 0121, South Africa

Pekka Niemelä, Faculty of Forestry, University of Joensuu, P.O.B. 111, FI-80101 Joensuu, Finland

Machiko Nohara, Entomological Laboratory, Graduate School of Bioresource and Bioenvironmental Sciences, Kyushu University, Fukuoka 812-8581, Japan

Takayuki Ohgushi, Center for Ecological Research, Kyoto University, Otsu, Shiga 520-2113, Japan

Kenichi Ozaki, Forestry and Forest Products Research Institute, 1 Matsunosato, Tsukuba, 305-8687, Japan

Luca Picciau, Department of Exploitation and Protection of the Agricultural and Forestry Resources, Entomology and Zoology Applied to the Environment "Carlo Vidano", Via Leonardo da Vinci 44, Grugliasco 10095, Italy

Carolyn C. Pike, University of Minncsota, Cloquet Forestry Center, 175 University Rd, Cloquet, MN 55720, USA

Peter W. Price, Department of Biological Sciences, Northern Arizona University, Flagstaff, Arizona 86011-5640, USA

Ambra Quacchia, Department of Exploitation and Protection of the Agricultural and Forestry Resources, Entomology and Zoology Applied to the Environment "Carlo Vidano", Via Leonardo da Vinci 44, Grugliasco 10095, Italy

Dan Quiring, Population Ecology Group, Faculty of Forestry and Environmental Management, University of New Brunswick, Fredericton, New Brunswick E3B 6C2, Canada

Zoë Randle, Centre for Ecology and Hydrology, CEH Dorset, Winfrith Technology Centre, Dorchester, DT2 8ZD, UK

Daniel J. Robison, North Carolina State University, Box 8008, Jordan Hall Room 3118, Raleigh, NC 27695-8008, USA

Heikki Roininen, Department of Biology, University of Joensuu, P.O.B. 111, FI-80101 Joensuu, Finland

Yasuaki Sakamoto, Hokkaido Research Center, Forestry and Forest Products Research Institute, Hitsujigaoka, Sapporo 062-8516, Japan

Shinsuke Sato, Laboratory of Ecology, Graduate School of Natural Science and Technology, Kanazawa University, Ishikawa 920-1192, Japan

Karsten Schönrogge, Centre for Ecology and Hydrology, CEH Dorset, Winfrith Technology Centre, Dorchester, DT2 8ZD, UK

Gabrijel Seljak, Chamber for Agriculture and Forestry of Slovenia, Agricultural and Forestry Institute Nova Gorica, Pri Hrastu 18, SI-5000 Nova Gorica, Slovenia

Chia-Chi Shen, Department of Entomology, National Chung Hsing University, Taichung 40227, Taiwan

Ei'ichi Shibata, Laboratory of Forest Protection, Graduate School of Bioagricultural Sciences, Nagoya University, Chikusa-ku, Nagoya 464-8601, Japan

Yukihiro Shimatani, Laboratory of River Engineering, Faculty of Engineering, Kyushu University, Fukuoka 812-8581, Japan

Graham N. Stone, Institute of Evolutionary Biology, The Kings Buildings, West Mains Road, Edinburgh, EH9 3JT, UK

Ken Tabuchi, JSPS Research Fellow, Hokkaido Research Center, Forestry and Forest Products Research Institute, 7 Hitsujigaoka, Toyohira, Sapporo 062-8516, Japan

Makoto Tokuda, National Institute of Advanced Industrial Science and Technology, 1-1-1 Higashi, Tsukuba, Ibaraki 305-8566, Japan

Gene-Sheng Tung, Division of Forest Protection, Taiwan Forestry Research Institute, 53 Nan-Hai Road, Taipei 10053, Taiwan

Nami Uechi, Okinawa Prefectural Agricultural Experiment Station, 4-222 Sakiyama-cho, Naha, Okinawa 903-0814, Japan

Akira Ueda, Hokkaido Research Center, Forestry and Forest Products Research Institute, 7 Hitsujigaoka, Toyohira, Sapporo 062-8516, Japan

Henri Vanhanen, Faculty of Forestry, University of Joensuu, P.O.B. 111, FI-80101 Joensuu, Finland

Timo O. Veteli, Faculty of Forestry, University of Joensuu, P.O.B. 111, FI-80101 Joensuu, Finland

Veli Vikberg, Liinalammintie 11 as. 6, 14200 Turenki, Finland

Risto Virtanen, Department of Biology, University of Oulu, P.O.B. 3000, FIN-90014 University of Oulu, Finland

Yu-Chu Weng, Department of Entomology, National Chung Hsing University, Taichung 40227, Taiwan

Man-Miao Yang, Department of Entomology, National Chung Hsing University, Taichung 40227, Taiwan

Kaori Yara, National Institute for Agro-Environmental Sciences, Tsukuba, Ibaraki 305-8604, Japan

Christopher T. Yarnes, Laboratory of Ecological Chemistry, Department of Biology, New Mexico State University, Las Cruces, New Mexico 88003, USA

Hiraku Yoshitake, Ito Laboratory, Department of General Systems Studies, Graduate School of Arts and Sciences, The University of Tokyo, 3-8-1 Komaba, Meguro-ku, Tokyo 153-8902, Japan

Junichi Yukawa, 1-5-12 Matsuzaki, Fukuoka 813-0035, Japan

Alexei Zinovjev, Zoological Institute of Russian Academy of Science, St Peterburg, Russia

1. Biodiversity and Community Structure

1 Latitudinal and Altitudinal Patterns in Species Richness and Mortality Factors of the Galling Sawflies on *Salix* Species in Japan

Heikki Roininen[1,2], Takayuki Ohgushi[2], Alexei Zinovjev[3], Risto Virtanen[4], Veli Vikberg[5], Kotaro Matsushita[6], Masahiro Nakamura[2,6], Peter W. Price[7], and Timo O. Veteli[8]

[1]Department of Biology, University of Joensuu, P.O.B. 111, FI-80101 Joensuu, Finland
[2]Center for Ecological Research, Kyoto University, Otsu, Shiga 520-2113, Japan
[3]Zoological Institute of Russian Academy of Science, St Peterburg, Russia
[4]Department of Biology, University of Oulu, P.O.B. 3000, FIN-90014 University of Oulu, Finland
[5]Liinalammintie 11 as. 6, 14200 Turenki, Finland
[6]The Institute of Low Temperature Science, Hokkaido University, Sapporo, Hokkaido 060-0819, Japan
[7]Department of Biological Sciences, Northern Arizona University, Flagstaff, Arizona 86011-5640, USA
[8]Faculty of Forestry, University of Joensuu, P.O.B. 111, FI-80101 Joensuu, Finland

Summary. Species richness of willow species and galling sawflies living on them were examined in latitudinal and altitudinal gradients in six Japanese river systems from Hokkaido to southern Honshu. Mortality factors of gallers including plant based mortality, parasitoids and inquilines during larval development were studied by dissecting sampled galls under a microscope. The association between environmental factors, mortality factors and local diversity of galling sawflies and their willow hosts were studied. Species richness of sawfly gallers and their host plants decreased towards the south. Species richness of gallers was lower in the delta areas at lower altitudes than at higher altitudes. Different mortality factors, plant based mortality, parasitoids or inquilines, showed no significant trends with latitude or altitude. Although some parasitoids showed a weak correlation with latitude and altitude, but overall survival of larvae was not correlated with latitude or altitude. Among sawfly gall types, *Pontania proxima*-type was distinct by having high plant-based mortality. The observed pattern of increasing diversity with increasing latitude is opposite to that in many other animals and plants. This pattern is unlikely explained by larval survival or different mortality factors since they showed no difference in latitudinal or altitudinal gradient. A possible explanation of the pattern may be the decreasing host plant richness with other host related factors, like increased habitat fragmentation and decreased abundance of host plants towards the south. In addition, leaf flush of host plants and egg laying of galling sawflies might be better synchronised in north with highly seasonal but predictable resource availability.

Key words. Parasitism, Inquiline, Plant based mortality, *Salix*, *Pontania*, *Eupontania*, *Phyllocolpa*, Elevation

1.1 Introduction

It is well known that species diversity in most groups of organisms increases towards the equator (e.g. Gaston and Spicer 2004; Giller 1984; Pianka 1966). There are few examples of the groups of organisms, which do not show any pattern or show reversed patterns of diversity in latitudinal gradients. The species diversity of sawflies and especially galling sawflies is an example of a reversed pattern. Price and Roininen (1993) demonstrated that the number of galling sawflies decreases from the north, at about 43°N, to the south in North America, while Kouki et al. (1994) have shown a similar latitudinal pattern in sawfly species in Eurasia. This pattern is unique for sawfly gallers since it has not been found in other gallers which increase up to about 34°N and then decline (Price et al. 1998).

Parasitoid assemblages per host or per groups of hosts (having different feeding habits) vary independently along climatic or latitudinal gradients. Ichneumonid richness declines towards the tropics in Australia (Gauld 1986). But in North America peak of ichneumonid richness is between 38 and 42 degrees north, and then declines south of that (Janzen 1981). Although there is lack of data for other parasitoid families, it is obvious that some families are common in the tropics, while some are not (Hespenheide 1979; Noyes 1989). At this time it is impossible to draw general conclusions about latitudinal pattern in species richness for insect parasitoids (Hawkins 1994). However, Hawkins (1990, 1994) has shown that parasitoid species per host species are not correlated with annual temperature or mean low temperature in endophytic feeders (including gallers) but are correlated with external feeders, so that number of parasitoids become smaller towards the tropics.

Some recent studies on sawfly gallers do not support Hawkins' generalisation that parasitoid assemblages of gallers are similar in numbers of species everywhere (Kopelke 1994; Roininen and Danell 1997; Roininen et al. 2002). The parasitoid and inquiline species in the communities of galling sawflies decrease towards the arctic areas. The most northern population of *Eupontania* Zinovjev gallers at northern Yamal peninsula has only one parasitoid (Roininen et al. 2002). On the contrary, very well studied *Eupontania* gallers in the middle of Europe have many parasitoid species (Kopelke 1994). Another typical pattern is that the number of species of inquiline parasitoids (eurytomid wasps, tephritid flies and curculionid weevils), which kill the sawfly host and utilise the gall tissue, decreases towards the north. Tephritids and weevils do not exist in the arctic area at all, and eurytomids are found in the subarctic area, although they are rare (Kopelke 1994; Roininen et al. 2002).

The total mortality caused by parasitoids in any insect groups including gallers does not indicate the existence of a climatic gradient (Hawkins 1994). *Eupontania* gallers support this generalisation (Kopelke 1994; Roininen et al. 2002), but *Euura mucronata* (Hartig) Man. (Chuchill), the bud galler, on willows showed decreased parasitism in arctic areas (Roininen and Danell 1997). Parasitoid assemblages of *Eupontania* gallers vary locally but the mortality caused by them is much more constant, independent of the species composition or numbers of parasitoids (Kopelke 1994).

In this study we address the following questions. Do any kind of patterns exist in species diversity of the leaf galling sawflies and their host plants in latitudinal or altitudinal gradients in Japan? Do different mortality factors of galling sawflies show any latitudinal or altitudinal gradients? Six river systems starting from the mountains and to sea shore were sampled, which show the maximum of 1250 km distance between sites and 1900 m difference in altitude.

1.2 Materials and Methods

1.2.1 Galling Sawflies and Willows

The galling sawflies (Hymenoptera; Tenthredinidae; Nematinae; Euurina) include four different genera: *Euura* Newman, *Pontania* A. Costa, *Eupontania* Zinovjev and *Phyllocolpa* Benson (Zinovjev and Vikberg 1999). All species of these genera form galls on different parts of Salicaceae plants. The shape of galls varies from open galls (leaf folds or rolls) to closed galls with many types. In this study the following gall types were included: folders, rollers, bud gallers, leaf blade galls (*proxima*-type), sausage galls (*dolichura*-type), pea-shaped galls on leaves (*viminalis*-type) and bean shaped galls (*vesicator*-type). Females lay eggs into the young and growing plant organs: leaves, buds or shoots. Species-specific galls are formed as a consequence of egg laying and developing larvae. Larvae develop inside the closed galls, rolls or folds, but in some leaf gallers external feeding can occur. It is a common habit for most rollers and folders (for more detailed biology of gallers see Kopelke 1982, 1986, 1991, 1999; Roininen et al. 2005; Zinovjev and Vikberg 1999). A total of 27 galler species or their morphotypes were identified in this study (Table 1).

Willows (*Salix* L.) are a diverse and a widespread genus in the Northern Hemisphere. Skvortsov (1999) lists 135 species for Europe and Russia with adjacent countries. Kimura (1989) records 36 willow species including *Chosenia* Pallas and *Toisusu* Kimura from Japan. Willows are especially typical plants in the mountain and arctic areas, 38% of species in Skvortsov's book are arctic or alpine species. Willow taxonomy is difficult and they are known to hybridize. In this study we apply Kimura's (1989) taxonomy used in the Japanese literature and give Skvortsov's (1999) suggested names in parenthesis. We also show the taxonomy suggested by Ohashi (2000) in Table 1.

We sampled galling sawflies along latitudinal and altitudinal gradients in Japan. The sampled river systems were as follows: River Ishikari on Hokkaido, River Akakawa in the Tsuruoka area, River Tainai in the Niigata area including the highest site at Mt. Bandai, River Tedori in the Kanazawa area, Shohnai and Yahagi rivers in the Nagoya area, and River Asahi in the Okayama area. The last five mentioned sites are in Honshu. These river systems represent a latitudinal gradient along the islands of Hokkaido and Honshu. Altitudinal classification was based on elevation and river morphology developed by Niiyama (1987). These elevation-related classes from lower elevation to high were delta zone, intermediate zone, alluvial fan zone and high valley zone (see the detailed description of zones in Niiyama 1987, 1989). We used this river morphology for classification, because it is ecologically more meaningful than the absolute elevation of sampling site. Absolute altitude of the sites was not the same in all river systems, but the order of classes was always identical. The data from the alpine zone were combined with the valley zone.

1.2.2 Sampling and Laboratory Analysis

At each river system we selected four sites, one in each river morphological class. In each sampling site we censused all willow species for the presence of galling sawflies. From each willow species at least 20 individuals (usually many more) were checked and sampled if galls were found. The total of dissected galls was 4258. Depending on the abundance of the galler species, from 78 to 434 galls were randomly sampled for the detailed study. Leaf rollers and folders were included in the species richness studies. In the studies of mortality factors and parasitoid assemblages, only species forming closed galls, *Eupontania* and *Pontania*, were included. Sampled galls were dissected under the microscope and the following characteristics of galls were measured: plant based mortality in the egg or larval stage (no evidence of predation was found), mortality by different parasitoid species, mortality by inquiline weevils or lepidopteran larvae.

We were able to classify parasitoids into the following 5 classes, which mostly have one dominant species: 1) *Pteromalus* sp., parasitoid killing small or medium sized larvae and emerges from gall in later summer. 2) *Bracon* sp. killing medium sized larvae, which have a characteristic of white silky cocoon. 3) Ichneumonid (*Scambus* sp.) exoparasitoid attacking old larvae in late autumn. Host larvae were paralyzed at the time of dissection of galls when most successfully developed sawfly larvae had left the galls. These parasitoids were so rare that data for them is not shown. 4) *Eurytoma* sp. (close to *E. aciculata*), which attacks young galls and kills the host and eats the gall tissue as a phytophagous parasitoid, overwinters in the gall. 5) *Eurytoma* sp. new, which is very small and living on the wall tissue of galls probably also parasitizing small weevil larvae. It was considered a parasitoid when it caused mortality of the galler. In many cases it can coexist with a sawfly larva without a lethal effect on it. If there were morphologically and ecologically similar parasitoid species we were not able to separate them. The identified parasitoid species belong to the same genera as found in other studies (e.g. Kopelke 1994; Price and Pschorn-Walcher 1988). Only the *Eurytoma* sp. new represents a new kind of parasitism/commensalism habit of life not found earlier because it was able to coexist with a sawfly larva.

1.2.3 Statistics

Standard non-parametric procedures were used to test trends in gall willow species richness. To test whether the mortality factors (grouped as collective parasitoid, inquilines, plant based mortality) and overall survival are related to biotic and/or environmental variables (*Salix* species, elevation (corresponds with classification of river structure), latitude, gall type, sawfly species) a reduced rank regression analysis (RDA) was run by using CANOCO Version 4 (ter Braak and Šmilauer 1998; see also ter Braak

1987 and ter Braak and Looman 1994). This analysis assumes a linear relationship between response variables and explanatory variables. *Salix* species, gall type and sawfly species were coded for the analysis as dummy (0/1) explanatory variables. In the RDA, species data were centered and standardized, and standardization was made also for samples to remove effect of variable sample sizes. The statistical significance between the explanatory variables and mortality factors was tested by using forward selection of variables and Monte Carlo permutation tests.

1.3 Results

1.3.1 Patterns in Sawfly and Willow Species

We found a total of 27 species or morphotypes of gallers representing seven different types of galls, of which 17 species have not been reported earlier (Abe and Togashi 1989; Price and Ohgushi 1995; Yukawa and Masuda 1996) from Japan on 16 willow species (Table 1).

Four (in Table 1) of the willow species did not have gallers in any of the study sites. The number of willow species and galling sawfly species declines towards the south (Fig. 1 and Table 2). In Hokkaido our sampling included 13 willow species and 21 morphotypes of galling sawflies but in Okayama 5 willow species and only 2 galling sawfly species. There were also differences in the relative proportion of galling sawfly species on willows; galling sawfly species per willow species and the number of willow species, which are known to have galling sawflies in Japan, declined toward the south (Fig. 1). In Hokkaido there were on average of 1.5 sawfly species per willow species, but in Okayama only 0.3. Species richness of willows and number of gallers per willow species were correlated positively in pooled samples from different latitudes (Fig. 2). Diversity of gall types was also highest in the north; in Hokkaido there were 5 different gall types, but in other sites only 1 to 3 different types.

We did not detect any altitudinal differences either in the total number of the willow species or the willow species that are known to have gallers (Fig. 3a; Friedman test; $H = 1.79$, $P > 0.05$ and $H = 1.00$, $P > 0.05$, respectively). However, high species number of gallers was more likely to be found at higher altitudes than the delta zones (Fig. 3b). The number of galling sawfly species per willow species with gallers somewhere in Japan was 2.5 times higher on average in other altitudes than on the delta area (Friedman test; $H = 8.90$, $P < 0.05$).

Table 1. The species of galling sawflies or their morphotypes and host plants included into the study. We used Kimura's (1989) taxonomy used in Japanese literature and give Skvortsov (1999) suggested names in parenthesis. If there is no name in parenthesis classification is identical

Galler	Willow species
Phyllocolpa "folder" sp.1	*Salix reinii*
Phyllocolpa "folder" sp.2	*S. rorida*
Phyllocolpa "folder" sp.3	*S. sachalinensis* (*S. udensis*[*])
Phyllocolpa "roller" sp.4	*S. pet-susu* (*S. schwerinii*[*])
Phyllocolpa "folder" sp.5	*Toisusu urbaniana* (*S. cardiophylla*[*])
Phyllocolpa "folder" sp.6	*S. jessoensis* (*S. pierotii*)
Phyllocolpa "folder" sp.7	*S. miyabeana*
Phyllocolpa "folder or roller" sp.8	*S. gracilistyla*
Phyllocolpa "folder" sp.9	*S. futura*
Phyllocolpa "folder" sp.10	*S. gilgiana*[#]
Eupontania "*viminalis*" sp.1	*S. reinii*
Eupontania "*viminalis*" sp.2[1]	*S. sachalinensis* (*S. udensis*[*])
Eupontania "*viminalis*" sp.3	*S. rorida*
Eupontania "*viminalis*" sp.4[2]	*S. gracilistyla*
Eupontania "*viminalis*" sp.5	*S. miyabeana*
Eupontania "*viminalis*" sp.6	*S. pet-susu* (*S. schwerinii*[*])
Eupontania "*viminalis*" sp.7	S. gilgiana
Pontania mirabilis	*Toisusu urbaniana* (*S. cardiophylla*[*])
Eupontania "*vesicator*" sp. close to *P. lapponica*	*S. pet-susu* (*S. schwerinii*[*])
Eupontania mandshurica	*S. jessoensis* (*S. pierotii*)
Eupontania amurensis	*S. miyabeana*
Eupontania "*vesicator*" sp.1[4]	*S. integra*
Eupontania "*vesicator*" sp.2[5]	*S. chaenomeloides*
Pontania "*proxima*" sp.1	*S. pet-susu* (*S. schwerinii*[*])
Pontania "*proxima*" sp.2[3]	*S. sachalinensis* (*S. udensis*[*])
Pontania "*proxima*" sp.3	*S. yezoalpina* (*S. nakamurana*[*]
Euura mucronata	*S. reinii*
Euura mucronata	*S. sachalinensis* (*S. udensis*[*])
Salix species with no galls in any sites	
S. pauciflora (*S. nummularia*[*])	
S. subfragilis (*S. triandra*[*])	
S. hultenii (*S. caprea*[*])	
S. bakko (*S. caprea*[*])	

Superscript numbers show the classification by Yukawa and Masuda (1996): [1]*Pontania* sp. I; [2]*Pontania* sp. K; [3]*Pontania* sp. H; [4]*Pontania* sp. J; and [5]*Pontania* sp. A. [#]Ohashi (2000) combined with *S. miyabeana*. [*]Enumerated also by Ohashi (2000).

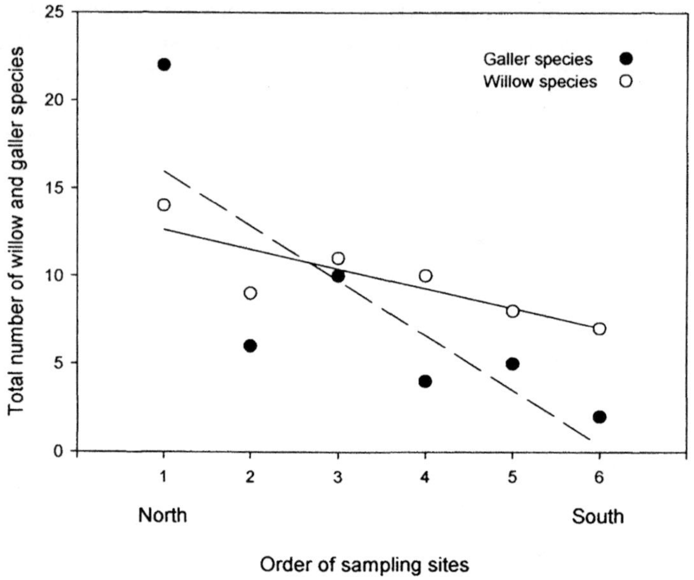

Fig. 1. Correlations between latitudinal order of sampling site and total number of galler species and number of willow species known to have galling sawflies in Japan (Spearman's rank correlation; for gallers, $r = -0.87$, $P < 0.019$, and for willows, $r = -0.83$, $P < 0.05$).

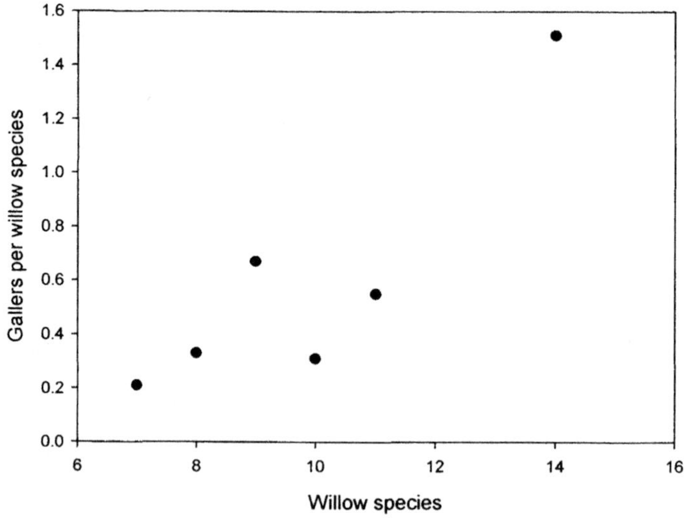

Fig. 2. Correlation between the number of gallers per willow species and the number of willow species (Pearson's correlation; $r = 0.88$, $P = 0.02$). Data are pooled by latitudinal sampling sites.

Table 2. Numbers of *Eupontania* (E), *Pontania* (P), *Phyllocolpa* (Ph) (includes all folders and rollers) and *Euura* (Eu) species or morphotypes found in different sites.

Site order	Locality	Altitude	River classification	# of willow Species	# of galling species E+P	Ph	Eu	Total number of gallers
1	Hokkaido, Daisetsu Lake, and Kurodake mountain	870-1964	Valley zone or alpine zone	8	8	5	1	14
1	Hokkaido, Kamikawa	300	Alluvial fan zone	9	7	5	0	12
1	Hokkaido, Shintutsukawa	100	Intermediate zone	5	4	3	0	7
1	Hokkaido, Oyafuru and Tobetsu	20	Delta	6	5	1	0	7
2	Tsuruoka, Gassan	800-1150	Valley zone	5	3	0	1	4
2	Tsuruoka, Kusibiki	100	Intermediate zone	6	4	1	0	5
2	Tsuruoka, Akakawa river from sea up to Road 7	<60	Delta zone	6	1	0	0	1
3	Fukushima pref., Bandai-san	1180-1300	Valley zone	8	1	2	1	4
3	Niigata, Oku-Tainai	200-320	Alluvial fan zone	10	1	4	0	4
3	Niigata, Tsuzumioka	80	Intermediate zone	7	3	2	0	5
3	Niigata, Nakajyo town	30	Delta zone	6	1	0	0	1
4	Kanazawa, Hakusan	1170	Valley zone	5	1	1	1	3
4	Kanazawa, Hiroshiyama	785	Alluvial fan zone	5	1	1	0	2
4	Kanazawa, Oguchi	330	Intermediate zone	9	0	2	0	2
4	Kanazawa, Tedori R. between 6 and 16 km from sea	60	Delta zone	7	0	0	0	0
5	Nagoya, Inabu Town	810-885	Valley zone	4	2	2	0	3
5	Nagoya, Inabu Town	500	Alluvial zone	3	0	1	0	1
5	Nagoya, Seto town	100	Intermediate zone	4	1	0	0	1
5	Nagoya, Kasugai Town	<60	Delta zone	3	0	0	0	0
6	Okayama, Kamogawa	200	Valley zone	2	0	0	0	0
6	Okayama, Takabe Town	120	Alluvial zone	4	1	0	0	1
6	Okayama, Shimoda and Nakamaki	70	Intermediate zone	5	1	0	0	1
6	Okayama, Yamaura	<20	Delta zone	2	0	0	0	0

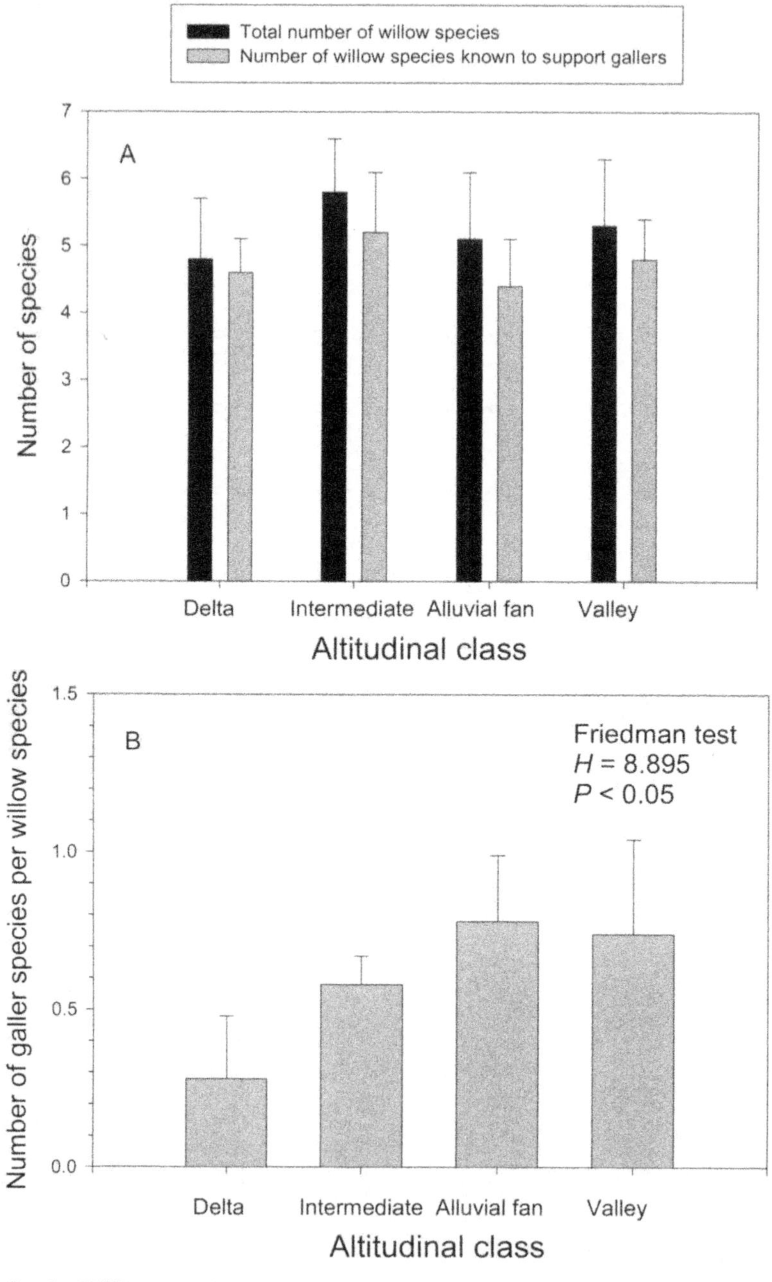

Fig. 3a, b. Willows and gallers by altitude (by river classification). **a** Numbers of willow species at different altitudes in two groups; all willow species and ones known to have gallers in Japan (Friedman's test for both cases; $P < 0.05$). **b** Number of galler species per willow species at different altitudinal classes.

Table 3. Survival and mortality factors in studied populations of of *Eupontania* and *Pontania* gallers.

Site	Sawfly sp. or morphotype	*Salix* sp.	River class	n	Survival (%)	Mortality factor (%)		
						Plant based	Parasitoids	Inquilines
Hokkaido	*P. proxima* sp.	*S. pet-susu*	Valley	106	33.0	46.2	9.4	11.3
Hokkaido	*P. mirabilis*	*T. urbaniana*	Valley	184	77.2	20.7	2.2	0.0
Hokkaido	*E. viminalis* sp	*S. gracilistyla*	Alluvial fan	187	81.8	15.5	1.6	1.1
Hokkaido	*E. vesicator* sp.	*S. jessoensis*	Alluvial fan	118	73.7	6.8	15.3	4.2
Hokkaido	*E. viminalis* sp	*S. rorida*	Alluvial fan	146	75.3	17.1	2.1	5.5
Hokkaido	*E. vesicator* sp.	*S. integra*	Delta	269	27.5	15.6	52.8	4.1
Hokkaido	*P. proxima* sp.	*S. sachalinensis*	Delta	160	43.1	51.3	4.4	1.3
Hokkaido	*E. viminalis* sp.	*S. sachalinensis*	Delta	196	52.0	19.4	17.3	11.2
Hokkaido	*E. amurensis*	*S. miyabeana*	Delta	236	47.0	9.7	33.1	11.2
Hokkaido	*E. viminalis* sp.	*S. miyabeana*	Delta	221	34.8	4.1	54.8	6.3
Tsuruoka	*E. vesicator* sp.	*S. integra*	Valley	210	79.5	6.2	11.0	3.3
Tsuruoka	*E. vesicator* sp.	*S. integra*	Alluvial fan	78	46.2	5.1	48.0	0.0
Tsuruoka	*E. vesicator* sp.	*S. jessoensis*	Alluvial fan	83	79.5	4.8	8.4	7.2
Niigata	*E. viminalis* sp.	*S. gilgiana*	Intermediate	254	55.1	10.2	20.1	14.6
Niigata	*E. viminalis* sp.	*S. gilgiana*	Delta	228	67.5	1.8	14.5	16.2
Niigata	*E. vesicator* sp.	*S. integra*	Valley	259	52.9	12.4	20.1	14.7
Niigata	*E. vesicator* sp.	*S. jessoensis*	Intermediate	81	54.3	0.0	23.5	22.2
Niigata	*E. viminalis* sp.	*S. sachalinensis*	Alluvial fan	236	68.6	4.2	13.1	14.0
Kanazawa	*E. viminalis* sp.	*S. sachalinensis*	Valley	143	53.8	0.0	36.4	9.8
Kanazawa	*E. viminalis* sp.	*S. sachalinensis*	Alluvial fan	131	53.4	9.9	27.5	9.2
Kanazawa	*E. viminalis* sp	*S. sachalinensis*	Intermediate	199	58.8	4.0	34.2	3.0
Nagoya	*E. vesicator* sp.	*S. integra*	Valley	434	49.5	7.4	38.5	4.6
Okayama	*E. vesicator* sp.	*S. chaenomeloides*	Intermediate	99	54.4	9.1	5.1	31.3

1.3.2 Patterns in Mortality Factors

Survival and mortality factors when divided into plant based mortality, parasitoids and inquilines, varied considerably within and among sites (Table 3), but in overall testing parasitoids caused significantly higher mortality than plant based factors or inquilines (ANOVA; $F_{2,68} = 5.84$, $P < 0.01$, Tukey HSD for multiple comparisons, $P < 0.05$ in both cases). Larval survival ranged from 27.5 to 81.8 percent, plant based mortality ranged from 0 to 51.3 percent, mortality caused by parasitoids ranged from 1.6 to 54.8 percent, and mortality caused by inquilines ranged from 0 to 31.3 percent among sampling sites (Table 3). The most remarkable within-site variation among galler species was a 2.3 times difference in survival between *P. proxima* sp. and *P. mirabilis* in the Hokkaido valley zone, and over 12 times difference in plant based mortality and parasitism between *P. proxima* sp. and *P. viminalis* sp. in the Hokkaido delta zone (Table 3).

The relationships between mortality factors to biotic and/or environmental variables (*Salix* species, elevation, latitude, gall type, sawfly species) was not strong according to the RDA analysis (permutation test; $P = 0.051$ for the significance of all canonical axes, trace = 0.73, F-ratio = 2.04 under the full model). Indeed, only gall type 1 (*Pontania proxima*) had high plant-based mortality (permutation test; $P = 0.026$). As gall type 1 turned out to have a very strong effect on the ordination, it was omitted in the remaining forward selection tests. In these tests, mortality factors were not significantly related either to presence of any single galler species, or to altitude or to latitude (P values for manual forward selection tests > 0.05).

1.4 Discussion

Our data show that the number of species of galling sawflies and their host plants increase towards the north. This is consistent with the observed increase in sawflies and especially galling sawflies, which have been found to be the exception to the general pattern (Kouki 1999; Kouki et al. 1994; Price and Roininen 1993) that species richness increases towards the south. Species richness of gallers increased on altitudinal gradients as well. Interestingly, our data suggest that species richness of sawflies declines much faster than their host willow species (Fig. 1, Table 2). Therefore the latitudinal gradient of species richness of galling sawflies could not just be a result of local species richness of host plants as suggested by Kouki et al. (1994). Probably, the reason for the steeper decline in the richness of sawflies than their host plants is related to the local abundance of host plant species. Willows are dominant plants in the arctic and alpine, and to some extent in boreal forest (Myklestad and Birks 1993; Skvortsov 1999) where sawfly richness also peaks.

Latitudinal and altitudinal gradients had no influence on mortality by parasitoids or inquilines. Our results are consistent with Hawkins (1994) that mortality caused by parasitoids did not show any climatic/latitudinal patterns in galling insects. Similarly, Roininen et al. (2002) found that mortality by parasitoids was not correlated with increasing climatic harshness of the environment in high latitudes, but they found that species richness of parasitoids decreased. There are no studies on altitudinal differences of galling insects but another endophytic group, miners, has shown opposite results to ours. For instance, Kato (1996) found that parasitoid assemblages of the honeysuckle leafminer are altitude-related; different parasitoid species caused the main part of mortality at different altitudes. Although Preszler and Boecklen (1996, and references therein) conclude that influence of parasitoid and predators on leafminers weaken with increasing elevation, our results do not support that. Sawfly gallers, which are known to have fewer parasitoid species than miners, do not show that clear pattern with altitude.

The mortality by parasitoids does not correlate with the latitudinal pattern in the species richness of galling sawflies. Therefore we should look for other possible factors responsible for the observed pattern of decreasing species richness of sawfly gallers towards the south. In the Japanese islands the dispersal of sawflies has been limited and it has probably been possible from the mainland only from the north through Sakhalin and the Kuril Islands. In this sense, higher species diversity in Hokkaido could result from more frequent dispersal. All the galler species, which colonized Hokkaido, may have not been able to disperse towards the south, although their host plants exist in more southern areas. The three willow species, *S. triandra, S. caprea, and S. nummularia*, do not have gallers in Japan, but are known to have them in continental Eurasia and Sakhalin (Zinovjev 1999). Most Japanese willows encounter their southern distributional limits in Japan (Skvortsov 1999). Therefore, dispersal may explain part of the patterns of distribution of galling sawflies in the Japanese islands, but less so in North America and Eurasia.

Although host plant diversity plays a key role explaining the diversity of willow-feeding sawfly gallers, we want to emphasize the other important factors in the relationships of galling sawflies and their host plants. All sawfly gallers, as well as most sawfly species, are specific in their egg-laying behaviour: eggs are laid in specific places usually inside the tissue of the plant, and most importantly into the plant tissue in a phenologically appropriate stage. This is because the formation of galls requires tissue in an early growing stage. In northern areas, the arctic and alpine zones, and boreal forest, leaf flush is well synchronized in most host plants, and takes place fast. A lot of resources are available but only for a short time period (Roininen 1991). For sawfly gallers, it may be easier to be synchronized with host plants in a highly predictable environment (see Yukawa 2000). Another reason for high species richness in boreal and arctic areas might be the abundance of willows and also their less fragmented habitats (Skvortsov 1999). Ishikawa (1983) found that the distribution of willow species in river systems became more fragmented from Hokkaido to central Honshu, and abundance of willows decreased accordingly. In addition, the distribution of willows in Japan is restricted to river systems (Kimura 1989), which increases habitat fragmentation.

1.5 Acknowledgements

We thank the Center for Ecological Research for a visiting professorship for HR and the personnel of CER for many kinds of help during the study period. For help with field work we thank Michihiro Ishihara, Naoto Kamata, Ken Shimizu, Hideki Ueno, Hironori Yasuda, Takao Itioka, Masahiro Nomura, Kenji Fujisaki and Yoko Inui. This study was partly supported by the Ministry of Education, Culture, Sports, Science and Technology Grant-in-Aid for Creative Basic Research (09NP1501) and Scientific Research (A-15207003) to TO, and the 21st Century COE Program (A2) to the Center for Ecological Research. HR was supported by the Finnish Academy (project no. 47574).

1.6 References

Abe M, Togashi I (1989) Symphyta (in Japanese). In: Hirashima Y (ed) A check list of Japanese insects. Entomological Laboratory, Faculty of Agriculture, Kyushu University, Fukuoka, pp 541-560

Gaston KJ, Spicer JI (2004) Biodiversity: an introduction. Blackwell Science, Oxford.

Gauld ID (1986) Latitudinal gradients in ichneumonid species richness in Australia. Ecological Entomology 11:155-161

Giller PS (1984) Community structure and the niche. Chapman and Hall, London.

Hawkins BA (1990) Global pattern of parasitoid assemblage size. Journal of Animal Ecology 59:57-72

Hawkins BA (1994) Patterns and process in host-parasitoid interactions. Cambridge University Press, Cambridge.

Hespenheide HA (1979) Are there fewer parasitoids in the tropics? American Naturalist 113:766-769

Ishikawa S (1983) Ecological studies on the floodplain vegetation in the Tohoku and Hokkaido districts, Japan. Ecological Review 20:73-114

Janzen DH (1981) The peak in North American ichneumonid species richness lies between 38 degree and 42 degree N. Ecology 62:532-537

Kato M (1996) Effects of parasitoid community structure upon the population dynamics of the honeysuckle leafminer, *Chromatomyia suikazurae* (Diptera: Agromyzidae). Researches on Population Ecology 38:27-40

Kimura Y (1989) Salicaceae (in Japanese). In: Satake Y, Hara H, Watari S, Tominari T (eds) Wild flowers of Japan, woody plant. Heibonsha, Tokyo, pp 39-58

Kopelke J-P (1982) Die gallenbildenden Pontania-Arten - ihre Sonderstellung unter den Blattwespen. Teil I: Gallenbildung, Entwicklung und Phänologie. Natur und Museum 112:356-365

Kopelke, J-P (1986) Zur Taxonomische und Biologie neuer Pontania-Arten der dolichura- Gruppe. Senckenbergiana Biologica 67:51-71.

Kopelke J-P (1991) Die Arten der viminalis-Gruppe, Gattung Pontania O. Costa 1859, Mittel- und Nordeuropas. (Insecta: Hymenoptera: Tenthredinidae). Senckenbergiana Biologica 71:65-128

Kopelke J-P (1994) Der Schmarotzerkomplex (Brutparasiten und Parasitoide) der gallenbildenden Pontania-Arten (Insecta: Hymenoptera: Tenthredinidae). Senckenbergiana Biologica 73:83-133

Kopelke J-P (1999) Gallenerzeugende Blattwespen Europas – Taxonomische Grundlagen, Biologie und Ökologie (Tenthredinidae: Nematinae: *Euura, Phyllocolpa, Pontania*). Courier Forschungsinstitut Senckenberg 212:1-183

Kouki J (1999) Latitudinal gradients in species richness in northern areas: some exceptional patterns. Ecological Bulletins 47:30-37

Kouki J, Niemelä P, Viitasaari M (1994) Reversed latitudinal gradient in species richness of sawflies (Hymenoptera, Symphyta). Annales Zoologici Fennici 31:83-88

Myklestad Å, Birks HJB (1993) A numerical analysis of the distribution patterns of Salix L. species in Europe. Journal of Biogeography 20:1-32

Niiyama K (1987) Distribution of salicaceous species and soil texture of habitats along the Ishikari river (in Japanese). Japanese Journal of Ecology 37:163-174

Niiyama K (1989) Distribution of *Chosenia arbutifolia* and soil texture of habitats along the Satsunai river (in Japanese). Japanese Journal of Ecology 39:173-182

Noyes JS (1989) The diversity of Hymenoptera in the tropics with special reference to Parasitica in Sulawesi. Ecological Entomology 14:197-207

Ohashi H (2000) A systematic enumeration of Japanese *Salix* (Salicaceae) (in Japanese). Journal of Japanese Botany 75:1-41

Pianka EP (1966) Latitudinal gradients in species diversity: A review of concepts. American Naturalist 100:33-46

Preszler RW, Boecklen WJ (1996) The influence of elevation on tri-trophic interactions: opposing gradients of top-down and bottom-up effects on a leaf-mining moth. Ecoscience 3:75-80

Price PW, Ohgushi T (1995) Preference and performance linkage in a *Phyllocolpa* sawfly on the willow, *Salix miyabeana*, on Hokkaido. Researches on Population Ecology 37:23-28

Price PW, Pschorn-Walcher H (1988) Are galling insects better protected against parasitoids than exposed feeders? a test using tenthredinid sawflies. Ecological Entomology 13:195-205

Price PW, Roininen H (1993) The adaptive radiation in gall induction. In: Wagner MR, Raffa KF (eds) Sawfly life history adaptations to woody plants. Academic Press, Orlando, pp 229-257

Price PW, Fernandes GW, Lara ACF, Brawn J, Barrios H, Wright MG, Ribeiro SP, Rothcliff N (1998) Global patterns in local number of insect galling species. Journal of Biogeography 25:581-591

Roininen H (1991) Temporal change in the location of egg-laying by a bud-galling sawfly, *Euura mucronata*, on growing shoots of *Salix cinerea*. Oecologia 87:265-269

Roininen H, Danell K (1997) Mortality factors and resource use of the bud-galling sawfly, *Euura mucronata* (Hartig), on willows (*Salix* spp.) in arctic Eurasia. Polar Biology 18:325-330.

Roininen H, Danell K, Zinovjev A, Vikberg V, Virtanen R (2002) Community structure, survival and mortality factors in Arctic populations of *Eupontania* leaf gallers. Polar Biology 25:605-611

Roininen H, Nyman T, Zinovjev A (2005) Biology, ecology, and evolution of gall inducing sawflies (Hymenoptera: Tenthredinidae and Xyelidae). In: Raman A, Schaefer CW, Withers TM (eds) Biology, ecology, and evolution of gall-inducing arthropods. Science Publishers, Enfield Plymouth, pp 467-494

Skvortsov AK (1999) Willows of Russia and adjacent countries: taxonomical and geographical revision. University of Joensuu, Joensuu.

ter Braak CJF (1987) Ordination. In: Jongman RHG, ter Braak CJF, van Tongeren OFR (eds) Data analysis in community and landscape ecology. Pudoc, Wageningen, pp 91-173

ter Braak CJF, Looman CWN (1994) Biplots in reduced rank regression. Biometric Journal 36:983-1003

ter Braak CJF, Šmilauer P (1998) CANOCO reference manual and user's guide to Canoco for Windows: software for canonical community ordination (version 4). Microcomputer Power, NY.

Yukawa J (2000) Synchronization of gallers with host plant phenology. Researches on Population Ecology 42:105-113

Yukawa J, Masuda H (1996) Insect and mite galls of Japan in colors (in Japanese, with English explanations for color plates). Zenkoku Nôson Kyôiku Kyôkai, Tokyo

Zinovjev AG (1999) Palearctic sawflies of the genus *Pontania* Costa (Hymenoptera: Tenthredinidae) and their host-plant specificity. Proceeding of an IUFRO Symposium in Mátrafüred, Hungary. USDA Forest Service. General Technical Report NC-199:204-225

Zinovjev AG, Vikberg V (1999) The sawflies of the *Pontania crassispina*-group with a key for the genera of the subtribe Euurina (Hymenoptera: Tenthredinidae, Nematinae). Entomologica Scandinavica 30:281-298

2 Species Richness of Eriophyid Mites on Finnish Trees and Shrubs

Pekka Niemelä[1], Heikki Roininen[2], Henri Vanhanen[1], and Timo O. Veteli[1]

[1]Faculty of Forestry, University of Joensuu, P.O.B. 111, FI-80101 Joensuu, Finland
[2]Department of Biology, University of Joensuu, P.O.B. 111, FI-80101 Joensuu, Finland

Summary. Host plant use and species richness of eriophyid mites (Arthropoda, Acarina, Eriophyiidae) on Finnish trees and shrubs was studied on the basis of published data. The number of eriophyid species ranged from 0 (*Picea*) to 15 (*Alnus*). Most of these mites were concentrated on two species-rich, host plant families, Betulacea and Salicacea. The two families harbor 42% of the total eriophyid fauna of Finnish trees and shrubs. However, *Tilia cordata*, the only species of the Tiliaceae family, had six species of eriophyids. We used the geographic range, average and total frequency (abundance) of the host plant, host plant height, leaf size and the number of host plant relatives (other plants in the same family) as explaining variables in a regression model. Species richness of eriophyid mites was best explained by the leaf size and number of relatives of host plants. These two factors explained 42 % of the variation in species richness. When conifers (which have a low number of eriophyid species) were excluded, the host plant abundance and leaf size explained 66 % of the variation in species richness. The results indicate that resource availability (both leaf size and abundance) is an important factor in increasing the probability of random colonization and adaptive radiation on eriophyid mites living on trees and shrubs.

Key words. Eriophyid mites, Biodiversity, Trees and shrubs, Host plant abundance, Leaf size

2.1 Introduction

The factors affecting the number of herbivore species on host plants was intensively studied in the late 70's and early 80's (e.g. Strong et al. 1984).

Herbivore species richness has attracted increasing attraction along with the rise in biodiversity research in the late 1980's, but it is now called biodiversity.

In most cases the best explaining factor for herbivore species richness has been the range or geographic distribution and/or abundance of the host plant, which has usually explained 20–60% of the variation in species richness (Blanche and Westoby 1996; Kelly and Southwood 1999; Roininen et al. 2005; Tahvanainen and Niemelä 1987). Plant characteristics such as size, structural complexity, time of occurrence in a particular area, number of habitats occupied and degree of taxonomic isolation, also explain part of the variation in insect species diversity (for reviews see Kelly and Southwood 1999; Strong et al. 1984; Tahvanainen and Niemelä 1987).

Most of the studies on herbivore species richness have concentrated on insects, like lepidopterans, and sawflies, some of which are important galling herbivores (Neuvonen and Niemelä 1981; Neuvonen and Niemelä 1983; Kelly and Southwood 1999). However, practically nothing is known about the factors affecting the species richness of eriophyid mites (Arthropoda, Acarina, Eriophyiidae) even though they are very common galling herbivores in both boreal and temperate forests.

Eriophyid mites are tiny (1/6–1/3 mm) herbivorous arthropods belonging to the family Eriophyidae (Liro and Roivainen 1951). Their bodies gradually taper toward the end and somewhat resemble a carrot in shape. They are yellow to pinkish white to purplish in colour, and wedge-shaped with the widest part of the body occurring just behind the head. They have two pairs of legs and two pairs of mandibles. No direct mating is known to happen between males and females, but rather fertilization occurs from the contact of females with sperm sacks laid down on the host by males. It is also assumed that parthenogenesis is common in some species, but evidence other than highly towards females biased sex-ratio is lacking. Up to 80 eggs per female may be laid over a month or so. No eggs are known to overwinter. Dispersal of these slow moving mites is by wind, water, birds, insects, or humans (Liro and Roivainen 1951).

The eriophyids are the most highly adapted of the plant-feeding mites and have evolved extremely intimate associations with their host plants (Krantz and Lindquist 1979). All species feed on leaves, although some attack leaves still in buds. However, several ecological niches exist on a leaf apparently, indicated by occupation of different species on upper or lower leaf surfaces, main veins or leaf lamina, leaf margins, buds, etc. with various forms of galls in each area. Heavily infested leaves take on a silvery or bronze appearance, depending on the species. Plants usually respond to eriophyid mite feeding by forming a tissue barrier around the feeding animal in the form of a gall in a species-specific manner. The species are easily identified by the type of gall and host plant. Species of eriophyid mites can be divided into bud-galling mites, leaf-galling mites, rust mites and true blister mites. Very high populations can cause early defoliation through abscission of the leaves.

The aim of this study is to evaluate the factors affecting the species richness of eriophyid mites on Finnish trees and shrubs.

2.2 Materials and Methods

2.2.1 Number of Eriophyid Species (*S*)

In the regression analysis described below, we used the number of eriophyid species inhabiting a particular tree or shrub species as the dependent variable. The number of eriophyid species was mainly obtained from the manual of Finnish eriophyid species (Liro and Roivainen 1951). Only Finnish records were included.

All the trees and shrubs surveyed in the third national forest inventory (NFI) (Kujala 1964) were included in the analysis. However, in this study we considered only the genus level, due to the inaccuracy of the food plant observations: most of them were reported only at the genus level.

2.2.2 Frequency of Host Plant (*ToF*)

Frequencies of host plants are mainly based on the results of the third NFI (Kujala 1964) from the same period as the manual on Eriophyid mites. Parallel research lines, situated at distances of 13–20 km, extend from south-west to north-east through Finland. The distribution and frequency of plants were depicted in the form of dot maps along the research lines. Dots of five different sizes were used to indicate the frequency: for common plants the occurrence on 1, 2–3, 4–5, 6–7 or 8–10 out of ten 0.1 ha study areas lying 1 km apart, and for less common plants the number of occurrences (1, 2–4, 5–7, 8–10 or over 10 occurrences) on a line 10 m wide and 10 km long. We counted the number of dots in each size class, multiplied the sums by the mean frequencies for the respective classes (for less common plants we used 12 as the mean of the largest, open class), and summed the products to obtain an estimate of total frequency (*ToF*) for each plant species. The sum for less common plants was divided by 10, because the studied area was ten times larger than that used for common plants (see above).

For rare plants not included in Kujala's (1964) study we obtained frequency estimates by multiplying the number of dots in Finland (only native occurrences) in Hultén's (1971) distribution maps by the ratio of the frequency estimate for *Ribes nigrum* in Kujala (1964) to Hultén's (1971) frequency estimate for *R. nigrum*. *R. nigrum* is the only species with these two kinds of data available, thus making the comparison possible. Scots pine (*Pinus sylvestris*) and Norway spruce (*Picea abies*) were also excluded from Kujala's maps. For these tree species we obtained frequency estimates by extrapolating from a regression equation relating the frequency estimates of Kujala to those in the appendix of Kalliola (1973).

The frequency estimate of a plant genus is the sum of species frequencies belonging to the same genus, except for the genus *Salix*, where the frequency is double the sum of those *Salix* species given in Kujala's (1964) maps. This is because only six species of *Salix* were included in Kujala's study, which is roughly the half of common *Salix* species found in Finland.

2.2.3 Range (*R*)

The range of plant species (*R*) is the number of those UTM squares (Universal Transverse Mercator, about 50 × 50 km) where the species is observed in Finland according to Atlas Florae Europae (Jalas and Suominen 1973, 1976) or Hultén (1971). The range of plant genus is the number of UTM squares occupied by all species in a given genus.

2.2.4 Average Frequency (*AvF*)

In order to carry out a more detailed analysis of the effect of range and abundance, the average frequency (*AvF*) was calculated by dividing the total frequency by the range in a similar way to Neuvonen and Niemelä (1981).

2.2.5 Height (*H*)

The height (m) of species (*H*) is the geometric mean of maximum and minimum values for the height given by Hiitonen and Poijärvi (1966) or Lid (1974). The height value of a genus is the height of the tallest species of the genus.

2.2.6 Leaf Size (*LS*)

The leaf size (*LS*, cm^2) was measured from herbarium samples using a planimeter. The value is an average of 3 to 10 leaf samples with an accuracy of 1 cm^2. The leaf size value of a genus is the leaf size of the species with the largest leaf area. We handled the leaflets of species such as *Sorbus* and *Fraxinus* as a separate leaf, due to the fact that a leaflet is a more relevant unit for phytophagous invertebrates.

2.2.7 Number of Relatives (*NoR*)

The number of relatives (*NoR*) is the number of native Finnish tree and shrub species according to Hämet-Ahti et al. (1977) belonging to the same family. The taxonomy follows Strassburger (1967). Very rare species (<5 sites in Finland) are excluded.

2.2.8 Faunal Similarity

The similarities of Finnish Eriophyid faunas of different tree and shrub genera were analysed with PC-ORD (PC-ORD, Multivariate Analysis of Ecological Data, Version 4) using Sorensen's Quotient of Similarity (*QS*: Sorensen 1948)

$$QS = 2c/(a + b), \quad (1.1)$$

in which c = number of species in common, a = number of species on host A, and b = number of species on host B, and group average clustering. We calculated similarities based on the host use of Eriophyid genera and not on the species level because only one of the Finnish Eriophyid species is polyphagous and therefore their host use is almost totally separated. Families with 2 or less eriophyid species were excluded from the analysis.

2.2.9 Data and Statistical Testing

From all of the above variables we generated the dataset shown in Table 1. Statistical tests were performed with SPSS for Windows-software (SPSS Inc. 2002, release 11.5.1). The normality of the data was checked with the Kolmogorov-Smirnov test before any further analysis. As far as all the host plants were concerned, S, *AvF*, *ToF* and *NoR* were not distributed normally and thus these variables were log-transformed. When we tested the normality of the data without coniferous species, only *AvF* and *ToF* needed transformation. We constructed the regression models in which each explanatory variable was treated as an independent variable. We then constructed the models by selecting two variables that significantly explained the highest amount of variance.

Table 1. Host plant characteristics and species number of eriophyid mites living on them. *S*, number of species of Eriophyid mites living on host plants; *R*, range of host plant; *AvF*, average frequency of host plants; *ToF*, total frequency of host plants; *H*, height of host plant; *LS*, leaf size of host plant; *NoR*, number of relatives of host plants belonging to the same family. For more accurate description of the variables, refer to the text

	S	*R*	*AvF*	*ToF*	*H*	*LS*	*NoR*
Picea (Pinaceae)	0	156	57.7	9006	2500	0	2
Pinus (Pinaceae)	1	163	60.3	9833	2000	1	2
Juniperus (Cupressaceae)	1	164	34.7	5699	122	0	0
Taxus (Taxaceae)	1	2	0.6	1	245	1	0
Quercus (Fagaceae)	1	16	0.6	10	2121	24	7
Corylus (Betulaceae)	6	30	0.6	18	424	54	7
Betula (Betulaceae)	10	164	68.5	11229	1500	10	7
Alnus (Betulaceae)	12	164	17.0	2787	1000	35	7
Ulmus (Ulmaceae)	6	23	0.6	14	1936	62	2
Myrica (Myricaceae)	2	60	0.8	50	55	2	0
Ribes (Grossulariaceae)	6	164	1.3	211	120	39	3
Prunus (Rosaceae)	4	155	0.7	107	600	17	16
Sorbus (Rosaceae)	5	164	30.4	4984	632	4	16
Rubus (Rosaceae)	1	141	8.3	1170	120	10	16
Rosa (Rosaceae)	2	143	0.4	57	158	2	16
Crataegus (Rosaceae)	2	7	0.6	4	387	11	16
Hippophae (Eleagnaceae)	2	20	0.6	11	141	4	0
Acer (Aceraceae)	5	36	0.7	26	1414	68	0
Rhamnus (Rhamnaceae)	2	7	0.6	4	345	10	2
Frangula (Rhamnaceae)	1	123	4.9	606	245	10	2
Populus (Salicaceae)	8	164	17.8	2912	1732	19	20
Salix (Salicaceae)	11	164	32.2	5280	545	15	20
Tilia (Tiliaceae)	6	66	0.5	30	1000	25	0
Daphne (Thymelaeaceae)	1	100	0.5	54	79	7	0
Viburnum (Caprifoliaceae)	1	103	0.3	29	245	28	2
Lonicera (Caprifoliaceae)	2	88	0.6	50	141	9	2
Fraxinus (Fraxaceae)	2	25	0.7	17	1732	5	0

2.3 Results

The greatest number of species of eriophyids was concentrated on two species-rich, host-plant families: Betulaceae and Salicaceae. These two families harbor 42% of the total eriophyid fauna of Finnish trees and shrubs (Table 1). Leaf size (*LS*) of the tree and shrub species is the best predictor of species richness (*S*) of eriophyid mites, explaining 27% of the variation alone (Table 2, Fig. 1). No other characteristics of the host plants were significant (Table 2, Fig. 1). However, the model that included both *LS* and *NoR* significantly explained 40% of the variation in species richness (Table 3). This was the best fit model for all the host plant species.

When the coniferous species, which have a very low number of eriophyid species (Table 1), were excluded from the data, three variables significantly explained the species richness of eriophyid mites: *AvF* (44%), *ToF* (39%) and *R* (22%)(Table 2, Fig. 2). The model including both *AvF* and *LS* significantly explained 66% of the variation in species richness (Table 3).

Table 2a, b. Measured characteristics of host plants in independent regression analysis. **a** All tree and shrub species (n = 27). **b** Deciduous trees and shrubs (n = 23). For variable names, see Table 1. **P* < 0.05; ***P* < 0.01

Variable entered		R^2	F
a)			
	Leaf Size	0.27	9.21**
	Number of Relatives	0.13	3.86
	Range	0.04	0.96
	Total Frequency	0.04	0.92
	Average Frequency	0.02	0.53
	Height	0.01	0.12
b)			
	Average Frequency	0.44	16.26**
	Total Frequency	0.39	13.35**
	Range	0.22	5.83*
	Height	0.13	3.04
	Leaf Size	0.13	3.00
	Number of Relatives	0.09	2.04

Table 3a, b. Summary table of the best fit multiple regression models for species richness of eriophyid mites. **a** All tree and shrub species (n = 27). **b** Deciduous trees and shrubs (n = 23). Dependent variable: Species richness of eriophyid mites. For abbreviations see Table 1. $^{**}P < 0.01$; $^{***}P < 0.001$

Model	R^2	F
a)		
$Y = 0.31 + 0.008LS + 0.213NoR$	0.40	7.85**
b)		
$Y = -2.09 + 3.582AvF + 0.086LS$	0.66	19.66***

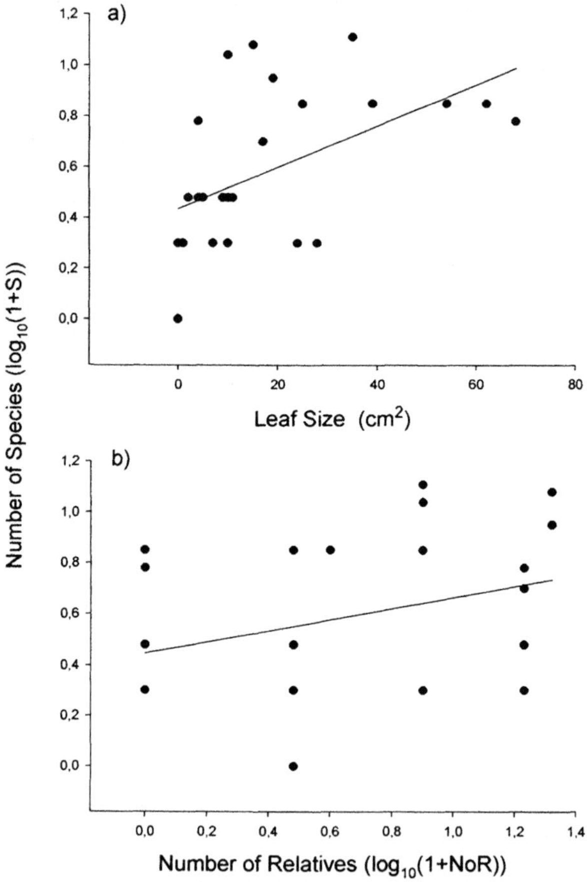

Fig. 1a, b. Correlation between the number of eriophyid species and **(a)** leaf size of host plant (*LS*, see text): $Y = 0.44 + 8.12LS$; $r = 0.52$ ($P < 0.01$); and **(b)** number of host plant relatives (*NoR*, see text): $Y = 0.45 + 0.22NoR$, $r = 0.36$ ($P < 0.01$) on all Finnish tree and shrub species.

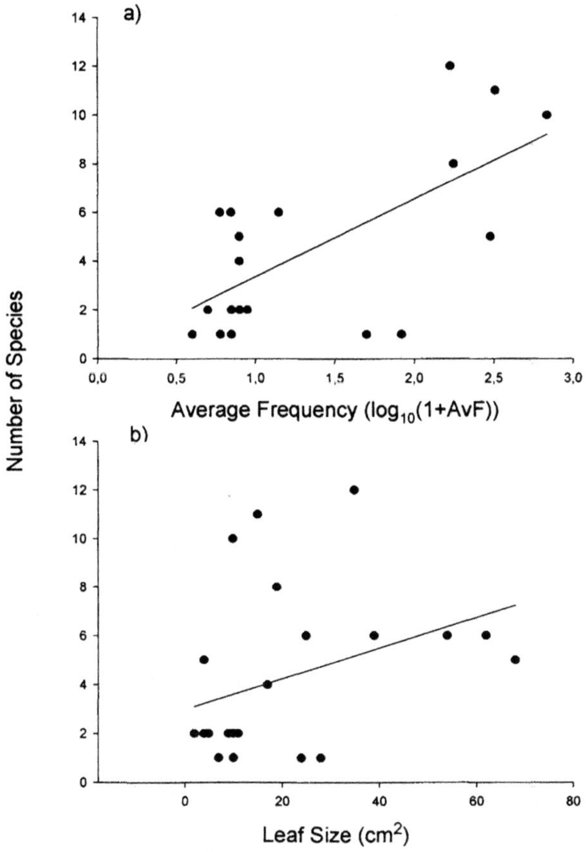

Fig. 2a, b. Correlation between the species richness of eriophyid mites and **(a)** average frequency of host plant (*AvF*, see text): $Y = 0.17 + 3.19AvF$, $r = 0.66$ ($P < 0.01$); and **(b)** leaf size of host plants (*LS*, see text): $Y = 2.98 + 0.06LS$, $r = 0.35$ ($P < 0.01$) on Finnish deciduous tree and shrub species.

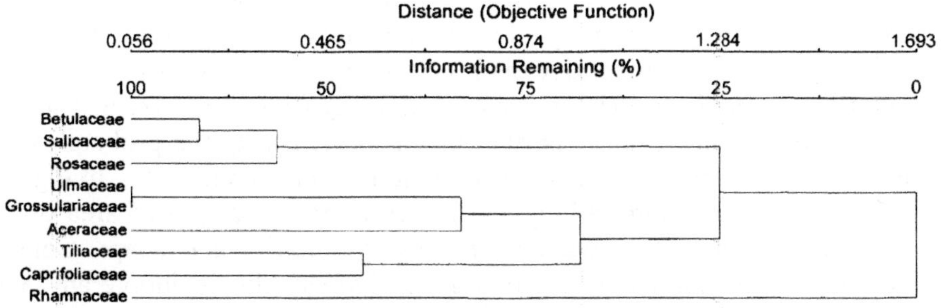

Fig. 3. Similarity of eriophyid species between Finnish tree and shrub genera.

In faunal similarity analysis, common and widespread plant families, Betulaceae, Salicaceae and Rosaceae, formed a clear cluster (Fig. 3). Another clear cluster was formed by Ulmaceae and Grossulariaceae. Rhamnaceae differed from other families by forming a single family-cluster.

2.4 Discussion

The families Betulaceae and Salicaceae had the highest total number of eriophyid mite species. Species richness is also high on Betulaceae when counted by the species in the family, but in Salicaceae it represents highest total numbers for all colonized host families. The exception is *Tilia cordata* (Tiliaceae), which is the only species in the family, but still had six eriophyid species. Plants belonging to families such as Betulaceae and Salicaceae are dominating trees and shrubs in boreal forests, indicating that abundant plants seem to have the highest number of species. The results of the regression analyses indicated that, when all the tree species and shrubs are taken into account, leaf size of the host plant explained 27% of the variation in species richness of eriophyid mites. The number of relatives of host plants best explained the residual variation of species richness, and increased the coefficient of determination to 40%. An even stronger pattern emerged when the coniferous species, which have very low number of eriophyid species, were excluded from the data. The average frequency explained 44% of the variation and the residual variation was best explained by leaf size, which improved the fitness of the model to 66%.

It could be argued that common plants are more carefully studied than rare plants, thus explaining the high number of species found on Betulaceae and Salicaceae. However, a similar analysis on lepidopterans indicated that underestimation of species richness on rare host plants did not change the overall pattern (Niemelä and Neuvonen 1983). Similarly, Kelly and Southwood (1999) found that in the very well known insect fauna of the British Islands the species richness was best explained by host abundance. They found the same pattern in many taxonomic groups of insects, indicating a lack of competitive exclusion among different taxa.

Our results indicate that resource availability (both leaf size and abundance of host plant) is an important factor in increasing the probability of random colonization and adaptive radiation on eriophyid mites living on trees and shrubs. Being galling sessile herbivores, a large leaf area obviously offers them more available niches to colonize and adapt. Consequently, they differ from free-living herbivores, like lepidopterans and sawflies where leaf size does not seem to be a significant factor explaining the species richness (Neuvonen and Niemelä 1981, 1983). We need more studies on other galling or relatively sessile herbivore groups in order to confirm this pattern.

Eriophyids usually are highly host specific, a factor that may limit niche exploitation and promote rapid speciation (Kranz and Lindquist 1979). The similarity analysis revealed clusters based on plant abundance (Betulaceae, Salicaceae, Rosaceae) and clusters based on the same host-plant habitat (Ulmaceae, Grossulariaceae, Tiliaceae, Caprifoliaceae). Thus, the similarity of host use does not mirror evolutionary history measured as taxonomic relatedness of the host plants very well, but evidently shows some evolutionary constraints. Host plant use in the genus *Callyntrotus* is restricted to *Lonicera*. However, there are several genera (*Eriophyes* subgenera *Aceria* and *Eriophyes*, *Phyllocoptes* subgenera *Vasates*) have species living on different plant families, indicating that radical steps in host use have occurred. Generally, plant taxonomy has been found to be a small but recognizable factor influencing the evolution and host swifts of herbivorous insects, especially in the case of sawflies (Kelly and Southwood 1999; Neuvonen and Niemelä 1983; Roininen et al. 2005). Although Fenton et al. (2000) found that the phylogeny of seven species of *Cecidophyopsis* mites living on *Ribes* species did not correlate with the phylogeny of *Ribes* hosts.

2.5 Acknowledgements

We thank the Finnish Centre of Excellence program funded by the Academy of Finland, project no. 64308. We are also grateful to John Derome for checking the language of this article.

2.6 References

Blanche KR, Westoby M (1995) The effect of the taxon and geographic range size of host eucalypt species on the species richness of gall-forming insects. Australian Journal of Ecology 21:332-335

Fenton B, Birch ANE, Malloch G, Lanham PG, Brennan RM (2000) Gall mite molecular phylogeny and its relationship to the evolution of plant host specificity. Experimental and Applied Acarology 24:831-861

Hiitonen I, Poijärvi A (1966) Koulu- ja retkeilykasvio. 15th ed. Otava, Helsinki

Hultén E. (1971) Atlas of the distribution of vascular plants in north-western Europe. 2nd ed. Generalstabens litografiska anstalts förlag, Stockholm

Hämet-Ahti L, Jalas J, Ulvinen T (1977) Suomen alkuperäiset ja vakiintuneet putkilokasvit 15th ed. Otava, Helsinki
Jalas J, Suominen J (1973) Atlas florae Europeae. Distribution of vascular plants in Europe. 2. Gymnospermae (Pinaceae to Ephedraceae). Maps 151-200. Suomalaisen kirjallisuuden kirjapaino, Helsinki
Jalas J, Suominen J (1976) Atlas florae Europeae. Distribution of vascular plants in Europe. 3. Gymnospermae (Salicaceae to Balanophoraceae). Maps 201-383. Suomalaisen kirjallisuuden kirjapaino Helsinki
Kalliola R (1973) Suomen kasvimaantiede. Werner Söderström, Porvoo-Helsinki
Kelly CK, Southwood TRE (1999) Species richness and resource availability: A phylogenetic analysis of insects associated with trees. Proceedings of the National Academy of Science, USA 96:8013-8016
Krantz GW, Lindquist EE (1979) Evolution of phytophagous mites (Acari). Annual Review of Entomology 24:121-158
Kujala V (1964) Metsä- ja suokasvilajien levinneisyys- ja yleisyyssuhteita Suomessa. Vuosina 1951-1953 suoritetun valtakunnan metsien III linja-arvioinnin tuloksia. Communicationes Instituti Forestalis Fenniae 59:1-137, maps 1-196
Lid J (1974) Norsk og Svensk Flora. 2nd ed. Det Norske Samlaget, Oslo
Liro JI, Roivainen H (1951) Suomen eläimet, Animalia Fennica 6, Äkämäpunkit, Eriophyidae. Werner Söderström, Porvoo-Helsinki
Neuvonen S, Niemelä P (1981) Species richness of Macrolepidoptera on Finnish deciduous trees and shrubs. Oecologia 51:364-370
Neuvonen S, Niemelä P (1983) Species richness and faunal similarity of arboreal insect herbivores. Oikos 40:452-459
Niemelä P, Neuvonen S (1983) Species richness of herbivores on hosts: how robust are patterns revealed by analysing published host plant lists? Annales Entomologici Fennici 49:95-99
Roininen H, Nyman T, Zinovjev A (2005) Biology, ecology, and evolution of gall inducing sawflies (Hymenoptera: Tenthredinidae and Xyelidae). In: Raman A, Schaefer CW, Withers TM (eds) Biology, ecology, and evolution of gall-inducing arthropods. Science Publishers, Enfield Plymouth, pp 467-494.
Sorensen T (1948) A method of establishing groups of equal amplitude in plant sociology based on similarity of species content and its application to analyses of vegetation of Danish commons. Biologiske Skrifter 5:1-34.
Strassburger E (1967) Textbook of botany. Longman, London
Strong DR, Lawton JH, Southwood R (1984) Insects on plants, community patterns and mechanisms. Blackwell Scientific Publications, Southhampton
Tahvanainen J, Niemelä P (1987) Biogeographical and evolutionary aspects of insect herbivory. Annales Zoologici Fennici 24:239–247

3 Diversity, Biology, and Nutritional Adaptation of Psyllids and their Galls in Taiwan

Man-Miao Yang[1], Ling-Hsiu Liao[1], Mei-Fiang Lou[1], Wei-Chung Chen[1], Shih Shu Huang[1], Gene-Sheng Tung[2], Yu-Chu Weng[1], and Chia-Chi Shen[1]

[1]Department of Entomology, National Chung Hsing University, Taichung 40227, Taiwan
[2]Division of Forest Protection, Taiwan Forestry Research Institute, 53 Nan-Hai Road, Taipei 10053, Taiwan

Summary. Psyllids, or jumping plant lice, are a group of sternorrhynchous Hemiptera. They are highly host specific sucking insects, and many of them form galls. In Taiwan, a revision and some supplemental works of psyllid taxonomy were done in the 1980s. In these publications, 18 species of psyllids were noted as being gall formers. Our survey, as of 1996 found a total of 98 types/species of psyllid galls, and many of them are new species/records. The host range of these gall forming psyllids belongs to a wide spectrum of vascular plants, consisting of 28 families, 45 genera, and 89 species. Galls are most abundant on Lauraceae and Myrtaceae. The psyllid galls as a whole are highly diverse in both gall shape and galling position, but they are mainly species specific. Life histories and the differentiation between gall tissue of several galling psyllids on *Machilus*, *Cinnamomum* (Lauraceae) and *Ficus* (Moraceae) are studied and compared. In these psyllid galls, different from normal plant tissue, phloem in the vesicular bundles is oriented closer to the larval chamber than the xylem in the enclosed type of galls. The psyllid species *Trioza shuiliensis* which induce globular galls on the leaves of *Machilus japonica* var. *kusanoi* were used as a model system for testing the nutritional hypothesis of gall adaptation. The results of this investigation support the nutritional sink hypothesis.

Key words. Psyllid, Gall, Diversity, Tissue differentiation, Nutritional adaptation

3.1 Galling Psyllids in Taiwan

Psyllids, or jumping plant lice, are a group of sternorrhynchous Hemiptera. They are highly host specific sucking insects, and many of them form galls. The fauna of Taiwanese psyllids was first established by foreign workers, especially by the Japanese (e.g. Kuwayama 1931) when they occupied Taiwan during the first half of the twentieth century. Exhaustive revisory works were later carried out by professor Chung Tu Yang and his students in the 1980s. The monograph on the psyllids of Taiwan by Yang (1984) described 108 adults and 81 nymphs and also provided reliable host plant records as well as some biological notes. Some earlier miscellaneous works (Yang and Tsai 1980a, b, c) and following supplementary works continued the contribution to the taxonomic revision (e.g. Fang 1990; Fang and Yang 1986; Fang et al. 1997; Lauterer et al. 1988; Yang and Fang 1986; Yang et al. 1986). Among these literatures, 19 species of psyllids were noted as being gall formers (Table 1). Our survey, which started in 1996, found a total of 120 types of psyllid galls when each particular shape of gall on the different parts of each plant species was counted as one type. Among these, we recognized 98 psyllid species with many of them being new species or new records. In this paper, we provide an overview of these galling psyllids in Taiwan based on the diversity of gall shapes, plant parts utilized, host plant species, and plant tissue differentiation. The tests of the nutritional hypothesis of gall adaptation are also discussed.

3.2 Host Spectrum of Psyllid Galls in Taiwan

The hosts of the Taiwanese gall-forming psyllids belong to a wide spectrum of vascular plants, consisting of 28 families, 45 genera and 89 species (Table 2). Galls are most abundant on Lauraceae, Myrtaceae, Moraceae, and Aquafoliaceae in terms of number of host species and number of gall types found.

3.3 Diversity of Galls Formed by Psyllids in Taiwan

The psyllid galls as a whole are highly diverse in gall shape and galling position, but they are specific for each species. Galls may be as simple as small pits on a leaf blade to several deep pits close together forming a complex large pit. It may show up as a slight curling of leaf edges to profuse rolling-up of certain areas of the edge of a leaf, forming a specific

Table 1. Formerly recorded gall-forming psyllids in Taiwan

Psyllid species	Host plant	Galling position and gall type	References
Cecidotrioza sozanica Boselli, 1930	*Daphniphyllum pentandrum* Hayata var. *pentandrum* (Daphniphyllaceae)	Deep pit galls on leaves	Yang 1984, p.234
Chineura alba Yang & Tsay, 1980	*Canarium album* (Lour.) Racusch. (Burseraceae)	Small pits on both sides of leaves	Yang and Tsai 1980a, p.66; Yang 1984, p.164
Eustenopsylla euryae Yang, 1984	*Eurya strigillosa* Hayata (Theaceae)	Leaf-folding	Yang 1984, p.202
Homotrioza beilschmiediae Yang, 1984	*Beilschmiedia erythrophloria* Hay. (Lauraceae) Hay. (Lauraceae)	Round galls on leaves	Yang 1984, p.225
Homotrioza epica Yang, 1984	*Symplocos morrisonicola* Hayata. (Symplocaceae)	Leaf-folding	Yang 1984, p.223
Neophacopteron euphoriae Yang, 1984	*Euphoria longana* Lam. (Sapindaceae)	Deep pit galls on young leaves	Yang 1984, p.167
Parastenospylla proboscidaria (Yu, 1956)	*Vaccinitum Wrightii* Gray (Vacciniaceae)	Leaf-rolling	Yang 1984, p.206
Parastenospylla vacciniae Yang, 1984	*Vaccinium randaiense* Hayata. (Vacciniaceae)	Leaf-rolling	Yang 1984, p.208
Pseudotrioza mallotcola Yang, 1984	*Mallotus philippinensis* (Lam.) Muell.-Arg. (Euphorbiaceae).	Round galls on leaves	Yang 1984, p.232
Psylla miultijuga Yang, 1984	*Pittosporum illicioides* Makino (Pittosporaceae)	Rolling galls on leaves	Yang 1984, p.101
Trioza (*Megatrioza*) *beilschmiediae* Yang, 1984	*Beilschmiedia erythrophloria* Hay. (Lauraceae) Hay. (Lauraceae)	Large galls on young leaves and branch top	Yang 1984, p.257
Trioza camphorae Sasaki, 1905	*Cinnamomum camphora* (Lauraceae)	Oval or roundish galls on surface of leaves	Kuwayama 1931, p.129
Trioza caseariae Yang, 1984	*Casearia membranacea* Hance (Flacourtiaceae)	Marginal fold galls on leaves	Yang 1984, p.265
Trioza elaeocarpi Yang, 1984	*Elaeocarpus sylvestris* (Lour.) Poiret. (Elaeocarpaceae)	Round galls on leaves	Yang 1984, p.277
Trioza euryae Yang, 1984	*Eurya japonica* Thunb. (Theaceae)	Leaf-folding	Yang 1984, P281
Trioza (*Megatrioza*) *neolitsae* Miyatake, 1965	*Mallotus philippinensis* (Lam.) Muell.-Arg. (Euphorbiaceae).	Circular gall on leaves	Yang 1984, p.251
Trioza neolitseacola Yang, 1984	*Neolitsea parvigemma* (Hay.) kaihira. (Lauraceae).	Pit galls on leaves	Yang 1984, p.286
Trioza shuiliensis Yang, 1984	*Machilus japonica* Sieb. & Zucc. var. *kusanoi* (Hyata) Liao	Round gall on leaves	Yang 1984, P.216
Trioza outeiensis Yang, 1984	*Eugenia* sp. (Myrtaceae)	Deep pit galls of leaves	Yang 1984, p.289

Table 2. The abundance of psyllid galls found on each host-plant family in Taiwan

Host family	No. of host species	No. of galls
Lauraceae	25	30
Myrtaceae	10	13
Moraceae	8	8
Aquifoliaceae	6	6
Theaceae	3	4
Proteaceae	3	3
Sapotaceae	3	3
Symplocaceae	3	3
Elaeocarpaceae	3	3
Flacourtiaceae	2	2
Lardizabalaceae	2	2
Fabaceae	2	2
Clusiaceae	2	2
Fagaceae	2	2
Sapindaceae	2	2
Euphorbiaceae	1	1
Araliaceae	1	1
Styracaceae	1	1
Daphniphyllaceae	1	1
Ebenaceae	1	1
Pittosporaceae	1	1
Sabiaceae	1	1
Asteraceae	1	1
Ulmaceae	1	1
Meliaceae	1	1
Caprifoliaceae	1	1
Actinidiaceae	1	1
Acanthaceae	1	1
Total	89 species 45 genera 28 families	98 types/species

purse-like gall, or it can appear as various enclosed type of galls, such as spherical, conical, and crown-like shapes (Yang 1999).

Almost all parts of the plant are utilized by galling psyllids. Galls are found on leaves, petioles, buds, stems, flowers, and fruits (Fig. 1). However, the leaves count as the major galling part (74.2%), followed by the buds (9.2%) and then the stems (8.3%).

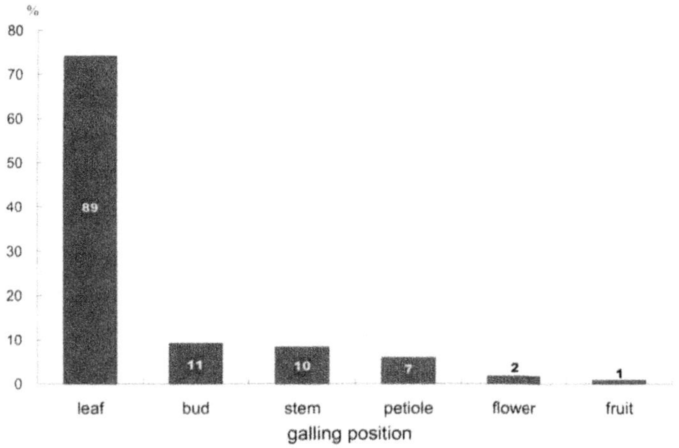

Fig. 1. Percentage of different gall types found on each plant part formed by galling psyllids in Taiwan. *The number on each bar* indicates the number of gall types found on each specific plant part.

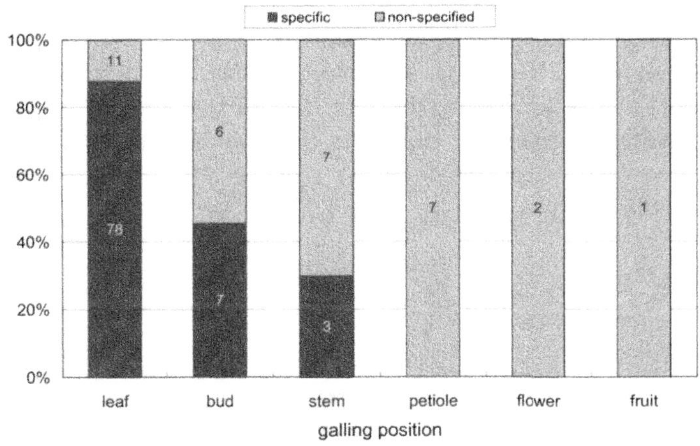

Fig. 2. Specificity of galling position for each plant part. *The number inside the bar* refers to the number of gall types.

The galling position is mostly species specific, with one species normally forming galls on one specific plant part. Eighty-six out of 98 galling species are specific to one plant part. If each of the 120 gall types were to be used as a basic unit to analyze the specificity of different galling parts, then leaf galls are the most common when compared to other galling positions (Fig. 2). Most of the non-specific galling occurs along the vesicular bundles, i.e. on the leaf basal part, major veins, petioles and stems.

3.4 Life History of Galling Psyllids and their Biology

Most of the galling psyllids in Taiwan are univoltine (Huang 2003; Liao 2003). The life history of *Trioza shuiliensis* (Fig. 3, 4A) forming globular galls on *Machilus japonica* var. *kusanoi* (Lauraceae) exemplifies this typical type. However, others are multivoline (Chen 2005; Huang 2003), and *Pauropsylla triozoptera* forming conical galls (Fig. 5A) on *Ficus ampelas* and *F. irisana* (Moraceae) represents this particular type of life strategy (Chen 2005). And, almost all of the different developmental stages of galls and nymphs can be found on one tree at any specific time.

3.5 Diversification of Gall Tissue and Nutritional Adaptation

Most psyllid galls are monolocular, but sometimes more than one individual may be found in one gall when first instar nymphs initiate galls nearby (Chen 2005; Huang 2003; Liao 2003). The differentiation of gall tissue varies, ranging from organoid, kataplasmatic (Fig. 5C, D) (Chen 2005), to prosoplasmatic (Fig. 5A, B) (Liao 2003) galls based on Küster's category (Küster 1911). It is interesting to note though that the phloem in the vesicular bundles is oriented closer to the larval chamber than to the

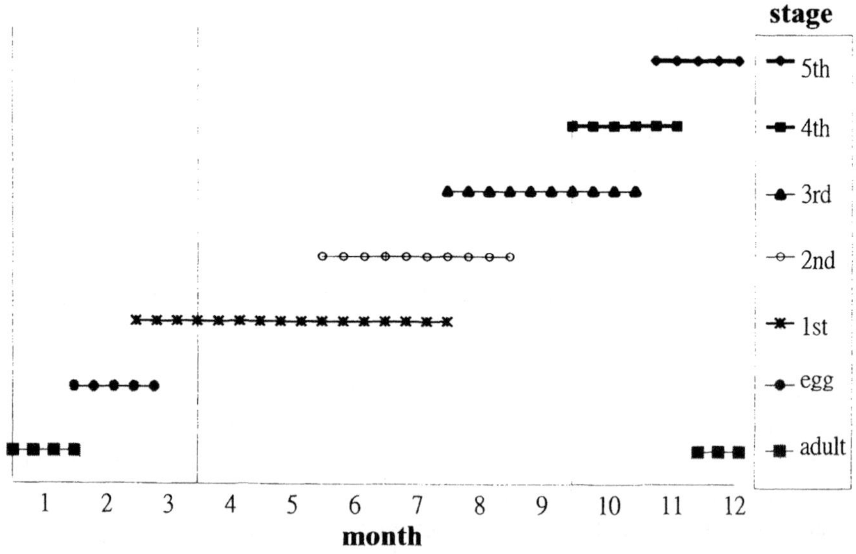

Fig. 3. Life history of psyllid *Trioza shuiliensis* (Yang) on *Machilus japonica* var. *kusanoi* (Hay.) Liao.

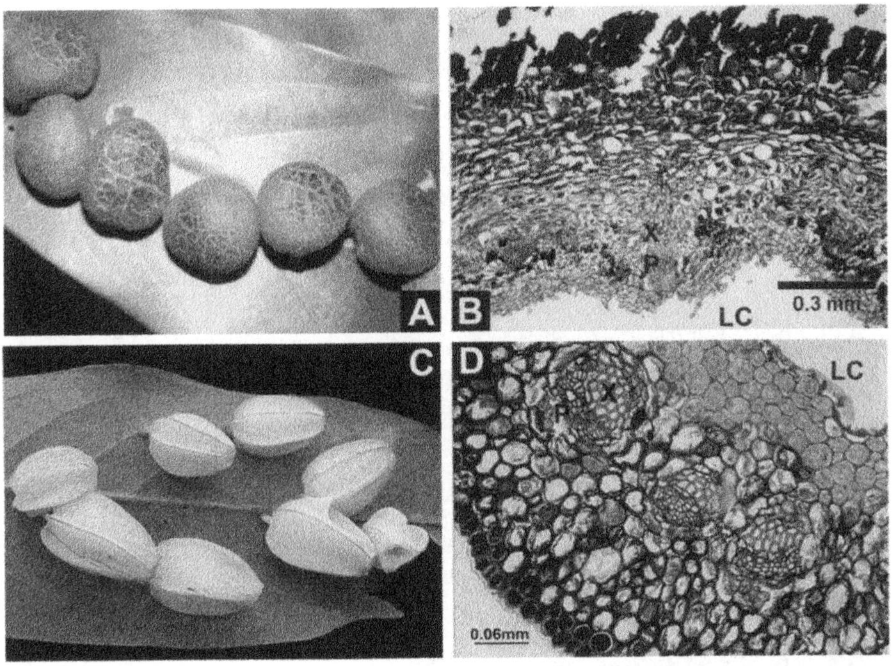

Fig. 4. The positions of xylem and phloem relative to the larval chamber in the galls of psyllids and cecidomyiids are different. **A** The mature galls of *Trioza shuiliensis* (Yang) on *Machilus japonica* var. *kusanoi* (Hay.) Liao. **B** The section of gall tissue of A. **C** The mature galls of *Daphnephila* cecidomyiids on *Machilus zuihoensis* Hayata. **D** The cross section of the gall of C. *X*, xylem; *P*, phloem; *LC*, larval chamber.

xylem, which is the reverse from the normal plant tissue, in the galls on *Machilus* (Fig. 4) (Liao 2003) and also other psyllid galls on different plant genus and families (Fig. 5) (Chen 2005; Tung 1997; Tung et al. 1998). Contrary to other cecidomyiid galls also on *Machilus*, the vesicular orientation remains the same as on usual plants (Weng 2003).

The psyllid species *Trioza shuiliensis* which induce globular galls (Fig. 4A) on the leaves of *Machilus japonica* var. *kusanoi* was selected as a model system for testing the nutritional hypothesis of gall adaptation (Liao 2003). The gall tissues did not show having the ability to work normally in photosynthesis based upon the analyses of the contents of chlorophylls and their derivatives. High contents of nutritional component were found in the gall tissue and they had to come from other tissues rather than the galls themselves. This result supports the hypothesis that gall tissue may create a sink to provide galling insects with more nutrition than free-living herbivores.

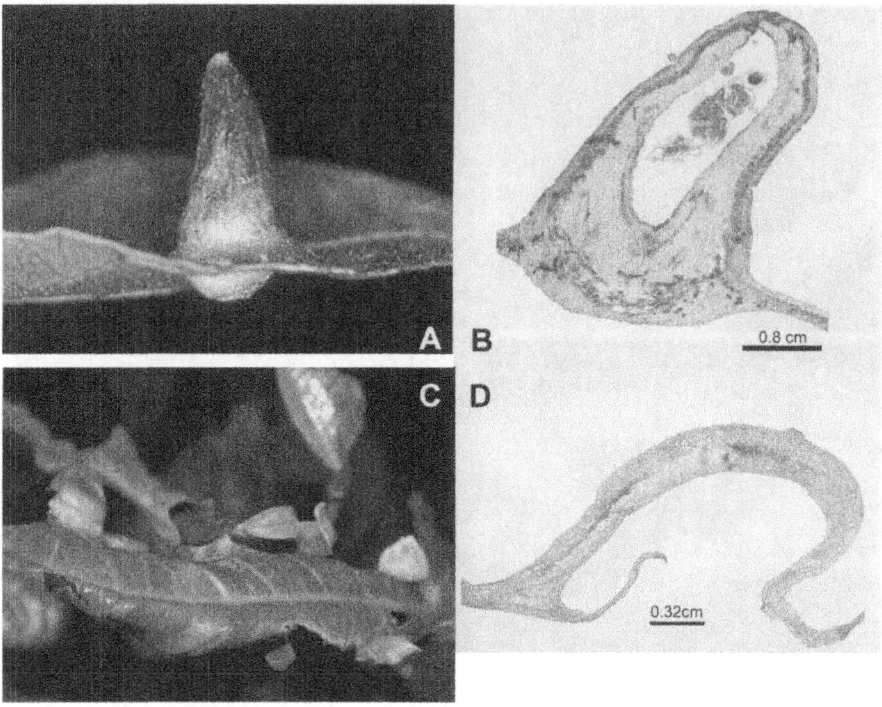

Fig. 5. Two psyllid galls represent different types of tissue differentiation. **A** The mature gall of *Pauropsylla triozoptera* Crawford on *Ficus ampelas* Burm. f. **B** A section of gall tissue of A represents the prosoplasmic gall. **C** The mature gall of *Dynopsylla pinnativena* (Enderlein) on *Ficus nervosa* Heyne. **D** A section of gall tissue of C exemplifies the kataplasmic gall.

3.6 Concluding Remarks

The diversity of psyllid galls in Taiwan is high, and shows not only in the number of galls found but also in the way they utilize plant resources. They provide an ideal system for further studies in gall evolution.

3.7 Acknowledgements

The authors are grateful to H. F. Yen, C. M. Wang, C. H. Ou, T. Y. Yang and many botanists of various institutes in Taiwan for their help in plant identification and the many friends who assisted with the collection of galls. We are indebted to anonymous reviewers for making helpful comments. This research was supported by the National Museum of Natural Science, and the National Science Council of Taiwan (NSC 93-2621-B005-005; NSC 87-2311-B-178-002).

3.8 References

Chen WC (2005) Gall-forming psyllids (Hemiptera: Psylloidea) on *Ficus* spp. in Taiwan and the anatomy of gall tissue (in Chinese). MS thesis, National Chung Hsing Unviersity, Taichung

Fang SJ (1990) Psylloidea of Taiwan supplement II (Homoptera). Journal of Taiwan Museum 43:103-117

Fang SJ, Yang CT (1986) Psylloidea of Taiwan (Homoptera: Sternorrhyncha) Supplement. Monograph of Taiwan Museum 6:119-176

Fang, SJ, Chang SH, Shih SP (1997) *Arytaina yangi* (Hemiptera, Psylloidea, Psyllidae), a new pest of *Cajanus cajan*. Journal of Agricultural Research, China 46:383-387

Huang SS (2003) Taxonomy and the comparison of gall-forming habits among pit galls induced by psyllids (Hemiptera: Psylloidea) on eleven host plants in Taiwan (in Chinese). MS thesis, National Chung Hsing Unviersity, Taichung

Küster E (1911) Die Gallen der Pflanzen. S. Hirzel, Leipzig

Kuwayama S (1931) A revision of the Psyllidae of Taiwan. Insecta Matsumurana 5:117-133

Lauterer P, Yang CT, Fang SJ (1988) Changes in the nomenclature of five species of psyllids from Taiwan (Homoptera: Psylloidea), with notes on the genus *Bactericera*. Journal of Taiwan Museum 41:71-74

Liao LH (2003) Nutritional adaptation of galling insects investigated by globular galls of *Trioza shuiliensis* (Yang) on *Machilus japonica* var. *kusanoi* (Hayata) Liao (in Chinese). MS thesis, National Chung Hsing Unviersity, Taichung

Tung GS (1997) Gall diversity of Lauraceae and the development of a psyllid gall in *Cinnamomum osmophloeum* Kaneh. MS thesis, National Taiwan University, Taipei

Tung GS, Yang MM, Yang PS (1998) The development of an oval-shaped psyllid gall in *Cinnamomum osmophloeum*. (Lauraceae). In: Csoka G, Mattson WJ, Stone GN, Price PW (eds) The biology of gall-inducing arthropods. General Technical Report NC-199. USDA Forest Service, North Central Research Station, St. Paul, pp 193-195

Weng YC (2003) Comparative biology of five types of cecidomyiid galls on *Machilus* in central Taiwan (in Chinese). MS thesis, National Chung Hsing Unviersity, Taichung

Yang CT (1984) Psyllidae of Taiwan. Taiwan Meseum special publication series, vol 3. Taiwan Museum,Taipei

Yang CT, Fang SJ (1986) A serious pest, *Heteropsylla cubana*, of Pacific Islands (Homoptera: Psylloidea: Pauropsyllidae). Journal of Taiwan Museum 39:59-62

Yang CT, Tsay CI (1980a) A new species of genus *Chinerua* (Homoptera, Psyllidae) from Taiwan. Proceedings of the National Science Council 4:65-68

Yang CT, Tsay CI (1980b) On the immature stages of nine psyllid type-species from Taiwan (Homoptera: Psyllidae). Bulletin of the Society of Entomology 15:285-308

Yang CT, Tsay CI (1980c) Immature stages of three species of genus *Epipsylla* (Homoptera: Psyllidae). Proceedings of the National Science Council 4:418-423

Yang MM (1999) An overview of gall-forming psyllids of Taiwan (Hemiptera: Psylloidea) (in Chinese). Proceedings of the application of insect identification in plant protection and quarantine. Special Issue of the Entomological Society of the Republic of China, 11:39-47

Yang MM, Yang CT, Chao JT (1986) Reproductive isolation and taxonomy of two Taiwanese *Paurocephala* species (Homoptera: Psylloidea). Monograph of Taiwan Museum 6:176-203

4 Trophic Shift in $\delta^{15}N$ and $\delta^{13}C$ through Galling Arthropod Communities: Estimates from *Quercus turbinella* and *Salix exigua*

Christopher T. Yarnes and William J. Boecklen

Laboratory of Ecological Chemistry, Department of Biology, New Mexico State University, Las Cruces, New Mexico 88003, USA

Summary. Galling arthropod communities have long been a model system for community ecologists, yet much remains to be explored concerning trophic interactions between hosts, herbivores, and natural enemies. While the utilization of stable isotope ratios can help to elucidate complex trophic interactions in such communities, estimates of trophic shift between community members are required before stable isotope analyses can be appropriately employed. In this chapter, we document the degree of trophic shift in carbon ($\delta^{13}C$) and nitrogen ($\delta^{15}N$) isotopes within galls and hosts of a cynipid gall wasp (Cynipidae) in *Quercus turbinella* (Fagaceae), and a gall midge (Cecidomyiidae) and sawfly (Tenthredinidae) on *Salix exigua* (Salicaceae). We found trophic shift in nitrogen isotopes to be reduced relative to estimates from other systems, while carbon isotopes were considerably enriched. In combination with our current results, we review estimates of trophic shift in gall communities and compare patterns of trophic shift across studies. We discuss physiological mechanisms that determine the distribution of stable isotopes throughout gall communities and their potential effect on estimates of trophic shift ($\delta^{13}C$ and $\delta^{15}N$).

Key words. Stable isotope, Gall midge, Gall wasp, Trophic shift

4.1 Introduction

Arthropod galls often harbor diverse, complex communities of parasitoids, hyperparasitoids, inquilines, and predators. These closed communities have long served as empirical models for community ecology, yet much remains to be explored surrounding the role of direct and indirect trophic interactions in structuring communities (Plantard et al. 1996; Price et al.

1980; Raman et al. 2005; Roininen et al. 1996; Stone et al. 1995; Washburn and Cornell 1981). The complex trophic interactions within galls may be elucidated through the use of stable-isotope techniques. Stable isotopes have proven useful in determining the structure of complex communities and the trophic position of species in other biological systems (Post 2002), as well as revealing important population-level aspects of organism nutrition (O'Brien et al. 2002). This has been particularly true for trophic systems that are not amenable to more traditional dietary or behavioral analysis (Blüthgen et al. 2003; Callaham et al. 2000). The use of stable isotopes in the examination of trophic interactions in other insect communities has been fruitful (i.e. Callaham et al. 2000; Markow et al. 2000; McNabb et al. 2001).

The successful application of stable isotopes to trophic interactions requires a valid a priori expectation of *trophic shift* (McCutchan et al. 2003; Post 2002; Vander Zanden and Rasmussen 2001) — patterns of consumer–diet fractionation (Δ; e.g., $\Delta = \delta^{13}C_{\text{consumer}} - \delta^{13}C_{\text{diet}}$) in carbon ($\Delta\delta^{13}C$) and nitrogen ($\Delta\delta^{15}N$) isotopes (DeNiro and Epstein 1978, 1981) between trophic levels. Nitrogen generally becomes steadily more enriched (greater relative amount of the heavier isotope, e.g., $^{15}N/^{14}N$) at higher trophic levels. The overall level of enrichment is determined by exogenous and endogenous variation in nitrogen assimilation and excretion (Steele and Daniel 1978). Carbon becomes only slightly enriched across trophic levels, and less reliably so than nitrogen; carbon isotope enrichment is primarily balanced by the ratio of respiration to growth (McCutchan et al. 2003).

Despite these generalities, the degree of trophic shift in nitrogen and carbon through food webs is known to be variable between systems due to the underlying physiological variation between organisms (McCutchan et al. 2003; Post 2002; Vanderclift and Ponsard 2003). Further, the trophic positioning of communities with high levels of omnivory, cannibalism, or parasitism is especially susceptible to error in assumptions concerning $\Delta\delta^{13}C$ and $\Delta\delta^{15}N$. It is important to obtain quality estimates of the isotopic baseline of the community (Vander Zanden and Rasmussen 2001), as determined by the primary consumers, and obtain estimates of trophic shift for secondary consumers. Additionally, progress depends on the documentation of patterns in the field and the integration with mechanistic physiological studies across a wide range of organisms (Gannes et al. 1997; Martínez del Rio and Wolf 2004), including endophagous insects and their parasitoids.

Patterns in the trophic shift of stable isotopes have only recently been reported for gall-formers or their associated parasitoids (Langellotto et al. 2005; Tooker and Hanks 2004; Yarnes et al. 2005). The establishment of an isotopic baseline of primary consumers (gall-formers) and quality estimates of trophic shift in gall-forming communities are critical. Global estimates of trophic shift based on metadata compiled across different taxa may be invalid for gall communities due to physiological differences between organisms (Martínez del Rio and Wolf 2004). In this chapter, we establish the baseline isotopic composition of an oak gall cynipid, *Andricus reticulatus* (Hymenoptera: Cynipidae), in the oak *Quercus turbinella* Greene, and for two separate taxa in *Salix exigua* Nuttall, *Rhabdophaga strobiloides* (Diptera: Cecidomyiidae) and *Euura exiguae* (Hymenoptera: Tenthredinidae). We also document patterns of trophic shift in carbon and nitrogen ($\Delta\delta^{13}C$, $\Delta\delta^{15}N$) for a torymid parasitoid in *R. strobiloides*. We compare the degree of $\Delta\delta^{15}N$ and $\Delta\delta^{13}C$ in these communities relative to global estimates constructed from a wide variety of ecosystems, and estimates from other gall-forming arthropods. We also discuss physiological characteristics of galling arthropods important to the analysis of trophic shift in gall communities.

4.2 Trophic Shift in Gall-forming Arthropods in *Quercus turbinella* and *Salix exigua*

4.2.1 Collection of Galls

On October 15th 2004, *A. reticulatus* galls were collected from *Q. turbinella* near Aguirre Springs Campground in the Organ Mountains, Dona Ana County, New Mexico U.S.A. Here *Q. turbinella* forms dunes of shrubs, 1–2 m. Six trees were haphazardly chosen from a small stand of *Q. turbinella* and examined for cynipid galls. All galled leaves were collected from each tree (range: 1–32 galls·species^{-1}·tree^{-2}). Galls were transported to the laboratory, placed in Petri dishes, and monitored for galler and parasitoid emergence.

On July 13th 2005, *E. exiguae* and *R. strobiloides* were collected from *S. exigua* clones along the Rio Grande south of the Picacho Street Bridge in northwestern Las Cruces, Dona Ana County, New Mexico U.S.A. *S. exigua* forms expansive clones 2–3 m in height and care was taken to sample from distinct clones. Ten clones bearing galls were examined for *E. exiguae* and *R. strobiloides* galls. Galls were placed on ice during transport and samples processed immediately. In *R. strobiloides*, an unknown torymid parasitoid (Torymidae) was found to emerge from mature bud galls on *S. exigua*.

4.2.2 Analysis of Trophic Shift in Gall Communities

Samples were prepared and analyzed through continuous-flow isotope-ratio mass spectrometry according to Yarnes et al. (2005). Results of the batches processed for this experiment yielded a level of precision of equal to ±0.2‰ for δ^{13}C and ±0.3‰ for δ^{15}N. Significance in the pair-wise trophic shift between plant tissues, gall tissues, and insects within trees was analyzed using Student's Paired *t*-Test. All analyses were carried out using SYSTAT Version 10.2 (© 2002, SSI, Richmond, CA U.S.A.). All reported estimates of trophic shift are accompanied by their respective standard error of the difference (± SE).

4.2.3 Results

The δ^{13}C composition of adult *A. reticulatus* was significantly enriched relative to gall tissue in *Q. turbinella* ($\delta^{13}C_{consumer-diet}$ = 4.3 ± 0.4‰) and leaves bearing galls ($\delta^{13}C_{consumer-diet}$ = 5.1 ± 0.7‰), while galls were slightly enriched (≈1‰; not significant) relative to galled leaves in *Q. turbinella* (Table 1). Galled leaf tissue was marginally depleted in δ^{13}C relative to gall tissue (Table 1). The $\delta^{15}N_{consumer-diet}$ between *A. reticulatus* and their galls on *Q. turbinella* was significantly enriched (2.3 ± 0.5‰); a similar level of enrichment was observed for $\delta^{15}N_{consumer-diet}$ between *A. reticulatus* and *Q. turbinella* leaves (Table 1). Galls and galled leaves were similar in their δ^{15}N composition (Table 1).

The δ^{13}C composition of larval gall-forming arthropods on *S. exigua* was depleted relative to inner gall tissues (*R. strobiloides*: $\delta^{13}C_{consumer-diet}$ = –0.2 ± 0.2‰; *E. exiguae*: $\delta^{13}C_{consumer-diet}$ = –0.5 ± 0.2‰), but enriched over

Table 1. Δ^{15}N ± SE, Δ^{13}C ± SE for adult cynipid gall wasps on *Quercus turbinella*

Species	Isotope	$\Delta_{herbivore-gall}$	$\Delta_{herbivore-leaf}$	$\Delta_{gall-leaf}$	$\Delta_{parasitoid-herbivore}$
Neuroterus sp. (Yarnes et al. 2005)	^{15}N	0.7 ± 0.2	0.5 ±0.3	–0.2 ± 0.2	2.1 ± 0.2
	^{13}C	4.4 ± 0.5	6.0 ± 0.5	1.6 ± 0.2	–0.3 ± 1.1
Andricus reticulatus	^{15}N	2.3 ± 0.5 t_3 = 5.099 P = 0.02	2.2 ± 0.4 t_3 = 6.116 P = 0.009	–0.1 ± 0.3 t_3 = -0.444 P = 0.687	n/a
	^{13}C	4.3 ± 0.4 t_3 = 11.052 P = 0.002	5.1 ± 0.7 t_3 = 7.677 P = 0.005	0.9 ± 0.5 t_3 = 1.908 P = 0.15	n/a

Table 2. $\Delta^{15}N \pm SE$, $\Delta^{13}C \pm SE$ for larval gall-forming arthropods on *Salix exigua*

Species	Isotope	$\Delta_{herbivore\text{–}inner}$	$\Delta_{herbivore\text{–}outer}$	$\Delta_{inner\text{–}outer}$	$\Delta_{herbivore\text{–}mean\ gall}$	$\Delta_{parasitoid\text{–}herbivore}$
Rhabdophaga strobiloides	^{15}N	1.1 ± 0.2 $t_9 = 5.94$ $P < 0.001$	1.0 ± 0.5 $t_{10} = 2.20$ $P = 0.05$	0.0 ± 0.7 $t_8 = 0.065$ $P = 0.95$	1.1 ± 0.2 $t_8 = 4.821$ $P = 0.001$	2.2 ± 0.5 $t_1 = -4.657$ $P = 0.135$
	^{13}C	-0.2 ± 0.2 $t_9 = -1.073$ $P = 0.311$	0.8 ± 0.2 $t_{11} = 4.397$ $P = 0.001$	0.9 ± 0.2 $t_8 = 6.059$ $P < 0.001$	0.2 ± 0.2 $t_8 = 0.247$ $P = 0.287$	1.2 ± 0.3 $t_1 = 4.460$ $P = 0.14$
Euura exiguae	^{15}N	1.3 ± 0.3 $t_4 = 4.104$ $P = 0.02$	1.7 ± 0.2 $t_4 = 8.102$ $p = 0.001$	0.4 ± 0.2 $t_4 = 1.789$ $P = 0.15$	1.5 ± 0.2 $t_4 = 6.278$ $P = 0.003$	n/a
	^{13}C	-0.5 ± 0.2 $t_4 = -2.478$ $p = 0.07$	-0.9 ± 0.3 $t_4 = -3.737$ $P = 0.02$	-0.4 ± 0.2 $t_4 = -2.227$ $P = 0.09$	0.7 ± 0.2 $t_4 = 3.410$ $P = 0.03$	n/a

outer gall tissue (*R. strobiloides*: $\delta^{13}C_{consumer\text{–}diet} = 0.8 \pm 0.2‰$; *E. exiguae*: $\delta^{13}C_{consumer\text{–}diet} = -0.9 \pm 0.3‰$). The $\delta^{13}C_{consumer\text{–}diet}$ between larvae and mean gall tissue (inner + outer) were not different in *R. strobiloides* ($\delta^{13}C_{consumer\text{–}diet} = 0.3 \pm 0.2‰$), but was significantly depleted in *E. exiguae* ($\delta^{13}C_{consumer\text{–}diet} = -0.7 \pm 0.2‰$). The $\delta^{13}C$ composition of inner gall tissue was not significantly different from that of outer gall tissue in *E. exiguae*, but was significantly enriched within galls of *R. strobiloides* (Table 2). The $\delta^{13}C_{consumer\text{–}diet}$ between the torymid parasitoid and *R. strobiloides* appeared to be enriched (1.2 ± 0.2‰; Table 2), however the small sample size (n = 2) precluded a robust statistical test. The $\delta^{15}N_{consumer\text{–}diet}$ between both larval *R. strobiloides* and *E. exiguae* and their inner and outer gall tissues in *S. exigua* were ≈1‰; the same relationship was true when estimating $\delta^{15}N_{consumer\text{–}diet}$ from mean gall $\delta^{15}N$ (Table 2). Torymid parasitoids appeared enriched in $\delta^{15}N$ relative to *R. strobiloides* ($\delta^{15}N_{consumer\text{–}diet} = 2.2 \pm 0.3‰$). The inner and outer gall tissues were similar in their $\delta^{15}N$ composition (Table 2).

4.3 Discussion

Several distinct patterns in the trophic shift within gall communities emerge when comparing the results from these species coupled with that of published data. Most striking may be the $\Delta\delta^{13}C$ between cynipids and their galls. In many systems $\Delta\delta^{13}C$ is much less variable (0–1‰; Table 3) across trophic levels than is $\Delta\delta^{15}N$. However, in rare instances $\Delta\delta^{13}C$ may

Table 3. Published estimates of $\Delta^{15}N$ ± SE, $\Delta^{13}C$ ± SE for larval gall-forming arthropods and their parasitoids in other host plants. Meta-data estimates of $\Delta^{15}N$ and $\Delta^{13}C$ are given by McCutchan et al. (2003)

Species	Isotope	$\Delta_{\text{herb–host}}$	$\Delta_{\text{P. californica–herb}}$	$\Delta_{\text{Tetrastichus–herb}}$	$\Delta_{\text{T. baccaridis–herb}}$	$\Delta_{\text{T. koebelei–herb}}$
Rhopalomyia californica (Langaletto et al. 2005[a])	^{15}N	3.3 ± 0.4	1.8 ± 0.3	0.4 ± 0.8	2.1 ± 0.4	2.3 ± 0.3
	^{13}C	1.1 ± 0.3	0.4 ± 0.2	1.6 ± 0.3	1.0 ± 0.3	1.2 ± 0.2
Species	Isotope	$\Delta_{\text{A. rufus–S. terebinthinaceum}}$	$\Delta_{\text{A. rufus–S. laciniatum}}$			
Antistrophus rufus (Tooker and Hanks 2004[b])	^{15}N	≈2.5	≈0.5			
	^{13}C	not reported	not reported			
Source of estimate	Isotope	$\Delta_{\text{consumer–diet}}$				
McCutchan et al. 2003	^{15}N	2.3 ± 0.2				
	^{13}C	0.5 ± 0.1				

[a]Trophic shift estimates are reported for *R. californica* ("herb") on its host, *Baccharis pilularis* and four parasitoids.
[b]Reported trophic shifts are rough estimates as they were reported through figures. They report estimates of $\Delta^{15}N$ for *A. rufus* in two host plants.

range to near 7‰ (Buckeye butterfly, *Junonea coenia*; McCutchan et al. 2003). The leaf-galling cynipids on *Q. turbinella* (*A. reticulatus* and *Neuroterus* sp.) exhibit relatively large $\Delta\delta^{13}C$ (4.27–6.02‰ over gall tissue, Table 1). While, $\Delta\delta^{13}C$ is determined by the balance of assimilation and respiration, it is unlikely that cynipids would require the abnormally high respiration rate for growth leading to such an elevated $\Delta\delta^{13}C$. Gleixner et al. (1998) found primary plant carbon storage compounds (sucrose, starch) in sink tissues to be enriched in $\delta^{13}C$ composition over their counterparts in source tissues owing to post-transport metabolic conversions of primary compounds. Galls are known to be metabolically intense and similar source-sink dynamics and patterns of post-transport metabolic activities within galls (Harper et al. 2004; Larson and Whitham 1991) would contribute to variation in $\delta^{13}C$ between different gall tissues. Notably, we do not see the same pattern in bud-galling Cecidomyiidae (Langellotto et al. 2005; Yarnes et al. 2005; Tables 2, 3). Nor do we see high estimates of $\Delta\delta^{13}C$ in stem-gallers representing other taxa (Tables 1, 3). At this time, an estimate of $\Delta\delta^{13}C$ near zero for all non-cynipid gall-formers may be appropriate. All the endoparasitoids of gall-forming arthropods examined so far (Tables 1, 2) exhibit only a marginal degree of $\Delta\delta^{13}C$ (−1‰ to 1‰) from their hosts, as expected from meta-data estimates (Table 3).

The trophic shift in nitrogen isotopes ($\Delta\delta^{15}N$) has been proposed as an index for diet protein quality for consumers (Martínez del Rio and Wolf 2004) as it represents the balance between nitrogen assimilation and excretion. As the quality and assimilation efficiency of diet protein increases, trophic shift decreases. The $\Delta\delta^{15}N$ between *A. reticulatus* and their galls in *Q. turbinella* was nearly equal to meta-data estimates (≈2.5‰; McCutchan et al. 2003) and similar to estimates of trophic shift in other gall-forming arthropods (Langellotto et al. 2005; Tooker and Hanks 2004; Table 3). However, the $\Delta\delta^{15}N$ in *A. reticulatus* on *Q. turbinella* is drastically different from that reported in another species of cynipid, *Neuroterus* sp., on *Q. turbinella* (Yarnes et al. 2005; Table 1). This result indicates that separate species of cynipids can exhibit different $\Delta\delta^{15}N$ on the same host plant. Further, Tooker and Hanks (2004) found that the $\Delta\delta^{15}N$ for *Antistrophus rufus* significantly differed across host plants (Table 3). These results may indicate differences in the physiological mechanisms of gall nitrogen supply across different galling arthropods, as well as differences in nutritional quality of different hosts. Clearly, ecological patterns of $\Delta\delta^{15}N$ in gall-formers are still in need of considerable study before researchers can rely upon published estimates. Literature surveys of predator-based studies point to a generalized 3‰ shift in $\delta^{15}N$ across trophic levels for secondary consumers (Post 2002). The lower observed trophic shift for a torymid parasitoid of *R. strobiloides* of ≈ 2‰ is consistent with the high nitrogen assimilation efficiency often observed in parasitoids (Greenblatt et al. 1982) and other parasitoids (Tables 1–3). While it appears reasonable to assume an estimate of ≈ 2‰ for $\Delta\ \delta^{15}N$ in parasitoids of gall-forming arthropods (Langellotto et al. 2005; Yarnes et al. 2005), additional estimates for gall-formers are necessary due to interspecific differences between gall-formers and on both the same and different host plants.

4.4 Analytical Considerations and Future Directions

The continued application of stable isotope analyses in gall-forming arthropod communities requires a number of analytical considerations. While significant differences in the trophic shift of isotopes exist within galls of most species, the magnitude of the trophic shift is typically reduced from that of meta-data estimates. This puts an increased demand upon analytical precision, particularly when measurements are performed in continuous-flow isotope ratio mass spectrometry (CF-IRMS). This may be further exacerbated by the sample size requirement of $^{15}N/^{14}N$ measurements in CF-IRMS (60 μg elemental N) when considering the body mass of many inhabitants of gall communities. While considerably more time consuming, the use of traditional dual-inlet techniques would provide better precision and reduced sample consumption requirements (typically a ten-fold reduction in sample gas requirements). It is our suggestion that dual-inlet techniques be strongly considered when available.

Recent technological developments in IRMS technologies now allow for the high-precision measurement of the isotope-ratios of individual compounds through the coupling of a gas chromatograph to an IRMS (GC-C-IRMS). In particular, this technique has been successfully applied to the measurement of isotope ratios in both amino acids and lipids (Teece and Fogel 2004). The application of GC-C-IRMS has the potential to elucidate long-standing physiological and biochemical aspects of cecidogenesis, as well as the nutrition of gall-forming arthropods. This may be particularly true for the role of lipids in the inner nutritive cells of cynipid galls (Bronner 1992, Harper et al. 2004) or the regulation of host amino acid production and supply by gall-formers (Hartley and Lawton 1992; Koyama et al. 2003).

The utility of stable isotopes goes far beyond any potential applications to the biology of gall-forming arthropods and their associates. Stable isotopes offer a risk-free alternative to the radioactive isotopes historically used in studies of cecidogenesis. Stable isotopes are safe both at the natural abundance level and in studies requiring the use of stable isotopic tracers. The broad capabilities and increasing availability of IRMS technologies leaves little room for the future consideration of radioactive tracers in ecological and biochemical studies of galls.

4.5 Conclusions

Significant differences in the trophic shift in carbon and nitrogen isotopes between gall community consumers and their hosts indicate that stable isotopes have the potential to provide considerable insight into the trophic interactions within gall communities. Additional information is still needed to set an isotopic baseline for $\Delta\ \delta^{15}N$ within gall communities. $\Delta\ \delta^{13}C$ appears to be fairly reliable across systems — with the exception of cynipid gall wasps. In particular, the examination of different gall tissues and compound-specific isotope studies may profoundly illuminate the physiological ecology of gall-forming insect larvae. The ultimate utility of stable isotopes to the study of food web ecology within galls will depend on the further documentation of patterns in trophic shift and the demonstration of applicability to more complex communities that include inquilines, predators, and hyperparasitoids.

4.6 Acknowledgments

The authors thank Melissa Marquez, Niki Rockwell, and Avis James for assistance in the laboratory and field. We thank Peter Price for help with *E. exiguae*. Funding for this work was kindly provided through NSF Grants # DEB-0129630 and DMS-0337789 to W.J. Boecklen and EPS-0132632 to New Mexico.

4.7 References

Blüthgen N, Gebauer G, Fiedler K (2003) Disentangling a rainforest food web using stable isotopes: dietary diversity in a species-rich ant community. Oecologia 137:426-435

Bronner R (1992) The role of nutritive cells in the nutrition of cynipids and cecidomyiids. In: Shorthouse JD, Rohfritsch O (eds) Biology of insect-induced galls. Oxford University Press, New York

Callaham Jr MA, Whiles MR, Meyer CK, Brock BL, Charlton RE (2000) Feeding ecology and emergence production of annual cicadas (Homoptera: Cicadidae) in tallgrass prairie. Oecologia 123:535-542

DeNiro MJ, Epstein S (1978) Influence of diet on the distribution of carbon isotopes in animals. Geochimica et Cosmochimica Acta 42:95-506

DeNiro MJ Epstein S (1981) Influence of diet on the distribution of nitrogen isotopes in animals. Geochimica et Cosmochimica Acta 45:341-351

Gannes LZ, O'Brien DM, Martínez del Rio C (1997) Stable isotopes in animal ecology: assumptions, caveats, and a call for more laboratory experiments. Ecology 78:1271-1276

Gleixner G, Scrimgeour C, Schmidt HL, Viola R (1998) Stable isotope distribution in the major metabolites of source and sink organs of *Solanum tuberosum* L.: a powerful tool in the study of metabolic partitioning in intact plants. Planta 207:241-245

Greenblatt JA, Barbosa P, Montgomery ME (1982) Host's diet effects on nitrogen utilization efficiency for two parasitoid species: *Brachymeria intermedia* and *Coccygomimus turionellae*. Physiological Entomology 7:263-267

Harper LJ, Schönrogge K, Lim KY, Francis P, Lichtenstein CP (2004) Cynipid galls: insect-induced modifications of plant development create novel plant organs. Plant, Cell and Environment 27:327-335

Hartley SE, Lawton JH (1992) Host plant manipulation by gall insects—a test of the nutrition hypothesis. Journal of Animal Ecology 61:113-119

Koyama Y, Yao I, Akimoto SI (2003) Aphid galls accumulate high concentrations of amino acids: a support for the nutrition hypothesis for gall formation. Entomologia Experimentalis et Applicata 113:35-44
Langellotto GA, Rosenheim JA, Williams MR (2005) Enhanced carbon enrichment in parasitoids (Hymenoptera): a stable isotope study. Annals of the Entomological Society of America 98:205-213
Larson KC, Whitham TG (1991) Manipulation of food resources by a gall-forming aphid: the physiology of sink-source interactions. Oecologia 88:15-21
Markow TA, Anwar S, Pfeiler E. (2000) Stable isotope ratios of carbon and nitrogen in natural populations of *Drosophila* species and their hosts. Functional Ecology 14:261-266
Martínez del Rio C, Wolf BO (2004) The interplay between a food's stoichiometry, digestion, and metabolism: using stable isotopes to find out what animals eat. In: Stark JM (ed) Physiological and ecological adaptations to feeding in vertebrates. Science Publishers, Enfield
McCutchan, Jr JH, Lewis Jr WM, Kendall C, McGrath CC (2003) Variation in trophic shift for stable isotope ratios of carbon, nitrogen, and sulfur. Oikos 102:378-390
McNabb DM, Halaj J, Wise DH (2001) Inferring trophic positions of generalist predators and their linkage to the detrital food web in agroecosystems: a stable isotope analysis. Pedobiologia 45:289-297
O'Brien DM, Fogel ML, Boggs CL (2002) Renewable and non-renewable resources: amino acid turnover and allocation to reproduction in Lepidoptera. Proceedings of the National Academy of Science, USA 99:4413-4418
Plantard O, Rasplus J-Y, Hochberg ME (1996) Resource partitioning in the parasitoid assemblage of the oak galler *Neuroterus quercusbaccarum* L. (Hymenoptera: Cynipidae). Acta Oecologica 17:1-15
Post DM (2002) Using stable isotopes to estimate trophic position: models, methods, and assumption. Ecology 83:703-718
Price PW, Bouton CE, Gross P, McPheron BA, Thompson JN, Weis A (1980) Interactions among three trophic levels: influence of plants on interactions between insect herbivores and natural enemies. Annual Review of Ecology and Systematics 11:41-65
Raman A, Schaefer CW, Withers TM (2005) Biology, ecology, and evolution of gall-inducing arthropods. Science Publishers, Enfield
Roininen H, Price PW, Tahvanainen J (1996) Bottom-up and top-down influences in the trophic system of a willow, a galling sawfly, parasitoids and inquilines. Oikos 77:44-50
Steele KW, Daniel RM (1978) Fractionation of nitrogen isotopes by animals: a further complication to the use of variation in the natural abundance of ^{15}N for tracer studies. Journal of Agricultural Science 90:7-9
Stone GN, Schönrogge K, Crawley MJ, Fraser S (1995) Geographic and between generation variation in the parasitoid communities associated with an invading gallwasp, *Andricus quercuscalicis* (Hymenoptera: Cynipidae). Oecologia 104:207-217

Teece MA, Fogel ML (2004) Preparation of ecological and biochemical samples for isotope analysis. In: de Groot PA (ed) Handbook of stable isotope analytical techniques, vol. I. Elsevier, Amsterdam

Tooker JF, Hanks LM (2004) Trophic position of the endophytic beetle *Mordellistena aethiops* Smith (Coleoptera: Mordellidae). Environmental Entomology 33:291-296

Vander Zanden MJ, Rasmussen JB (2001) Variation in delta N-15 and delta C-13 trophic fractionation: implications for aquatic food web studies. Limnology and Oceanography 46:2061-2066

Vanderclift MA, Ponsard S (2003) Sources of variation in consumer-diet $\delta^{15}N$ enrichment: a meta-analysis. Oecologia 136:169-182

Washburn JO, Cornell HV (1981) Parasitoids, patches and phenology—their possible role in the local extinction of a cynipid gall wasp population. Ecology 62:1597-1607

Yarnes CT, Rockwell JN, Boecklen WJ (2005) Trophic shift in $\delta^{15}N$ and $\delta^{13}C$ through a cynipid gall wasp community (*Neuroterus* sp.) in *Quercus turbinella*. Environmental Entomology 34:1471-1476

5 Temporal Variation in the Structure of a Gall Wasp Assemblage along a Genetic Cline of *Quercus crispula* (Fagaceae)

Masato Ito

JSPS Research Fellow, Hokkaido Research Center, Forestry and Forest Products Research Institute, 7 Hitsujigaoka, Toyohira, Sapporo, Hokkaido 062-8516, Japan

Summary. I examined temporal variations in the structure of a gall wasp assemblage along a genetic cline among 12 half-sib families (HSFs) of the host oak *Quercus crispula*, by comparing patterns in the species composition, species richness, and abundance of component species of the assemblage in 2 years. Three of 14 gall sorts were dominant in both years (common gall wasps), and 3 dominated in only one of the years (opportunistic gall wasps). Species composition reflected the genetic cline among HSFs, and differed between years. However, the relative similarity of species composition across HSFs did not vary between years. The mean species richness differed among HSFs, but the relative richness across HSFs did not differ between years. Two of the common gall wasps showed similar, significant responses to the genetic cline between years, whereas the opportunistic gall wasps showed relatively generalist responses. Thus, the overall structure of the assemblage along the genetic cline was characterized mainly by the common gall wasps, whereas the temporal variation was caused additively by the opportunistic gall wasps. The lack of HSF versus year interactions suggests that the effects of plant genetics on this assemblage were not masked by environmental interactions.

Key words. Cynipidae, Fagaceae, Genetics–environment interaction, Plant–herbivore relationship, Species composition

5.1 Introduction

Studies of hybrid plants have suggested that susceptibility to herbivore attack changes along a genetic cline in the host plants that is caused by hybridization (Floate and Whitham 1993; Fritz et al. 1994). Consequently, the structure of a herbivore assemblage may change along the genetic cline of host plants (Whitham et al. 1994). Gall wasps (Hymenoptera: Cynipidae) dominate the gall-inducing fauna on oaks, *Quercus* spp. (Fagaceae) (Roskam 1992). Oaks are known to hybridize within species groups (e.g., Wigston 1974), and the host plant genetics are known to affect the success or failure of gall formation (Anderson et al. 1989; Glynn and Larsson 1994; Larsson and Strong 1992). Thus, the oak–gall wasp system offers a suitable system for exploring the responses of multiple herbivore species to such a genetic cline in the host plant.

The ambient environment also affects host plant traits (e.g., Price and Clancy 1986; Sipura and Tahvanainen 2000), and some studies have shown interaction effects on herbivore abundance between plant genetics and environments (Graham et al. 2001; Maddox and Root 1987; Stiling and Rossi 1996). Such interactions suggest that the structure of a herbivore assemblage that is based on plant genetics would vary among environments, and would thus differ among sites and years. These issues can be treated by transplant experiments combined with observations over a period of time (Fritz 1990; Graham et al. 2001).

Quercus crispula Blume is known to hybridize with *Quercus dentata* Thunberg (Ishida et al. 2003; Orita et al. 1991). Ito and Ozaki (2005) used half-sib families (HSFs) of *Q. crispula*, and detected a genetic cline among the HSFs caused by a history of hybridization with *Q. dentata*. They also showed that the species composition of the gall wasp assemblage changed along the genetic cline within *Q. crispula*. However, it remains uncertain whether the structure of the gall wasp assemblage along the oak genetic cline is affected by the ambient environment in any oak–gall wasp system (Aguilar and Boecklen 1992; Boecklen and Spellenberg 1990; Ito and Ozaki 2005; Moorehead et al. 1993).

In the present study, I focused on temporal variations in the structure of the gall wasp assemblage along the genetic cline of *Q. crispula*, and looked for interaction effects between plant genetics and the ambient environment. I tested hypotheses: (1) that the species composition of the gall wasp assemblage across HSFs of *Q. crispula* would vary between years, and if it varied, that the variation would be caused by (2) the difference in the species richness of the assemblage or (3) the difference in the response of the component species between years.

5.2 Study System and Method

Studies were carried out in a young *Q. crispula* plantation (7 years old in

2002) at the Hokkaido Research Center, Forestry and Forest Products Research Institute (43°10'N, 141°30'E). The Institute planted 11 HSFs in 20 blocks in a randomized complete block design, and one HSF (EI-8) in 19 of the 20 blocks (Table 1; see Ito and Ozaki [2005] for more detailed design). Hybrids of *Q. crispula* and *Q. dentata* have intermediate leaf morphology (Ishida et al. 2003; Orita et al. 1991). I used canonical discriminant function analysis based on leaf morphology to arrange the 12 HSFs along a cline from typical *Q. crispula* to resemblance of *Q. dentata* (Ito and Ozaki 2005). The canonical variate in this analysis was used as the index of hybridization with *Q. dentata* for each HSF (Table 1).

I identified gall wasps based on the gall morphology according to Yukawa and Masuda (1996), and used the number of gall sorts as an indicator of the species richness. I examined all branches of each tree and recorded gall sorts that were present in 2002 and 2004. I analyzed the species composition of the gall wasp assemblage based on the presence or absence of galls on each tree, and thus defined the abundance of a gall wasp species as the proportion of trees with galls of each sort in each HSF.

I analyzed the similarity in species composition of the gall wasp assem-

Table 1. Index of hybridization with *Q. dentata* and mean stem length (± SE) of 12 half-sib families

Half-sib family	Index of hybridization[a]	Stem length (cm)	
		2002	2004
KO	−1.82	137.8 ± 13.3	184.8 ± 15.6
SN-8	−1.31	128.4 ± 9.0	164.4 ± 11.2
TA	−1.11	165.1 ± 10.0	209.8 ± 12.1
EI-9	−0.37	128.7 ± 7.7	154.7 ± 8.6
EI-7	−0.26	131.3 ± 4.8	158.4 ± 6.3
EI-5	−0.12	99.6 ± 7.2	125.9 ± 9.8
EI-6	0.20	134.5 ± 7.9	164.3 ± 10.8
EI-8	0.41	116.1 ± 10.9	145.4 ± 13.5
OM-11	0.76	110.4 ± 7.0	137.8 ± 9.9
EM-11	0.92	127.2 ± 5.5	164.9 ± 8.5
OM-3	0.96	75.3 ± 5.2	89.7 ± 7.9
OM-5	1.75	83.0 ± 5.4	95.1 ± 7.1

[a]Canonical variate by a discriminant function analysis of leaf morphology; the smallest and largest values represent typical *Q. crispula* and resemblance of another oak *Q. dentata*, respectively (Ito and Ozaki 2005).

Table 2. Gall wasps recorded in the present study and the number of trees bearing their galls

Gall wasp species	Gall sort code[a]	Total no. of galled trees	
		2002	2004
BRG (Agamic generation of *Andricus symbioticus* Kovalev)		14	17
BG1 (Agamic generation of *Andricus mukaigawae* [Mukaigawa])		3	1
BG2 (Sexual generation of *Aphelonyx glanduliferae* Mukaigawa)		1	0
BG3 (Undescribed species)	C-065, 066?	35	27
BG4 (Undescribed species)	C-186	8	28
LG1 (Sexual generation of *A. symbioticus*)		12	14
LG2 (Sexual generation of *Andricus moriokae* Monzen)		1	3
LG3 (Sexual generation of *A. mukaigawae*)		5	7
LG4 (Undescribed species)	C-064?	6	32
LG5 (Undescribed species)	C-141	35	39
LG6 (Undescribed species)	C-136	55	38
LG7 (Undescribed species)	C-068	1	20
LG8 (Undescribed species)	C-139	1	0
LG9 (Undescribed species)	C-146, 191?	7	179

BRG, branch galler; BG, bud galler; LG, leaf galler.
[a]Provided only for undescribed species according to Yukawa and Masuda (1996).

blage among HSFs and years using non-metric multidimensional scaling (NMS) with a Bray–Curtis distance measure in a three-dimensional solution, and tested the difference in the species composition between HSF groups in respective years using multi-response permutation procedures (MRPP) with a Bray–Curtis distance measure. In these analyses, the proportion of trees was arcsine-root-transformed. I compared the relative similarity of the species composition across HSFs between years using Pearson's correlation coefficient based on the scores of the HSFs on each NMS axis.

Tree size differed among HSFs (Table 1), indicating that the presence of gall wasps could have been affected by differences in sampling effort. To test the effects of HSF, block position, and year on the number of gall sorts, I performed an analysis of covariance (ANCOVA) using stem length as the covariate, as this is a valid indicator of sampling effort for each tree (Ito and Ozaki 2005). The number of gall sorts was log-transformed.

I tested the difference in the response of gall wasps to HSFs between years by means of stepwise logistic regression for each gall sort, using HSF, year, and their interaction as the explanatory variables. Block position was not considered, because its effect was not significant for any gall sort. I conducted this analysis for dominant gall sorts, which I defined as galls that occurred on more than 24 trees (>10% of all trees) in each year.

I performed the NMS and MRPP analyses using PC-ORD version 4.20 (MjM software, Oregon, USA) and the other analyses using SPSS version 9.0J (SPSS Inc., Chicago, USA).

5.3 Results

I recorded a total of 14 gall sorts (Table 2). Bud galler 3 (BG3), leaf galler 5 (LG5), and leaf galler 6 (LG6) were dominant species in both years, whereas bud galler 4 (BG4), leaf galler 4 (LG4), and leaf galler 9 (LG9) were dominant only in 2004. The first axis of the NMS of the similarity of species composition explained 49.5% of the variation among HSFs and years, the second axis explained 35.0%, and the third axis explained 8.8% (*Stress* = 8.04). As a result, I expressed the ordination in two dimensions (Fig. 1). In each year, species composition varied among HSFs and the series of species composition nearly reflected the variation in the index of hybridization among HSFs (Fig. 1). HSFs were significantly divided into two species groups based on the year (MRPP, $A = 0.23$, $P < 0.001$). The scores of the HSFs on each NMS axis in 2002 were positively correlated with those in 2004 (Pearson's correlation coefficient, $r = 0.99$, $P < 0.001$

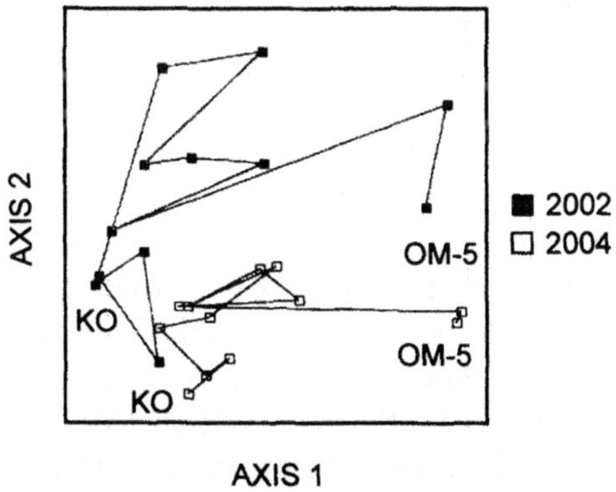

Fig. 1. Results of the NMS ordination of similarity in the species composition of gall wasps among 12 half-sib families in 2002 and 2004. Half-sib families are connected with lines in order of the magnitude of the index of hybridization.

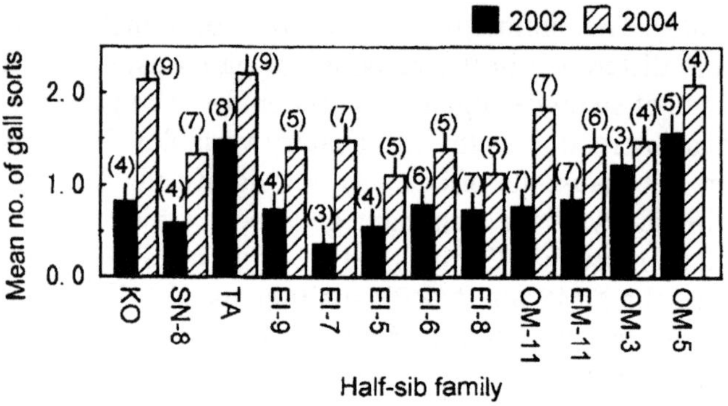

Fig. 2. Mean number of gall sorts per tree (+ SE) in each half-sib family in 2002 and 2004. The numbers in parentheses represent the total number of gall sorts. Mean values were adjusted for stem length. Half-sib families are arranged from left to right in order of increasing index of hybridization.

on the first axis; $r = 0.76$, $P = 0.004$ on the second axis).

Stem length was a significant covariate for the number of gall sorts (ANCOVA, $F_{1, 415} = 37.66$, $P < 0.001$). Interactions of the explanatory variables with the covariate were not significant (HSF, $F_{11, 354} = 0.92$; year, $F_{1, 354} = 0.01$; block, $F_{19, 354} = 1.44$; HSF × year, $F_{11, 354} = 0.44$; year × block, $F_{19, 354} = 1.29$; $P > 0.1$ in all cases). The mean number of gall sorts differed among HSFs ($F_{11, 415} = 6.05$, $P < 0.001$) and between years ($F_{1, 415} = 98.04$, $P < 0.001$), but the effect of block position ($F_{19, 415} = 1.27$, $P = 0.202$) and its interaction with year ($F_{19, 415} = 1.61$, $P = 0.050$) were not significant. The interaction between HSF and year was also not significant ($F_{11, 415} = 1.26$, $P = 0.245$). The mean number of gall sorts did not monotonically increase or decrease as a function of the index of hybridization, and was larger in 2004 than in 2002 in all HSFs (Fig. 2).

Interaction between HSF and year was eliminated from the models or not significant for the abundance of all dominant gall wasps (Table 3). The difference in abundance between years was highly significant for LG9, which was predominant in 2004, and less strongly significant for LG6, which was dominant in both years (Table 3). BG3 and LG4 appeared most frequently in HSFs with larger indices of hybridization (Fig. 3), but the effect of HSF was not significant for LG4 (Table 3). In contrast, LG5 tended to appear in HSFs with smaller indices of hybridization (Fig. 3, Table 3). The abundance of LG6 and LG9 was lower in OM-3 and OM-5, both of which have the largest indices of hybridization and the shortest stems (i.e., the lowest sampling efforts), but they showed weak overall trends with respect to this index (Fig. 3, Table 3). The abundance of BG4 did not differ significantly among HSFs (Fig. 3, Table 3).

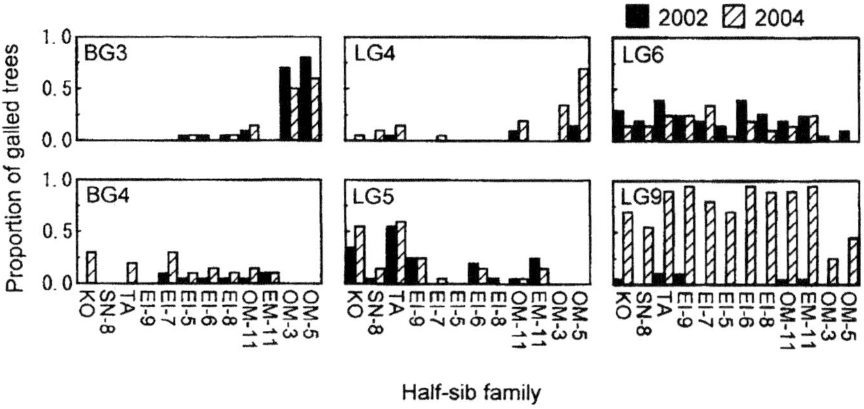

Fig. 3. Proportion of trees galled by dominant gall wasps. Half-sib families are arranged from left to right in order of increasing index of hybridization. *BG*, bud galler; *LG*, leaf galler.

Table 3. Results of stepwise logistic regressions testing the effects of half-sib family and year on the presence or absence of the galls of dominant gall wasps

Gall wasp species	Variable	Wald χ^2	*P*
BG3	HSF	64.13	< 0.001**
BG4	HSF	3.07	0.990
	HSF × Year	5.02	0.930
LG4	HSF	8.75	0.645
	HSF × Year	12.35	0.338
LG5	HSF	51.84	< 0.001**
LG6	HSF	20.76	0.036*
	Year	4.04	0.045*
LG9	HSF	47.83	< 0.001**
	Year	116.50	< 0.001**

BG, bud galler; LG, leaf galler; HSF, half-sib family.
d.f. = 11 for HSF, 1 for Year, and 11 for HSF × Year.
$*P < 0.05$; $**P < 0.01$.

5.4 Discussion

In each year, the species composition of the gall wasp assemblage differed among HSFs and reflected the genetic cline in *Q. crispula* that resulted from hybridization with *Q. dentata* (Table 1, Fig. 1). These results are consistent with those of studies of other hybrid oak–gall wasp systems (Aguilar and Boecklen 1992; Boecklen and Spellenberg 1990) and other hybrid plant–herbivore systems (Floate et al. 1996; Ishida et al. 2003). However, the mean species richness did not change monotonically along the genetic cline (Fig. 2), and the patterns of abundance along the genetic cline differed among component species (Fig. 3). The bimodal response of species richness would reflect the contrasting responses of some species to the genetic cline (Ito and Ozaki 2005). These results differ from those of previous studies of other oak–gall wasp systems, in which consistent trends were found along a genetic cline from parental oaks to their hybrids (Aguilar and Boecklen 1992; Boecklen and Spellenberg 1990). This difference may result from differences in the genetic range tested, because the present study dealt with the variation in a single oak species. Another explanation is that a genetic gap in the cline may have prevented the establishment of gall wasps from other oak species (Floate and Whitham 1993; Graham et al. 2001), because *Q. dentata* did not grow at the study site.

The structure of the gall wasp assemblage along the genetic cline of *Q. crispula* varied between the 2 years (Fig. 1), probably due to variations in the ambient biotic and/or abiotic environments. However, the ordination scores on each NMS axis in one year were positively correlated with those on the same axis in the other year (Fig. 1), indicating that the relative species composition across HSFs did not fluctuate between years. This can be explained by the lack of significant interaction effects between *Q. crispula* genetics (HSF) and year in determining the structure of the assemblage.

There were no significant interaction effects between HSF and year in terms of the abundance of gall wasps that dominated in the 2 years (BG3, LG5, and LG6; hereafter referred as "common gall wasps"), indicating that these gall wasps responded to HSF similarly in both years (Table 3, Fig. 3). Two of the common gall wasps responded clearly to HSF, whereas gall wasps that dominated only in 2004 (BG4, LG4, and LG9; hereafter referred as "opportunistic gall wasps") showed relatively generalist responses to HSF (Fig. 3). Consequently, the relative species richness of gall wasps across HSFs did not change between years, as indicated by the lack of significant interaction effect between HSF and year on the number of gall sorts (Fig. 2). Therefore, the structure of the gall wasp assemblage along the genetic cline of *Q. crispula* is likely to be characterized mainly by common gall wasps, whereas the variation in the structure will arise additively by including the effects of the opportunistic gall wasps.

Some studies have shown interaction effects between plant genotype and environment on herbivore abundance (Graham et al. 2001; Maddox and Root 1987; Stiling and Rossi 1996). These authors emphasized the strong effects of environment on genetically based plant resistance to herbivores. On the other hand, Fritz (1990) demonstrated no significant interactions of plant genotype with site and with year in most of the galling sawflies on arroyo willow, although plant genotype, site, and year individually tended to affect their abundance. Thus, the present results support Fritz's statement that the effects of plant genetics on herbivore assemblage cannot be masked by interactions with the environment even when interactions occur.

The dynamics of plant–herbivore assemblage relationships may reflect interactions among multiple herbivore species. If the susceptibility of a plant to a herbivore species is related to its susceptibility to other herbivore species, interactions of herbivores with the plant will be "diffuse"; that is, a herbivore species can affect the interactions between its host plant and other herbivore species (Janzen 1980). If the susceptibility to a herbivore is independent of the susceptibility to another herbivore, then interactions of herbivores with the plant will be "pairwise" (Hougen-Eitzman and Rausher 1994). In the present study, the abundance patterns of common gall wasps along the genetic cline were not altered by the presence of opportunistic gall wasps (Fig. 3). Therefore, these gall wasps are unlikely to interact with each other. Moreover, LG6, a common gall wasp, showed no clear response to HSF (Fig. 3), suggesting that its abundance was not affected by the other gall wasps that dominated on particular HSFs. Thus, there are unlikely to be any interactions among many gall wasps via plant genetics in the *Q. crispula*–gall wasp system.

5.5 Acknowledgments

I thank the Arboretum and Nursery Office, Hokkaido Research Center, Forestry and Forest Products Research Institute, for the permission to use the study site, and the members of the Laboratory of Forest Protection, Nagoya University, for their helpful suggestions.

5.6 References

Aguilar JM, Boecklen WJ (1992) Patterns of herbivory in the *Quercus grisea* × *Quercus gambelii* species complex. Oikos 64:498-504

Anderson SS, McCrea KD, Abrahamson WG, Hartzel LM (1989) Host genotype choice by the ball gallmaker *Eurosta solidaginis* (Diptera: Tephritidae). Ecology 70:1048-1054

Boecklen WJ, Spellenberg R (1990) Structure of herbivore communities in two oak (*Quercus* spp.) hybrid zones. Oecologia 85:92-100

Floate KD, Whitham TG (1993) The "hybrid bridge" hypothesis: host sifting via plant hybrid swarms. The American Naturalist 141:651-662

Floate KD, Fernandes GW, Nilsson JA (1996) Distinguishing intrapopulational categories of plants by their insect faunas: galls on rabbitbrush. Oecologia 105:221-229

Fritz RS (1990) Effects of genetic and environmental variation on resistance of willow to sawflies. Oecologia 82:325-332

Fritz RS, Nichols-Orians CM, Brunsfeld SJ (1994) Interspecific hybridization of plants and resistance to herbivores: hypotheses, genetics, and variable responses in a diverse herbivore community. Oecologia 97:106-117

Glynn C, Larsson S (1994) Gall initiation success and fecundity of *Dasineura marginemtorquens* on variable *Salix viminalis* host plants. Entomologia Experimentalis et Applicata 73:11-17

Graham JH, McArthur ED, Freeman DC (2001) Narrow hybrid zone between two subspecies of big sagebrush (*Arstemisia tridentata*: Asteraceae). XII. Galls on sagebrush in a reciprocal transplant garden. Oecologia 126:239-246

Hougen-Eitzman D, Rausher MD (1994) Interactions between herbivorous insects and plant-insect coevolution. Evolution 143:677-697

Ishida TA, Hattori K, Sato H, Kimura MT (2003) Differentiation and hybridization between *Quercus crispula* and *Q. dentata* (Fagaceae): insights from morphological traits, amplified fragment length polymorphism markers, and leafminer composition. American Journal of Botany 90:769-776

Ito M, Ozaki K (2005) Response of a gall wasp community to genetic variation in the host plant *Quercus crispula*: a test using half-sib families. Acta Oecologica 27:17-24

Janzen DH (1980) When is it coevolution? Evolution 34:611-612

Larsson S, Strong DR (1992) Oviposition choice and larval survival of *Dasineura marginemtorquens* (Diptera: Cecidomyiidae) on resistant and susceptible *Salix viminalis*. Ecological Entomology 17:227-232

Maddox GD, Root RB (1987) Resistance to 16 diverse species of herbivorous insects within a population of goldenrod, *Solidago altissima*: genetic variation and heritability. Oecologia 72:8-14

Moorehead JR, Taper ML, Case TJ (1993) Utilization of oak hosts by a monophagous gall wasp: how little host character is sufficient? Oecologia 95:385-392

Orita H, Koono K, Okuyama K, Eiga S (1991) On genetic characters of leaf forms in *Quercus* (in Japanese). Transactions of Hokkaido Branch of the Japanese Forest Society 39:44-46

Price PW, Clancy KM (1986) Multiple effects of precipitation on *Salix lasiolepis* and populations of the stem-galling sawfly, *Euura lasiolepis*. Ecological Research 1:1-14

Roskam JC (1992) Evolution of the gall-inducing guild. In: Shorthouse JD, Rohfritsch O (eds) Biology of insect-induced galls. Oxford University Press, New York, pp 34-49

Sipura M, Tahvanainen J (2000) Shading enhances the quality of willow leaves to leaf beetles—but does it matter? Oikos 91:550-558

Stiling P, Rossi AM (1996) Complex effects of genotype and environment on insect herbivores and their enemies. Ecology 77:2212-2218

Whitham TG, Morrow PA, Potts BM (1994) Plant hybrid zones as centers of biodiversity: the herbivore community of two endemic Tasmanian eucalypts. Oecologia 97:481-490

Wigston DL (1974) Cytology and genetics of oaks. In: Morris MG, Perring FH (eds) The British oak. Classey, Berkshire, pp 27-50

Yukawa J, Masuda H (1996) Insect and mite galls of Japan in colors (in Japanese with English explanations for color plates). Zenkoku Noson Kyoiku Kyokai, Tokyo

6. Effects of Floods on the Survival and Species Component of *Rhopalomyia* Gall Midges (Diptera: Cecidomyiidae) Associated with *Artemisia princeps* (Asteraceae) Growing in a Dry Riverbed in Japan

Tomoko Ganaha[1], Nami Uechi[2], Machiko Nohara[1], Junichi Yukawa[3], and Yukihiro Shimatani[4]

[1]Entomological Laboratory, Graduate School of Bioresource and Bioenvironmental Sciences, Kyushu University, Fukuoka 812-8581, Japan
[2]Okinawa Prefectural Agricultural Experiment Station, 4-222 Sakiyama-cho, Naha, Okinawa 903-0814, Japan
[3]Kyushu University, Fukuoka 812-8581, Japan
[4]Laboratory of River Engineering, Faculty of Engineering, Kyushu University, Fukuoka 812-8581, Japan

Summary. The Kitagawa River flows through Miyazaki Prefecture, Kyushu, Japan. Its frequent floods disturb insect and plant communities in the dry riverbed. The species composition of *Rhopalomyia* gall midges on *Artemisia princeps* after the great flood in 2004 was apparently different from that at normal water level in 2001–2002. Running water with a velocity of 3.0 m/s removed *A. princeps* leaves at a relatively high rate when the plants had been submerged in water for more than several days. At the time of the flood, the river attained a water level of 15.7 m and the velocity of running water was about 3.4 m/s. This situation lasted for three days. Under such conditions, many *A. princeps* leaves were removed, together with midge galls. Submergence of galls caused the death of midge larvae and pupae, when it lasts for more than two days. The stem galler, *R. struma*, would have more chance to survive in the hard stem galls than the other leaf gallers under two-day submerged conditions. Besides, the stem galls would not readily detach from the plants even when they are submerged in the running water with a high velocity. Such differences in galling traits between the *Rhopalomyia* species are considered to have changed the species composition of the gall midges.

Key words. floods, *Rhopalomyia*, gall midge, Cecidomyiidae, dry riverbed, *Artemisia princeps*, species composition

6.1 Introduction

Precipitation affects the survival of insects directly at the individual level (e.g., Wakisaka et al. 1991; Kamata and Igarashi 1994; Tanzubil et al. 2000) or indirectly through different trophic levels (e.g., Lensing et al. 2005; Nakamura et al. 2005). Although the effects of precipitation have been studied from various points of view, the effects of floods on survival have seldom been compared among congeneric species that belong to the same guild. To fill these gaps, we selected, as target insects, some gall midge species of the genus *Rhopalomyia* (Diptera: Cecidomyiidae). They belong to the same guild as gall inducers on *Artemisia princeps* Pampan (Asteraceae), which is one of the most dominant plants in dry riverbeds in Japan. At least 11 *Rhopalomyia* species occur on five *Artemisia* species in Japan and their galls are species-specific, being diverse in shape, structure and galled part (Yukawa and Masuda 1996). We assumed that floods might influence different gall midge species differently because gall shape, structure, and galled organ might be related to their survival under flood conditions, leading to changes in the species composition in the dry riverbeds.

We surveyed species component of *Rhopalomyia* gall midges on the dry riverbed and riverbanks, and around the river edge of the Kitagawa River that flows through Miyazaki Prefecture, Kyushu, Japan. During the last decade, heavy rains with 200–500 mm/48 hours caused floods of this river twice in 1997 and 2004 with water volume exceeding 4500 m^3/s, of which velocity was estimated to be more than 3 m/s (The Kitagawa Data obtained from the Nobeoka Office of River and National Highway, Japan Ministry of Land, Infrastructure and Transport). We also examined the survival of midge larvae under submerged conditions and tested whether running water with a current of 3 m/s could detach *A. princeps* leaves from the stems. Based on these investigations, we discuss the effects of floods on the survival and species component of *Rhopalomyia* gall midges.

Fig. 1a, b, c, d, e, f, g. Galls induced by *Rhopalomyia* species. **a** Leaf galls by *R. cinerarius*. **b** Stem galls by *R. giraldii*. **c** A terminal bud gall by *R. iwatensis*. **d** Stem galls by *R. struma*. **e** An axillary bud gall by *R. shinjii*. **f** Leaf galls by *R. yomogicola*. **g** A leaf gall by *Rhopalomyia* sp. 1.

6.2 Materials and Methods

6.2.1 Field Survey

In 2000–2004, the species of *Rhopalomyia* on *A. princeps* plants were surveyed eight times at different habitats mainly in Susa Area (32°38′35″N, 131°41′22″E), Nobeoka City, Miyazaki Prefecture, Japan and partly in its vicinity, Hyono, Matono, and Honmura Areas, Kitagawa Town of the same Prefecture. The habitats were divided into the following three categories; the dry riverbed of the Kitagawa River, the riverbanks, and the river edge because these habitats could be differently affected by different water levels. At each habitat, the abundance of respective species was surveyed by recording the number of different sorts of gall on randomly selected 90–705 plants. We identified *Rhopalomyia* species by the shape of galls (Fig. 1a–g). Then, the number of *Artemisia* plants that bore different sorts of midge gall was compared with chi-square test between the three habitats and between gall midge species. In particular, a field survey on 4 November 2004, 10 days after a great flood, provided species composition data to compare with those at normal water level in 2001–2002.

6.2.2 Effects of Running Water on the Removal of Submerged *Artemisia princeps* Leaves

Preliminary tests in the Laboratory of River Engineering, Kyushu University indicated that no fresh leaves of *A. princeps* detached from the plants in artificial running water with a current of 2.3–2.5 m/s. Then, we tested, in a natural stream that runs continuously and faster than artificial running water, if the running water with a current of 3.0 m/s can remove the leaves that have been submerged in water beforehand. In August 2005, 60 plants (22 to 36 cm in height) of *A. princeps* were dug from the Hakozaki Campus of Kyushu University (33°37'35"N, 130°25'42"E), Fukuoka City, Japan, and the number of leaves attached to each plant was recorded. These plants with roots were kept standing in buckets (height 55.5 cm; diameter 49.5 cm) and submerged in groundwater, which was renewed every other day and kept at 20.5–23.5°C under laboratory conditions. After 3, 4, 5, 6, 7, and 8 days, respectively, 10–20 of these plants were taken out of the buckets and submerged, for 60 seconds, in running water with a current of about 3.0 m/s. The duration of submergence was determined as 60 seconds because we noted, in the preliminary experiments, that in most cases leaves were removed soon after submergence in the running water. The number of leaves detached was recorded to estimate the survival of submerged leaves under flood conditions. A strong current was located in a stream in the suburb of Fukuoka City. The water that shot out in a narrow stream from the side of a low sand-trap dam (about 90 cm in height) was guided onto a U-shaped plastic gutter (182 cm in length, 66 cm in width, and 23 cm in depth) that leaned against the dam sidewall at about 30°. We measured the current velocity by recording the movement of a floating object on the current with a video camera. Analysis of the movement indicated that the velocity was about 3.0 m/s, which was about the maximum velocity of the Kitagawa River at the time of the floods in October 2004 (The Kitagawa Data).

6.2.3 Effects of Submergence on the Survival of *Rhopalomyia* Gall Midges

In August 2005, *Artemisia* plants with galls containing midge larvae or pupae were submerged in buckets, using the methods described earlier, in order to examine the effects of submergence on the survival of *Rhopalomyia* larvae and pupae. *R. cinerarius* Monzen (leaf galler), *R. yomogicola* (Matsumura) (leaf galler), and *R. struma* Monzen (stem galler) were used in this examination. After one, two, and three days, respectively, 19–116 individuals were taken out of the submerged galls and observed the movement of larvae and pupae in 75% ethanol under a binocular microscope to judge whether they were alive or dead.

Table 1. *Rhopalomyia* gall midges found on *Artemisia princeps* growing on the dry riverbed of the Kitagawa River, the riverbanks, and the riverfront at Susa, Nobeoka City, Miyazaki Prefecture, Japan and at Hyono, Matono, and Honmura Areas, Kitagawa Town in the same Prefecture

Gall midge	Galled part	2000 Nov.	2001 Nov.	2002 Nov.	2003 Oct.	2004 Nov.
R. cinerarius Monzen	Leaf	O	O	O	O	-
R. iwatensis Shinji	Terminal bud	O	O	O	-	-
R. shinji Gagné*	Axillary bud	-	O	O	-	-
R. struma Monzen*	Stem	O	O	O	O	O
R. yomogicola (Matsumura)	Leaf	O	O	O	O	O
Rhopalomyia sp. 1	Leaf	-	O	-	O	O
Rhopalomyia sp. 2	Leaf vein	-	O	-	-	-
Total number of species		4	7	5	4	3

* *R. shinjii* and *R. struma* were treated as a single species in the world catalog (Gagné 2004), but they are different species (Yukawa and Masuda 1996).
O, the species was found; -, the species was not found.

6.3 Results

6.3.1 Species Composition of *Rhopalomyia* Gall Midges

During the field surveys, five named and two unidentified species of *Rhopalomyia* were found on *A. princeps* at the census fields in Susa, Nobeoka City, Miyazaki Pref., Japan (Table 1). They included three leaf gallers, two bud gallers, one stem galler, and one leaf vein galler. Because the 2000 and 2001 data showed that more species appeared in November than in March, May, and June, we concentrated our field surveys in October or November thereafter. Usually four or more species were recorded in the years from 2000 to 2003, but species number reduced to three after the great flood in October 2004 (Table 1). *R. struma* and *R. yomogicola* were found throughout the surveys, while *Rhopalomyia* sp. 2 and *R. shinjii* were found only once or twice, respectively.

The number of *Rhopalomyia* species found in the surveys did not differ very much between the three habitats, whilst population densities exhibited distinct differences between some species and between different habitats (Tables 2–4). Gall midge densities were so low in 2003 that their densities could not be compared statistically, hence the data is not shown in this paper. On the dry riverbed and riverbank, the population density of *R. struma* was significantly highest among the congeners in 2001 and 2004, and subequal to that of *R. yomogicola* in 2002 (Tables 2–4). In the river edge, the density of *R. struma* was highest in 2001 and subequal to that of *R. yomogicola* in 2002 and 2004. *R. cinerarius* was found four times throughout the surveys (Table 1), but its density was very low, and it was frequently absent in two or all three habitats (Tables 2–4).

Table 2. Relative abundance of *Rhopalomyia* gall midges in November 2001 at different habitats in Susa

Gall midge	Habitat			Total
	Dry riverbed	Riverbank	River edge	
Plants examined	595	373	705	1673
R. cinerarius	0 (0.00)	0 (0.00)	8 (1.13)^{x}	8 (0.48)^{x}
R. iwatensis	1 (0.17)^{x}	2 (0.54)^{x}	0 (0.00)	3 (0.18)^{xy}
R. shinjii	1 (0.17)^{x}	0 (0.00)	0 (0.00)	1 (0.06)^{y}
R. struma	34 (5.71)^{ay}	15 (4.02)^{aby}	15 (2.13)^{bx}	64 (3.83)^{z}
R. yomogicola	6 (1.01)^{x}	4 (1.07)^{x}	0 (0.00)	10 (0.60)^{x}
Total	42 (7.06)^{a}	21 (5.63)^{ab}	23 (3.26)^{b}	86 (5.14)

Numerals indicate the number of gall-bearing plants and percentages in parentheses.
Different letters in the same line (a, b, c) or column (x, y, z) indicate a significant difference ($\chi^2 > \chi^2_{0.025}$).

Table 3. Relative abundance of *Rhopalomyia* gall midges in November 2002 at different habitats in Susa and its vicinity

Gall midge	Habitat			Total
	Dry riverbed	Riverbank	River edge	
Plants examined	300	224	150	674
R. cinerarius	0 (0.00)	2 (0.89)^{x}	0 (0.00)	2 (0.29)^{x}
R. iwatensis	0 (0.00)	0 (0.00)	0 (0.00)	0 (0.00)
R. shinjii	0 (0.00)	0 (0.00)	0 (0.00)	0 (0.00)
R. struma	2 (0.67)^{ax}	9 (4.02)^{by}	4 (2.67)^{abx}	15 (2.23)^{y}
R. yomogicola	8 (2.67)^{ay}	12 (5.36)^{ayz}	6 (4.00)^{ax}	26 (3.86)^{yz}
Total*	10 (3.33)^{a}	20 (8.93)^{b}	10 (6.67)^{ab}	40 (5.93)

Numerals indicate the number of gall-bearing plants and percentages in parentheses.
Different letters in the same line (a, b, c) or column (x, y, z) indicate a significant difference ($\chi^2 > \chi^2_{0.025}$).
* Some plants bore more than one sort of gall.

Table 4. Relative abundance of *Rhopalomyia* gall midges in November 2004 at different habitats in Susa

Gall midge	Habitat			Total
	Dry riverbed	Riverbank	River edge	
Plants examined	161	145	156	462
R. cinerarius	0 (0.00)	0 (0.00)	0 (0.00)	0 (0.00)
R. iwatensis	0 (0.00)	0 (0.00)	0 (0.00)	0 (0.00)
R. shinjii	0 (0.00)	0 (0.00)	0 (0.00)	0 (0.00)
R. struma	16 (9.94)^{ax}	39 (26.90)^{bx}	21 (13.46)^{ax}	76 (16.45)^{x}
R. yomogicola	5 (3.11)^{ay}	24 (16.55)^{by}	25 (16.03)^{bx}	54 (11.69)^{y}
Rhopalomyia sp. 1	6 (3.73)^{ay}	11 (7.59)^{bz}	17 (10.90)^{bx}	34 (7.36)^{z}
Total*	25 (15.53)^{a}	69 (47.59)^{b}	55 (35.26)^{c}	149 (35.50)

Numerals indicate the number of gall-bearing plants and percentages in parentheses.
Different letters in the same line (a, b, c) or column (x, y, z) indicate a significant difference ($\chi^2 > \chi^2_{0.025}$).
* Some plants bore more than one sort of gall.

6.3.2 Effects of Running Water on the Survival of *Artemisia princeps* Leaves

None of *A. princeps* leaves detached in the running water when they had been submerged for less than four days, but some leaves detached from the plants that had been submerged for four or more days (Table 5). In particular, the leaves submerged for more than six days detached at a significantly higher rate than those submerged for less than six days. The proportion of leaves that had been shredded into small pieces was not clearly related to the duration of submergence, whereas the proportion of the total number of leaves detached and shredded was significantly higher in the plants submerged for more than five days than in the others (Table 5).

6.3.3 Effects of Submergence on the Survival of *Rhopalomyia* Gall Midges

Because survival rate was not significantly different between larvae and pupae in the three *Rhopalomyia* species examined, their data were combined for comparison among the species. The survival rate of *R. struma* was significantly higher than that of two other species within one and two days after submergence (Table 6). The survival rate of *R. cinerarius* was significantly higher than that of *R. yomogicola* within one day, but no significant difference existed between them after the two-day submergence.

Table 5. Effects of running water on the detachment of submerged *A. princeps* leaves

Duration of submergence	Plants examined	Leaves examined	Leaves detached	Leaves shredded	Total leaves affected
3 days	10	180	0 (0.0)	3 (1.7)[a]	3 (1.7)[a]
4 days	10	164	2 (1.2)[a]	7 (4.3)[ab]	9 (5.5)[a]
5 days	10	133	2 (1.5)[a]	1 (0.8)[a]	3 (2.3)[a]
6 days	10	117	3 (2.6)[ab]	12 (10.3)[c]	15 (12.8)[b]
7–8 days	20	321	22 (6.9)[b]	23 (7.2)[bc]	45 (14.0)[b]

Numerals indicate the number of plants or leaves and percentages in parentheses. Different letters in the same column (a, b, c) indicate a significant difference ($\chi^2 > \chi^2_{0.05}$).

Table 6. Comparison among the three *Rhopalomyia* species of the survival rate of immature stages (larvae and pupae were taken together) after one, two, and three days submergence in dead water

Duration of	*R. struma*		*R. cinerarius*		*R. yomogicola*	
submergence	n	Survived	n	Survived	n	Survived
1 day	116	88 (75.9)[ax]	34	13 (38.2)[b]	56	8 (14.3)[c]
2 days	70	33 (47.1)[ay]	30	5 (16.7)[b]	51	8 (15.6)[b]
3 days	72	0 (0.0)	22	0 (0.0)	19	0 (0.0)

Numerals indicate the number of survived individuals and percentages in parentheses.
Different letters in the same line (a, b, c) or column (x, y, z) indicate a significant difference ($\chi^2 > \chi^2_{0.05}$).

6.4 Discussion

The species composition of *Rhopalomyia* gall midge after the great flood on 4 November 2004 was apparently different from that at normal water level in 2001–2002. The relative abundance of *R. struma* to *R. yomogicola* became greater (Table 4) and no galls of *R. cinerarius* were found. This means that the flood changed the species component by differently influencing different gall midge species. Running water with a velocity of 3.0 m/s removed *A. princeps* leaves at a relatively high rate when the plants had been submerged in water for more than several days (Table 5). At the time of the 2004 flood, the Kitagawa River attained an extremely high water level of 15.7 m and the velocity of running water was estimated to be about 3.4 m/s (The Kitagawa Data). In addition to the high velocity, various sorts of driftage might directly crush the plants. This situation lasted for three days (The Kitagawa Data). Under such conditions, many *A. prin-*

ceps leaves were probably removed, together with midge galls if they bore them. Submergence of galls may have caused the death of midge larvae and pupae, if it lasts for more than two days. All these factors led to the decrease of population density of the leaf gallers.

Before and after a flood, the duration of submergence may vary from place to place because of different water levels determined by the topography of dry riverbeds. The stem galler, *R. struma*, would have more chance to survive in some places where the duration is shorter than at other places because its larvae and pupae living in the hard stem galls survived at a higher rate than the other species under two-day submerged conditions (Table 6). Besides, the stem galls would not readily detach from the plants even when they are submerged in the running water with a high velocity. Such differences in galling traits between the *Rhopalomyia* species are considered to determine the species composition of the *Rhopalomyia* gall midges.

Except for the great flood in October, *A. princeps* in the dry riverbed was submerged under a water level more than 8.0 m for at least 50 days in 2004 (The Kitagawa Data) due to copious rain, which was not so heavy as the rain that caused the great flood. The high water level did not last more than four days but the leaf gallers might be affected by the submergence. In particular, if the water attains high level in winter, the effect would be more severe for every species because they overwinter on the ground in the withered leaf or stem galls, which could be carried away from the dry riverbed. However, the population density usually recovers in October and November (Table 1) because all of the species are multivoltine (Yukawa and Masuda 1996).

Thus, floods certainly kill gall midges, leading directly to the reduction in gall midge density. However, population density of respective species is determined not only by floods but also by various kinds of biotic and abiotic factors. Therefore, we need more detailed life table data to evaluate the population densities of multivoltine gall midges and compare them between successive years.

Rhopalomyia gall midges are commonly seen on *Artemisia* in a dry riverbed although it is frequently disturbed by floods. We have seen that some species recolonized the dry riverbed sooner or later in the subsequent generation after a flood. Disturbance may provide *Rhopalomyia* gall midges with many oviposition sites on regrowths from damaged *Artemisia* and promote the recolonization of the dry riverbed by the gall midges, as has been noted in the plant vigor hypothesis (e.g., Vieira et al. 1996).

6.5 Acknowledgements

We thank Dr. K. M. Harris for reading through an early draft of this paper and Prof. S. Sugio and Mr. M. Takahira for the information on the Kitagawa River. We are indebted to the Nobeoka Office of River and National Highway, Japan Ministry of Land, Infrastructure and Transport, and the Nobeoka Office for Public Works, Miyazaki Prefecture for various data of the Kitagawa River. Our thanks are also due to Dr. S. Kamitani and some students in the Entomological Laboratory, Kyushu University for their assistance in the field surveys and to Mr. D. Yamaguchi, Mr. T. Higuchi, Dr. K. Paku, and Mr. N. Ikematsu for their help in the running water experiment. TG, NU, and MN thank Prof. O. Tadauchi for his support in various ways. This study was supported by the Foundation for Riverfront Improvement and Restoration, Tokyo, Japan. This is a contribution from the Entomological Laboratory, Faculty of Agriculture, Kyushu University, Fukuoka (series 6, No. 18).

6.6 References

Gagné RJ (2004) A catalog of the Cecidomyiidae (Diptera) of the world. Memoirs of the Entomological Society of Washington 25:1-408

Kamata N, Igarashi Y (1994) Influence of rainfall on feeding behavior, growth, and mortality of larvae of the beech caterpillar, *Quadricarcarifera punctatella* (Motchulsky) (Lepidoptera: Notodontidae). Journal of Applied Entomology 118:347-353

Lensing JR, Todd S, Wise DH (2005) The impact of altered precipitation on spatial stratification and activity-densities of springtails (Collembola) and spiders (Araneae). Ecological Entomology 30:194-200

Nakamura M, Utsumi S, Miki T, Ohgushi T (2005) Flood initiates bottom-up cascades in a tri-trophic system: host plant regrowth increases densities of a leaf beetle and its predators. Journal of Animal Ecology 74:683-691

Tanzubil PB, McCaffery AR, Mensah GWK (2000) Diapause termination in the millet stem borer, *Coniesta ignefusalis* (Lepidoptera: Pyralidae) in Ghana as affected by photoperiod and moisture. Bulletin of Entomological Research 90:89-95

Vieira EM, Andrade I, Price PW (1996) Fire effects on a *Palocourea rigida* (Rubiaceae) gall midge – a test of the plant vigor hypothesis. Biotropica 28:210-217

Wakisaka S, Tsukuda R, Nakasuji F (1991) Effects of natural enemies, rainfall, temperature and host plants on survival and reproduction of the diamondback moth. In: Talekar NS (ed) Diamondback moth management: proceedings of the second international workshop. The World Vegetable Center, Tainan, pp 15-26

Yukawa J, Masuda H (1996) Insect and mite galls of Japan in colors (in Japanese with English explanation for color plates). Zenkoku Nôson Kyôiku Kyôkai, Tokyo

7 Guild Structure of Gall Midges on *Fagus crenata* in Relation to Snow Gradient: Present Status and Prediction of Future Status as a Result of Global Warming

Naoto Kamata[1], Shinsuke Sato[1], and Jiro Kodani[2]

[1] Laboratory of Ecology, Graduate School of Natural Science and Technology, Kanazawa University, Ishikawa 920-1192, Japan
[2] Ishikawa Forest Experiment Station, Sannnomiya, Hakusan, Ishikawa 920-2114, Japan

Summary. Twenty six species of gall midges (Diptera: Cecidomyiidae) induce leaf galls on *Fagus crenata*. Because the adult life span of gall midges is very short, they must emerge and oviposit in the short period of a specific stage of budburst. *Fagus crenata* is mainly distributed in regions with heavy snowfall. Snow cover prevents the emergence of gall midges that overwinter on the ground as immature stages. Therefore the time of snow melt in relation to that of budburst is likely to be an important factor determining the success of gall midges. The species number and density of the *Fagus* gall midges tend to be higher in intermediate snowfall areas, in which snow covers the ground surface throughout the winter but the time of snow melt is earlier than that of budburst. The gall midge fauna is known to be poor in *F. crenata* forests with little snowfall because of desiccation during the winter. As a result of global warming, the distribution range of *F. crenata* will shift to regions with higher elevation and/or higher latitude although the rate of this vegetation shift is considered to be slower than that of temperature change. We hypothesize that the *Fagus* gall midge fauna will become richer in the short term because *F. crenata* forests with intermediate snowfall will increase by the global warming. However, in the long term, the gall midge fauna will become poorer following the retrenchment of *F. crenata* forests.

Key words. Snow melt, Budburst, Synchrony, Gall midge, Beech leaf

7.1 Gall Midge on *Fagus crenata* and Snow

Snow plays an important role in maintaining moisture and stabilizing ground surface temperature during winter. Synchronization between bud-burst and adult emergence is critical for the females to lay their eggs successfully on appropriate host buds, in particular for such short lived insects as gall midges (Diptera: Cecidomyiidae) (Yukawa 2000). Gall midges that overwinter on the ground as immature stages are naturally expected to be most abundant in regions with intermediate amount of snowfall, in which snow covers the ground throughout winter and disappears before budburst. In contrast, heavy snow coverage may delay the emergence of gall midges and cause asynchrony with host budburst. Thus, we have been suspecting that the abundance and distribution of the gall midges are greatly affected by snow accumulation, which is now predicted to be reduced gradually as a result of global warming.

In Japan, 26 sorts of midge gall are known to occur on the leaves of *Fagus crenata* Blume (Fagaceae) (Sato and Yukawa 2001; Yukawa and Masuda 1996). These galls have been considered to be induced by different gall midge species, respectively, because galls are fundamentally species-specific in shape and structure (Yukawa and Masuda 1996). Therefore, in this paper we regard each midge gall as a representative of gall midge species, although most of them have not been identified yet. Gall midges on beech foliage are appropriate objects for studying insect guilds because many species coexist on a single plant species and the density of respective species is easily determined by counting galls collected by litter traps deployed near the forest floor throughout the seasons. By putting this convenience to field survey, we studied the guild structure of gall midges on *F. crenata* in relation to snow-coverage gradient.

7.2 Overwintering Patterns of Gall Midges on *F. crenata*

Three different overwintering patterns have been described for gall midges on *F. crenata* (Yukawa and Masuda 1996):

(1) Species that overwinter on the ground (Type G).
- (1-1) Species that overwinter on the ground in the larval stage (Subtype Gl).
- (1-2) Species that overwinter on the ground in the pupal stage (Subtype Gp).

(2) Species that overwinter inside buds in the larval stage (Type B).
- (2-1) Galls fall early in the season (Subtype Be).

(2-2) Galls fall late in the season (Subtype Bl).
(2-3) Adults emerge before gall fall (Subtype Bn).

Table 1 shows the life history patterns of 26 species of gall midges on *F. crenata*. Gall midges belonging to Type G are likely to be strongly influenced by snow and desiccation. In contrast, those with the life history trait of overwintering in host buds in the larval stage are not influenced by phenological asynchrony caused by heavy snow.

7.3 Gall Midge Fauna in Relation to Snow

7.3.1 Study Sites and the Estimation of Gall Midge Density

We established seven study sites in natural *F. crenata* forests in Ishikawa Prefecture, Central Japan. The area is famous for its heavy snowfall in winter because of the wet NW wind that emanates from the Japan Sea. The average snow depth ranges from 98 to 258 cm (Japan Meteorological Agency 2002). Density of gall midges was determined by the number of galls collected in five rectangular litter traps (1 m×1 m) set in the understory of each study site throughout the seasons. Collecting intervals were between 2 weeks and 1 month.

7.3.2 Species Diversity and Abundance of Gall Midges

Twenty-four gall midge species out of the 26 were collected during the four-year survey in the seven study plots. The number of gall midge species collected at each site throughout the survey ranged from 7 to 23. Gall density in each year at each plot ranged from 11.8 to 4044.2/m^2.

Annual means of maximum snow depth did not show a negative relationship with and the number of gall midge species (Fig. 1a), but negatively correlated with the gall density (Fig. 1b).

7.3.3 Gall Midge Fauna in Heavy Snowfall Regions

Five gall midge species were abundant (yearly mean gall density >10/m^2) even in heavy snowfall regions (Table 1). Two of the five species belonged to Type B and the three others to Type G. At first, snow around *Fagus* tree trunks began to melt. Even in heavy snowfall regions, the ground around the tree trunks was exposed synchronizing with the time of budburst (Fig. 2). Local populations of Type G species seem to have been sustained by some individuals that overwinter close to the trees. Only two gall midge species were more abundant in the sites with heavy snowfall than in intermediate snowfall sites nearby although their densities were not so high. These results suggest that snow cover at the time of budburst strongly limits gall midge density to the low level.

Table 1. Twenty-six gall midge species on *Fagus crenata* in Japan and their life history patterns

Japanese common name of the midge galls	Gall midge species[a]	Patterns[b]	Habitat specification[c]
Buna-ha-akagetamafushi	NI	Be	Heavy snowfall area
Buna-ha-magetamafushi	NI	Be	
Buna-ha-togetsunofushi	NI	Be	
Buna-ha-futotsunofushi	NI	Be	
Buna-ha-futokotsunofushi	NI	Bl	
Buna-ha-nagatsunofushi	NI	Bn	
Buna-hamyaku-kobufushi	NI	Bn	Heavy snowfall area
Buna-ha-kometsubufushi	NI	Gl	Heavy snowfall area
Buna-ha-marutsunofushi	NI	Gl	
Buna-ha-fukurefushi	NI	Gl	Dry forest
Buna-ha-kaigarafushiura	*Hartigiola faggalli*	Gl	Heavy snowfall area
Buna-ha-kaigarafushiomote	*Hartigiola faggalli*	Gl	Dry forest
Buna-haura-kobufushi	*Janetiella infrafoli*	Gl	
Buna-ha-tamafushi	NI	Gl	Heavy snowfall area
Buna-ha-kibatsunofushi	NI	Gl	
Buna-ha-hishigatafushi	NI	Gl	Dry forest
Buna-ha-marutamafushi	NI	Gp	
Buna-haberi-tamafushi	NI	Gp	
Buna-ha-tsunofushi	NI	Gp	
Buna-ha-kotsunofushi	NI	Gp	
Buna-haberi-hosofushi	NI	Gp	
Buna-hasuji-togaritamafushi	NI	Gp	
Buna-hasuji-donngurifushi	NI	Gp	
Buna-ha-hekomikotsunofushi	NI	UN	
Buna-haura-kefushi	NI	UN	
Buna-ha-ootsunofushi	NI	UN	

[a]Gall midge species: NI, not identified yet
[b]Patterns: UN, unknown; other abbrebiations are shown in texts.
[c]Habitat specification indicates speceis that were abundant in heavy snowfall areas in Ishikawa Prefecture and/or in dry beech forest in Kyushu.

7.4 Gall Midge Fauna on *F. crenata* in Dry-winter Areas

At Mt. Sefuri, located at the border between Fukuoka and Saga Prefectures, Kyushu, winter precipitation (December–February) was 285 mm and maximum snow depth was 8 cm for 30-year average (1971–2000) (Japan Meteorological Agency 2002). These climate values in the beech forests in Ishikawa Prefecture ranged 1031–1463 mm for winter precipitation and 98–256 cm for the maximum snow depth (Japan Meteorological Agency 2002). Hence, we judged beech forests in Mt. Sefuri as dry beech forests. According to the surveys on Mt. Sefuri (S. Sato and J. Yukawa, unpublished data), the species richness of gall midges was low and only four Type G1 species were dominant although their density was relatively low (Table 1). Thus, the *Fagus* gall midge fauna in the dry beech forests was poorer than in heavy and intermediate snowfall regions.

7.5 Influences of Global Warming on Gall Midge Fauna on *F. crenata*

7.5.1 Influence of Global Warming on *F. crenata* Forests in Japan

Distribution of *F. crenata* forests is determined by four climatic factors (summer precipitation, winter precipitation, minimum temperature of the coldest month, and warmth index) (Matsui et al. 2004a). Snow cover is known to be an important determinant of the distribution of *F. crenata* in Japan. *F. crenata* cannot grow in cool regions with little snowfall because this species is less tolerant of cold than is *Quercus crispula* Blume (Fagaceae), which otherwise occupies a similar ecological niche. As a result of global warming, the distribution range of *F. crenata* is forecasted to shift to regions with higher elevation and/or higher latitude and to shrink (Matsui et al. 2004b). Areas in which *F. crenata* can survive with high probability (>0.5) will decrease by 91%, retreating from the southwest, shrinking in central regions, and expanding toward the northeast beyond the current northern limit. However, the expansion of distribution range in woody plants is generally slower than the change of temperature because it takes a long time to start reproduction and the distance of seed dispersal can be a limiting factor.

In the short-term, the distribution range of *F. crenata* will not change greatly. *F. crenata* will disappear gradually from the southernmost and lowest regions of its current distribution range. The rate of range expansion to more northern and higher elevation areas would be smaller than the rate of disappearance (Kamata 2005).

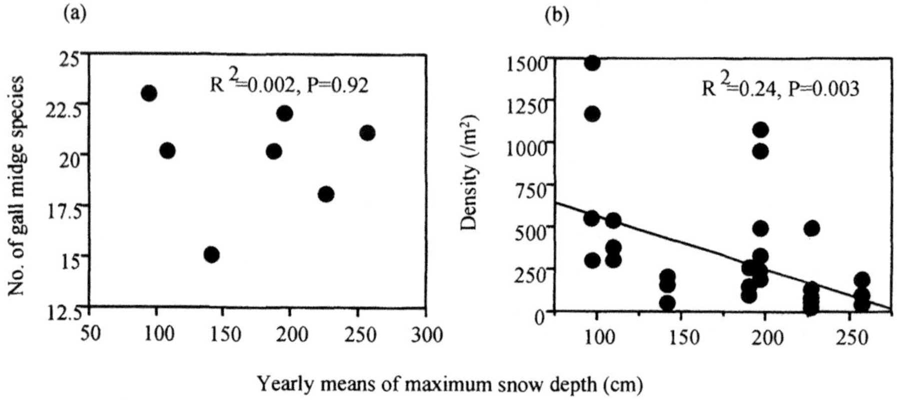

Fig. 1. Relationships of annual means of maximum snow depth with **(a)** the number of gall midge species and **(b)** gall density in seven *F. crenata* forest stands.

Fig. 2. Snow after budburst in a *F. crenata* stand in a heavy snowfall region. Litter on snow indicates scales and female flowers of *F. crenata*. The ground near the tree trunks started to be exposed at the time of budburst.

7.5.2 Influence of Global Warming in Phenology

As a result of global warming, snowfall would decrease and both snow melt and budburst would occur earlier, of which the former would start much earlier than the latter (Fig. 3). In the short term, the *Fagus* gall midge fauna will become richer in large areas because areas that presently have heavy snowfall will change to areas with intermediate snowfall, in which snow will disappear before budburst. In the long term, however, the gall midge fauna on *F. crenata* will become poorer following the retrenchment of *F. crenata* forests (Fig. 4).

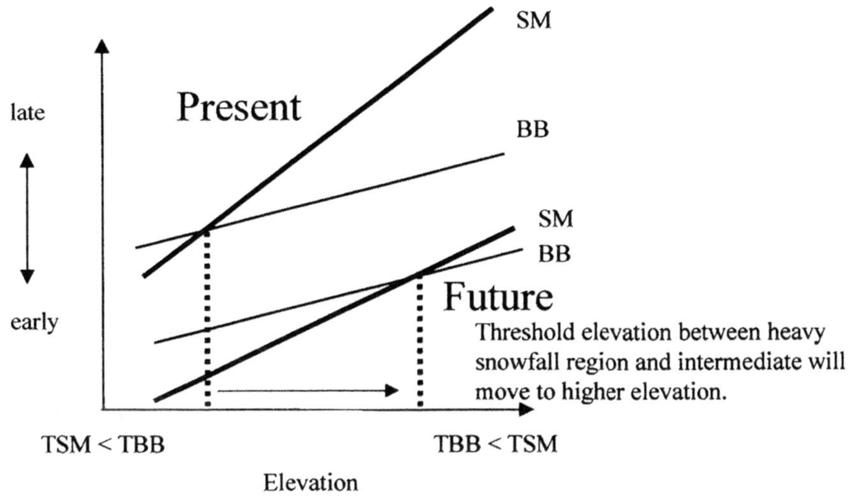

Fig. 3. Changes in threshold elevation between areas with heavy and intermediate snowfalls by global warming (modified from Kamata 2005). The time of budburst (TBB) will become earlier than at present. The time of snow melt (TSM) will also change but more drastically than TBB because TSM is influenced by both snowfall and speed of snow melting. As a result, the threshold elevation will rise more greatly than would be estimated on the basis of temperature change alone.

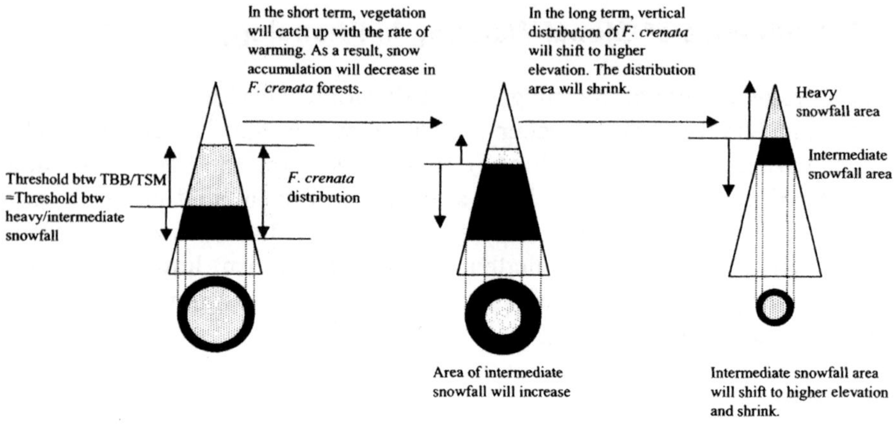

Fig. 4. Prediction of vertical shift of *F. crenata* distribution and the *Fagus* gall midges (modified from Kamata 2005). Area with intermediate snowfall, which is a suitable habitat for the gall midges, will increase as a result of global warming in the short term. However, the distribution of *F. crenata* will shift to higher elevation and the area of suitable habitat will decrease greatly in the long term.

7.6 Conclusion

At present, the gall midge fauna on *F. crenata* is most abundant in areas with intermediate snowfall, in which snow covers the ground in winter and melts before the budburst of *F. crenata*. In heavy snowfall regions, snow cover delays the emergence of gall midges that overwinter on the ground as immature stages, causing asynchrony of the emergence and oviposition with the time of budburst. In dry beech forests, mortality caused by desiccation in winter seems to limit the diversity of gall midges. The distribution range of *F. crenata* forests is forecasted to shift northward/upward and to shrink as a result of predicted global warming. Because plants cannot respond to these changes rapidly, in the short term, *F. crenata* forests with intermediate snowfall, which is preferable to the gall midges, will increase. However, in the long term, the gall midge fauna will become poorer following the retrenchment of *F. crenata* forests.

7.7 Acknowledgement

This study was partly supported by a grant-in-aid to Prof. Junichi Yukawa (No. 11308021) and 21st-century COE Program to Prof. Kazuichi Hayakawa (No. 1440101) both from MEXT. Associate Prof. Satoshi Kamitani (Kyushu University) kindly helped us for climate data. We would like to express our sincere thanks to all of these.

7.8 References

Japan Meteorological Agency (2002) Mesh climate data for 1971–2000 (CD-ROM). Japan Meteorological Business Support Center, Tokyo

Kamata N (2005) Diverse world of forest insects in Japan: ecology, evolution, and conservation. Tokai University Press, Hatano

Matsui T, Yagihashi T, Nakaya T, Tanaka N, Taoda H (2004a) Climatic controls on distribution of *Fagus crenata* forests in Japan. Journal of Vegetation Science 15:57-66

Matsui T, Yagihashi T, Nakaya T, Taoda H, Yoshinaga S, Daimaru H, Tanaka N (2004b) Probability distributions, vulnerability and sensitivity in *Fagus crenata* forests following predicted climate changes in Japan. Journal of Vegetation Science 15:605-614

Sato S, Yukawa J (2001) Absence record of *Fagus* gall midges (Diptera: Cecidomyiidae) on Ulleung Island, Korea and in North America. Esakia 41:17-25

Yukawa J (2000) Synchronization of gallers with host plant phenology. Population Ecology 42:105-113

Yukawa J, Masuda H (1996) Insect and mite galls of Japan in colors (in Japanese, with English explanation for color plates). Zenkoku Nôson Kyôiku Kyôkai, Tokyo

2. Biological Control and Galling Arthropods

8 Early Parasitoid Recruitment in Invading Cynipid Galls

Karsten Schönrogge[1], Seiichi Moriya[2], George Melika[3], Zoë Randle[1], Tracey Begg[4], Alexandre Aebi[4], and Graham N. Stone[4]

[1]Centre for Ecology and Hydrology, CEH Dorset, Winfrith Technology Centre, Dorchester, DT2 8ZD, UK
[2]National Agricultural Research Center, Tsukuba, Ibaraki 305-8666, Japan
[3]Systematic Parasitoid Laboratory, Vas County Plant Protection and Soil Conservation Service, Kelcz-Adelffy St. 6, Kőszeg 9730, Hungary
[4]Institute of Evolutionary Biology, The Kings Buildings, West Mains Road, Edinburgh, EH9 3JT, UK

Summary. Biological invasions are widely seen as the biggest threat to biodiversity next to the loss of habitats. One aspect of considerable interest is the recruitment of natural enemies after the establishment of the invading species and how such enemies link invaders to native communities. However, not all invaders are invasive. Eight cynipid species originating in south-eastern Europe invaded Britain over the last 200 years. Presently they cause no economic concern or have any detectable detrimental effect on the native cynipid fauna. Since their invasions have been allowed to progress without intervention, they provide an excellent opportunity to study the recruitment of natural enemies and their integration into native communities. In contrast, the invasion of Japan by *Dryocosmus kuriphilus* from China caused great economic concern, because considerable damage to its host trees, *Castanea* spp. a valuable fruit tree in Japan and elsewhere in the world, is caused by high infestation rates. Here we review the early recruitment of parasitoids to the alien species in the UK and *D. kuriphilus* in Japan, their role in the invaders population dynamics, and how they link the invaders to native cynipid communities.

Key words. Biological invasions, Biological control, Parasitoid recruitment, Community structures, Gall attributes

8.1 Introduction

Among cynipid gall wasps there are at least three independent examples of biological invasions that have resulted from human activity. Two cases represent introductions of individual species that are significant because they affect economically important host plants. Introduced European cork oak, *Quercus suber*, in California has been colonised by a European cynipid, *Plagiotrochus suberi* (Bailey and Stange 1966), and Chestnut (*Castanea*) in Japan, North America and Europe has been colonised by *Dryocosmus kuriphilus*, a native of China and Korea (Brussino et al. 2002; Moriya et al. 1989a; Payne 1978). The most significant example of cynipid range expansion is associated with human dispersal of *Quercus cerris* in Europe. This section Cerris oak is native to Italy, the Balkans and Asia Minor, and is the host for one or both generations of a wide diversity of cynipids (Melika et al. 2000; Stone et al. 2001). The natural distributions of all European section Cerris oaks correspond closely to glacial refugia for oaks during the Pleistocene ice ages, and following the retreat of the ice sheets only oaks in the section Quercus (particularly *Quercus robur* and *Quercus petraea*) were able to escape the refugia and recolonise northern Europe (Stone et al. 2001). As a result, no cynipids, dependent on *Q. cerris* (or any other section Cerris oak) for one or both generations in their lifecycle, occur naturally in northern Europe.

Over the last 400 years *Q. cerris* has been planted widely north and west of its native range (Stone and Sunnucks 1993), creating a mosaic of *Q. cerris* patches within the natural distribution of section *Quercus* oaks. In contrast to the situations with *D. kuriphilus* and *P. suberi* there are no geographical barriers between the native and invaded range for cynipid gall wasps using *Q. cerris*. As a result at least 10 species have subsequently invaded north-western Europe, including eight host-alternating *Andricus* species (*Andricus aries*, *Andricus corruptrix*, *Andricus gemmeus*, *Andricus grossulariae*, *Andricus kollari*, *Andricus lignicolus*, *Andricus lucidus* and *Andricus quercuscalicis*) and two species currently thought to be wholly dependent on *Q. cerris* (*Aphelonyx cerricola* and *Neuroterus saliens*) (Stone et al. 2002). *A. aries*, *A. corruptrix*, *A. grossulariae*, *A. lignicolus*, *A. lucidus*, *A. quercuscalicis* and *A. cerricola* all reached Britain between 1950 and 2000, 2000 km from the nearest natural *Q. cerris* stands, apparently without direct human assistance. *A. kollari* has spread naturally across most of northern Europe, but was also deliberately introduced into Britain from the eastern Mediterranean in the first half of the nineteenth century (Askew 1984; Stone et al. 2001). Range expanding cynipids have proven to be valuable model systems for studies on the recruitment of communities of natural enemies (Stone and Schönrogge 2003; Stone et al. 2002).

We first review the studies on parasitoid recruitment to the galls of *A. quercuscalicis,* the most extensively studied species in Europe. Secondly we consider the invasions of *A. corruptrix*, *A. kollari*, *A. lignicolus*, and *A. quercuscalicis* and how parasitoid species link the aliens to the native cynipid community, asking why these communities are invasible. Lastly, we will draw on evidence from these gall wasp invasions including that of *D. kuriphilus* in Japan to assess the role of parasitoid recruitment in the invasion process.

8.2 Parasitoid Recruitment to the Galls of *Andricus quercuscalicis*

A. quercuscalicis is a host-alternating invader in Western Europe (described above), establishing populations along a 2000 km invasion route from the Balkans to Britain (Schönrogge et al. 1995). First records outside the native range date as far back as 1631 from the southeast of Germany. Since then the species was described from the area of Berlin in 1787, and Bejerinck in the Netherlands described its life-cycle and the switch from *Q. cerris* to *Q. robur* between generations in 1882 (Schönrogge 1994 and therein). Claridge (1962) first recorded the species in Britain in 1958 and in 1986 the galls were first recorded from Ireland (Schönrogge 1994 and therein). These and some additional natural historic records provide a rough timeline of the spread of *A. quercuscalicis* across Europe. With the above dates and data on the parasitoid assemblage richness from the asexual galls from sites along the invasion route, it is possible to estimate the residence time of *A. quercuscalicis* in the regions described by Schönrogge et al. (1995). A simple regression of residence time against the residuals of a log-log regression of parasitoid species richness against sample size (correcting for variable sample sizes) explains 91% of the variability in parasitoids species richness across the invaded range ($F_{1,3} = 30.5$, $P < 0.05$). Thus parasitoid species richness is increasing with residence time and, possibly more significantly, the parasitoid assemblages associated with the asexual galls of *A. quercuscalicis* in the invaded range were subsets of those known from the native range (Schönrogge et al. 1995).

After *A. quercuscalicis* arrived in Britain the galls of the asexual generation remained virtually parasitoid free for about 20 years. Despite extensive rearings the first parasitoid records (*Torymus cyanaeus* and *Mesopolobus amaenus*) were only obtained in the late 1970's. *T. cyanaeus* has never been reared again from *A. quercuscalicis*, while *M. amaenus* attacked inquilines soon after their appearance in *A. quercuscalicis* galls, but has since remained rare. In this regard it seems to represent an early successional species in community development in the U.K. Curiously, though native to Europe this parasitoid has not been found in rearing of galls from the European continent (Hails et al. 1990; Schönrogge et al. 1995). Other opportunistic parasitoid species included ichneumonid parasitoids, *Mastrus castaneus* and *Gelis formicarius*, and the diapriid *Spilomicrus stigmaticalis* (Hails et al. 1990). The only inquiline species recorded during the 1980s was *Synergus pallicornis* and attack rates for both inquilines and parasitoids were generally low (<10%). However, subsequent recruitment of species such as *Mesopolobus sericeus*, *Sycophila biguttata, Eupelmus urozonus* represents the addition of taxa that are known to attack hosts inside cynipid galls, and were later found to attack hosts in the asexual galls of *A. quercuscalicis* in its native range, but were also all part of the native British parasitoid fauna (Askew 1961; Hails et al. 1990; Schönrogge et al. 1995).

Between 1990 and 1995 there was a sharp increase in the abundance of some parasitoid species and others were recorded for the first time. Both were closely associated with the recruitment of inquiline *Synergus* spp., particularly *Synergus gallaepomiformis*, to the asexual galls of *A. quercuscalicis* in the south-east of Britain. Both the geographical pattern of inquiline infestation and parasitoid recruitment followed the invasion route of the host (Schönrogge et al. 1996). Thus the recruitment process observed across Europe was mirrored on a smaller scale within Britain. The fact that parasitoid assemblages in the invaded range were almost perfect subsets of those recorded from the native range leaves in our view two non-exclusive hypotheses: (a) The recruited parasitoid species are from the native community and are pre-adapted to exploit the new host at the species level (native recruitment hypothesis), or (b) the parasitoid species attacking the host in its invaded range are "strains" that have pursued the host from the native range and represent invading genotypes (pursuit hypothesis).

Recruitment of parasitoid species to sexual generation galls of *A. quercuscalicis* was similar to that of the asexual galls in that assemblages recorded from Britain represented a subset of those recorded from the continent (Stone et al. 1995). However, we are not aware of time lags between the arrival in a new range and the recruitment of parasitoids to this generation. Given the time lag of parasitoid recruitment to the agamic generation on a tree species native to the UK, and since the sexual generation galls are formed on an introduced tree, *Q. cerris*, this is perhaps surprising, but all the parasitoid species attacking these galls also attack the sexual generation galls of *A. kollari* which arrived and spread throughout Britain 100 years before *A. quercuscalicis*. Thus a shift in the host searching behaviour of the parasitoid species between tree species could have happened even before *A. quercuscalicis* was in the country.

The convergence of the parasitoid faunas in the invaded and native ranges, despite differing environmental conditions and resident cynipid faunas, suggests a strong link between gall attributes (including host tree species, phenology, the plant organ galled, gall morphologies) and parasitoid community composition, species richness and abundance.

The galls induced by the two generations of *A. quercuscalicis* differ in several respects. The sexual generation gall is thin walled, 1–2 mm long, and develops very rapidly on the catkins of *Q. cerris*. In contrast, the asexual generation gall has a thick woody wall, reaches a diameter of up to 20 mm, and develops over several months on the acorns of *Q. robur*. These differences have two major consequences for the associated communities: (a) The asexual generation galls develop through a clear sequence of structural stages, and the parasitoids attacking this generation form a successional series associated with increasing host size from small parasitoid species with short ovipositors to larger species with long ovipositors (Schönrogge et al. 1995). In contrast, the rapid development of the sexual generation galls prevents such temporal structuring of parasitoid attack. (b) The asexual generation galls are attacked by several inquiline *Synergus* species, but even mature sexual generation galls appear to be too small and perhaps develop too rapidly to be suitable for inquiline development. Despite the major differences in host gall properties, in its native range the two generations of *A. quercuscalicis* support equally rich assemblages: 12 species in the sexual generation (all parasitoids) and 13 in the asexual generation (10 parasitoids, 3 inquilines). However, in the asexual generation only 4 of the 10 parasitoids feed predominantly on the gall wasp larva (the rest attacking inquilines), whereas in the sexual generation all of the parasitoid species attack the gall wasp or other parasitoids (Schönrogge et al. 1995; Stone et al. 1995). Only a single parasitoid is common to the communities of both generations, but plays a different role in each. In the sexual generation gall, the parasitoid *Cecidostiba fungosa* attacks the gall wasp larva. In the asexual generation this parasitoid attacks only inquilines in the outer wall of the gall, probably because at the time *C. fungosa* attacks, the gall wasp larva is beyond the reach of its short ovipositor. A final difference between the two communities is that across the range of *A. quercuscalicis* the mortality inflicted by parasitoid attack is far higher in the sexual generation (20–49%) than in the asexual generation (5–15%) (Hails and Crawley 1991, 1992; Schönrogge et al. 1995; Stone et al. 1995). Although the reasons for this difference are probably complex, it is tempting to suggest that the thin-walled, sexual generation galls are more vulnerable to attack by a rich assemblage of small parasitoids with short-ovipositors than the asexual generation galls. The differences in size and phenology between the galls induced by the two generations of *A. quercuscalicis* are shared with other cynipid species (Melika et al. 2000) and although detailed studies of the parasitoid assemblages associated with both generations of a cynipid species are rare (Askew 1961, 1980), strong differences between host generations are probably a common feature of oak cynipid communities.

8.3 Invaded Cynipid Communities in Britain

Indirect interactions through having shared natural enemies, and in particular through apparent competition (Holt 1977), are thought to be common and strong among endophytic insects because the sessile life style during their larval stages makes direct competitive interactions less likely. Yet indirect interactions through the food plant have also been reported (Sitch et al. 1988; Whitham 1978).

Eight communities of cynipid galls in Britain were studied intensively in 1994 and 1995 that included 1–4 of four alien species (*A. corruptrix*, *A. kollari*, *A. lignicolus*, and *A. quercuscalicis*) (Schönrogge et al. 1998, 2000). However, although Schönrogge and Crawley (2000) used equivalent methods to those applied in similar studies on aphid and leaf-miner systems in which apparent competition appeared to play a major role (Morris et al. 2004; Müller et al. 1999), they found no indication that apparent competition was important in shaping cynipid communities. They observed strong impacts of the aliens on the parasitoid populations, i.e. host shifts, changes in population sex-ratios and, in one community, satiation effects where host densities alternated between years and parasitoid abundance appeared to be limited by the low host densities (Schönrogge and Crawley 2000; Schönrogge et al. 1999), yet the indirect interactions among hosts were always weak.

The analysis of the quantified webs established for the eight communities also indicated that locally the parasitoid species appeared much more specialized than their species status as generalists would suggest. Where parasitoid species were shared among hosts, with few exceptions (e.g. *E. urozonus*, *Eurytoma brunniventris*), more than 90% of adults emerged from only one type of gall and generally less than 5% from any single alternative host. Based on current data it is not possible to distinguish whether local specialization is genetically fixed or whether host switches occur. However, it has been suggested that increased host fidelity promotes the stability of host–parasitoid communities that include generalist parasitoid species (Hastings and Godfray 1999).

The main feature of all eight communities studied by Schönrogge and Crawley (2000) is the isolation of the parasitoid assemblages associated with the galls of the sexual generations of the alien species on *Q. cerris*. These assemblages include four pteromalid species, *Mesopolobus dubius*, *Mesopolobus fuscipes*, *Mesopolobus tibialis* and *Mesopolobus xanthocerus*. As indicated above, all four *Mesopolobus* species have been known from the British fauna before the arrival of *A. quercuscalicis* and appear to have switched to the invading host while attacks on native galls are at least exceedingly rare. With the exception of *M. tibialis* (and probably *M. dubius*) the species are univoltine and only a few individuals of *M. tibialis* were reared from galls other than those on *Q. cerris*. As a consequence, there is no parasitoid species that is shared between any native cynipid and any of the invaders that would link those through both generations, but with shared natural enemy attack in only alternate generations, apparent competition would not be expected to occur.

We would suggest that this isolation and the degree of specialization within local parasitoid populations allow the coexistence of such a species rich fauna. Also the local community structure of galls and parasitoids determines part of the niche space into which an invading species may fit.

However, the absence of a tight linkage between the population dynamics of hosts and parasitoids does not mean that parasitoid attack can not affect the host populations. With attack rates sometimes found to be higher than 90%, they do affect gall densities and species that share parasitoid attack, such as the sexual generations of the alien species on *Q. cerris*, are likely to affect each others mortality rates within that generation. However, other sources of mortality such as bird predation that shows a more density dependent pattern of attack on this same set of galls (Schönrogge et al. 1999) are more likely to exert a degree of population regulation or even control.

8.4 Parasitoid Recruitment by *Dryocosmus kuriphilus* in Japan

D. kuriphilus is a univoltine cynipid gall wasp that induces multilocular galls on chestnut trees, *Castanea* spp. After its accidental introduction from China to Japan in 1941 (Shiraga 1951) it spread rapidly and was recorded by the late 1950's throughout the range of its host trees (Moriya et al. 2003). During a study from 1978 to 1981 at the Fruit Tree Research

Station at Yatabe, Japan, 10 species of parasitoids were recorded from the invasive galls and 5 of them were reared regularly between years (Ôtake et al. 1982). All but one of the 10 species are thought to have more than one generation each year, which means that their population dynamics were not closely tight to *D. kuriphilus* and they were not expected to exert any population control on the invading gall wasp. *Torymus beneficus* is univoltine and its phenology was reasonably matched with *D. kuriphilus* yet by 1981 *D. kuriphilus* populations remained uncontrolled. In 1982 a closely related parasitoid species from China, *Torymus sinensis*, was released and within 5 years the galling rate per chestnut shoot decreased spectacularly from about 43% to only 3%. Since the tolerable injury level was estimated at 30% (Gyoutoku and Uemura 1985; Moriya et al. 1989b), this represents economic control of the pest.

Like *T. beneficus*, *T. sinensis* is univoltine and its phenology well synchronized with *D. kuriphilus* and it is not clear why one species has very little effect on the host population, while the other is a more efficient control agent. Two features in the parasitoids biology have been suggested to explain the difference. Although *T. beneficus* seems to be univoltine, precise phenology data showed the occurrence of two emergence peaks (referred to as early and late emergence peaks) separated by 4 weeks during a year. In contrast, *T. sinensis* emerges in one emergence period between the two emergence peaks of *T. beneficus* and appears to be better synchronized with *D. kuriphilus*. This very fine phenological difference (1–2 weeks) may explain the higher efficiency of *T. sinensis* to control *D. kuriphilus* populations (Moriya et al. 1989a). Another explanation lies in the fact that *T. beneficus* has a shorter ovipositor than *T. sinensis*. *T. beneficus* does in fact attack only relatively small galls, i.e. the galls can outgrow the parasitoid, and so represent a refuge from attack, while *T. sinensis* can overcome this defense mechanism and attack larger galls. (Ôtake 1980; Ôtake et al. 1982).

Since the invasions of Japan, *D. kuriphilus* has invaded the United States and Europe. Details of these later invasions and prospects of *D. kuriphilus* in Europe are discussed in the chapter by Aebi et al. in this volume.

8.5 Concluding Remarks

Both the alien species in Britain and *D. kuriphilus* in Japan recruited largely multivoltine generalist parasitoids and at least for the asexual galls of *A. quercuscalicis* in Britain there was a considerable time lag (about 20 years) until a consistent parasitoid assemblage was established. While univoltine parasitoid species are rarer than multivoltine species within the assemblages, they appear to have shifted hosts almost entirely to the alien species. This is true for the three *Mesopolobus* species that attack the sexual galls of *A. quercuscalicis* and *T. beneficus* attacking *D. kuriphilus*. As a consequence the parasitoid assemblages associated with the sexual galls of *A. quercuscalicis* and the other species that make galls on *Q. cerris* studied today are isolated from those of native species. The apparent lack of tightly coupled host–parasitoid pairings is one aspect that allows the co-existence of cynipids in species rich communities and is one aspect that makes native cynipid communities invasible. Also, the differences between *T. sinensis* and *T. beneficus* in their biologies and the impact they have on *D. kuriphilus* population dynamics demonstrate that host–parasitoid dynamics need to be finely tuned for the parasitoid to exert population control.

8.6 References

Askew RR (1961) On the biology of the inhabitants of oak galls of Cynipidae (Hymenoptera) in Britain. Transactions of the Society for British Entomology 14:237-268

Askew RR (1980) The diversity of insect communities in leaf-mines and plant galls. Journal of Animal Ecology 49:817-829

Askew RR (1984) The biology of gall-wasps. In: Anathakrishnan TN (ed) The biology of galling insects. Oxford and IBH publishing, New Delhi, pp 223-271

Bailey SF, Stange LA (1966) The twig wasp of cork oak. Journal of Economic Entomology 59:663-668

Brussino G, Bosio G, Baudino M, Giordano R, Ramello F, Melika G (2002) Il cinipide galligeno *Dryocosmus kuriphilus* Yasumatsu: un pericoloso insetto esotico per il castagno europeo. L'Informatore Agrario 37:59-61

Claridge MF (1962) *Andricus quercuscalicis* (Burgsdorf) in Britain. The Entomologist 95:60-61

Gyoutoku Y, Uemura M (1985) Ecology and biological control of the chestnut gall wasp, *Dryocosmus kuriphilus* Yasumatsu (Hym.: Cynipidae). 1. Damage and parasitization in Kumamoto Prefecture (in Japanese). Proceedings of the Association for Plant Protection of Kyushu 31:213-215

Hails RS, Crawley MJ (1991) The population dynamics of an alien insect: *Andricus quercuscalicis* (Hymenoptera: Cynipidae). Journal of Animal Ecology 60:545-562

Hails RS, Crawley MJ (1992) Spatial density dependence in populations of a cynipid gall-former *Andricus quercuscalicis*. Journal of Animal Ecology 61:567-584.

Hails RS, Askew RR, Notton DG (1990) The Parasitoids and inquilines of the agamic generation of *Andricus quercuscalicis* (Hym.; Cynipidae) in Britain. The Entomologist 109:165-172

Hastings A, Godfray HCJ (1999) Learning, host fidelity, and the stability of host-parasitoid communities. American Naturalist 153:295-301

Holt RD (1977) Predation, apparent competition, and the structure of prey communities. Theoretical Population Biology 12:197-229

Melika G, Csóka G, Pujade-Villar J (2000) Check-list of oak gall wasps of Hungary, with some taxonomic notes (Hymenoptera: Cynipidae, Cynipinae, Cynipini). Annales Historico Naturales Musei Nationalis Hungarici 92:265-296

Moriya S, Inoue K, Mabuchi M (1989a) The use of *Torymus sinensis* to control chestnut gall-wasp, *Dryocosmus kuriphilus*, in Japan. Technical Bulletin of the Food and Fertilizer Technology Center 118:1-12

Moriya S, Inoue K, Ôtake A, Shiga M, Mabuchi M (1989b) Decline of the chestnut gall-wasp population, *Dryocosmus kuriphilus* Yasumatsu (Hymenoptera: Cynipidae) after the establishment of *Torymus sinensis* Kamijo (Hymenoptera: Torymidae). Applied Entomology and Zoology 24:231-233

Moriya S, Shiga M, Adachi I (2003) Classical biological control of the chestnut gall wasp in Japan. In: VanDriesche RG (ed) Proceedings of the 1st international symposium on biological control of arthropods. USDA Forest Service, Washington, pp 407-415

Morris RJ, Lewis OT, Godfray HCJ (2004) Experimental evidence for apparent competition in a tropical forest food web. Nature 428:310-313

Müller CB, Adriaanse ICT, Belshaw R, Godfray HCJ (1999) The structure of an aphid–parasitoid community. Journal of Animal Ecology 68:346-370

Ôtake A (1980) Chestnut gall-wasp, *Dryocosmus kuriphilus* Yasumatsu (Hymenoptera: Cynipidae): a preliminary study on trend of adult emergence and some other ecological aspects related to the final stage of its life cycle. Applied Entomology and Zoology 15:96-105

Ôtake A, Shiga M, Moriya S (1982) A study on parasitism of the chestnut gall wasp, *Dryocosmus kuriphilus* Yasumatsu (Hymenoptera: Cynipidae) by parasitoids indigenous to Japan. Bulletin of the Fruit Tree Research Station A9:177-192

Payne JA (1978). Oriental chestnut gall- wasp: new nut pest in North America. In: Macdonald WL, Cech FC, Luchok J, Smith C (eds) Proceeding of the American chestnut symposium. West Virginia University Press, Morgantown, pp 86-88

Schönrogge K (1994) Dynamics of the guild structure in the parasitoids and inquilines of an alien gall wasp, *Andricus quercuscalicis* Burgsdorf. PhD Thesis, Imperial College University of London, London

Schönrogge K, Crawley MJ (2000) Quantitative webs as a means of assessing the impact of alien insects. Journal of Animal Ecology 69:841-868

Schönrogge K, Stone GN, Crawley MJ (1995) Spatial and temporal variation in guild structure—parasitoids and inquilines of *Andricus quercuscalicis* (Hymenoptera, Cynipidae) in its native and alien ranges. Oikos 72:51-60

Schönrogge K, Stone GN, Crawley, MJ (1996) Alien herbivores and native parasitoids: rapid development of guild structure in an invading gall wasp *Andricus quercuscalicis* (Hymenoptera: Cynipidae). Ecological Entomology 21:71-80

Schönrogge K, Walker P, Crawley MJ (1998) Invaders on the move: parasitism in the sexual galls of four alien gall wasps in Britain (Hymenoptera: Cynipidae). Proceedings of the Royal Society of London Series B—Biological Sciences 265:1643-1650

Schönrogge K, Walker P, Crawley MJ (1999) Complex life cycles in *Andricus kollari* (Hymenoptera, Cynipidae) and their impact on associated parasitoid and inquiline species. Oikos 84:293-301

Schönrogge K, Walker P, Crawley MJ (2000) Parasitoid and inquiline attack in the galls of four alien, cynipid gall wasps: host switches and the effect on parasitoid sex ratios. Ecological Entomology 25:208-219

Shiraga T (1951) Problems and control of the chestnut gall wasp (in Japanese). Nogyô oyobi Engei (Agriculture and Horticulture) 26:167-170

Sitch TA, Grewcock DA, Gilbert FS (1988) Factors affecting components of the fitness in a gall-making wasp (*Cynips divisa* Hartig). Oecologia 76:371-375

Stone GN, Schönrogge K (2003) The adaptive significance of insect gall morphology. Trends in Ecology and Evolution 18:512-522

Stone GN, Sunnucks P (1993) Genetic consequences of an invasion through a patchy environment—the cynipid gallwasp *Andricus quercuscalicis* (Hymenoptera, Cynipidae). Molecular Ecology 2:251-268

Stone GN, Schönrogge K, Crawley MJ, Fraser S (1995) Geographic and between generation variation in the parasitoid communities associated with an invading gallwasp, *Andricus quercuscalicis* (Hymenoptera: Cynipidae). Oecologia 104:207-217

Stone G, Atkinson R, Rokas A, Csóka G, Nieves-Aldrey JL (2001) Differential success in northwards range expansion between ecotypes of the marble gallwasp *Andricus kollari*: a tale of two lifecycles. Molecular Ecology 10:761-778

Stone GN, Schönrogge K, Atkinson RJ, Bellido D, Pujade-Villar J (2002) The population biology of oak gall wasps (Hymenoptera: Cynipidae). Annual Review of Entomology 47:633-668

Whitham TG (1978) Habitat selection by *Pemphigus* aphids in response to resource limitation and competition. Ecology 59:1164-1176

9 Parasitoid Recruitment to the Globally Invasive Chestnut Gall Wasp *Dryocosmus kuriphilus*

Alexandre Aebi[1], Karsten Schönrogge[2], George Melika[3], Alberto Alma[4], Giovanni Bosio[5], Ambra Quacchia[4], Luca Picciau[4], Yoshihisa Abe[6], Seichii Moriya[7], Kaori Yara[8], Gabrijel Seljak[9], Graham N. Stone[1]

[1]Institute of Evolutionary Biology, The Kings Buildings, West Mains Road, Edinburgh, EH9 3JT, UK
[2]Centre for Ecology and Hydrology, CEH Dorset, Winfrith Technology Centre, Dorchester, DT2 8ZD, UK
[3]Systematic Parasitoid Laboratory, Vas County Plant Protection and Soil Conservation Service, Kelcs-Adelffy St. 6, Köszeg 9730, Hungary
[4]Department of Exploitation and Protection of the Agricultural and Forestry Resources, Entomology and Zoology applied to the Environment "Carlo Vidano", Via Leonardo da Vinci 44, Grugliasco 10095, Italy
[5]Phytosanitary Service, Regione Piemonte, Via Livorno 60, Torino 10144, Italy
[6]Laboratory of Applied Entomology, Graduate School of Agriculture, Kyoto Prefectural University, Kyoto 606-8522, Japan
[7]National Agricultural Research Center, Tsukuba, Ibaraki 305-8666, Japan
[8]National Institute for Agro-Environmental Sciences, Tsukuba, Ibaraki 305-8604, Japan
[9]Chamber for Agriculture and Forestry of Slovenia, Agricultural and Forestry Institute Nova Gorica, Pri Hrastu 18, SI-5000 Nova Gorica, Slovenia

Summary. The chestnut gall wasp *Dryocosmus kuriphilus* is a global pest of chestnut (*Castanea*). Established as a pest in the mid 20th century in Japan, Korea and the USA, this species has now reached Europe. Successful deployment of a biocontrol agent, *Torymus sinensis*, in Japan has led to its early release in Italy. Here we provide the first overview of the natural enemies associated with *D. kuriphilus* in its native and invaded ranges, and discuss general patterns in community development. We then use what is known about European oak gall wasp communities to predict possible future developments for *D. kuriphilus*, and possible interactions between parasitoid communities attacking hosts on chestnut and oaks.

Key words. Biological invasions, Biological control, Parasitoid recruitment, Community structure, *Dryocosmus kuriphilus*

9.1 Introduction

Because of human transportation, non-native species have become integral components of many ecosystems worldwide, threatening ecosystem integrity, biodiversity, agriculture and human health. Understanding the causes and consequences of biological invasions (invasion biology) represents an increasingly important challenge for ecologists and evolutionary biologists. The chestnut gall wasp, *Dryocosmus kuriphilus* Yasumatsu (Hymenoptera, Cynipidae), is the most important global insect pest of chestnut (*Castanea* spp., Fagaceae) (Brussino et al. 2002; Moriya et al. 1990; Murakami et al. 1995; Payne et al. 1983). This gall wasp is a member of the tribe Cynipini, most of which induce galls on oaks (*Quercus*) and is the only member of this tribe to attack chestnut (Stone et al. 2002).

Attack by *D. kuriphilus* reduces fruit yield by 50–75% (Payne et al. 1983), and heavy attack reduces tree vigor and wood production (Kato and Hijii 1997) and can kill the tree (Moriya et al. 2003). Frequent exchange of cultivars between chestnut growers and a parthenogenetic reproductive strategy (Nohara 1956) have made *D. kuriphilus* a successful invader in Asia, North America and Europe. Japan was the first region invaded by *D. kuriphilus*, and here the promise of future biological control was demonstrated by successful reduction of *D. kuriphilus* populations using an introduced Chinese parasitoid, *Torymus sinensis*. *T. sinensis* has just been released in an attempt to control the recent *D. kuriphilus* invasion, in Italy, representing the first recorded establishment of this pest in Europe.

Preliminary surveys of the natural enemies attacking *D. kuriphilus* in Italy show that it is already attacked by a suite of chalcid parasitoids characteristic of native European oak cynipid gall wasp communities on oak (reviewed in Stone et al. 2002). Overlap in the natural enemies attacking the invader and native communities raises the possibility of natural biological control, and of impacts on native faunas of introduction of *T. sinensis*. Here we review the global history of invasion and recruitment of natural enemies to *D. kuriphilus*, and discuss possible future development of the community centered on *D. kuriphilus* in Europe.

9.2 The Invasion History of *D. kuriphilus*

D. kuriphilus emerged as a pest in the mid-twentieth century. Although first formally recorded from Japan in 1958 (Murakami et al. 1980), outbreaks in China in 1941 and 1959 suggest that *D. kuriphilus* is native to this region (Moriya et al. 1990; Murakami et al. 1980). The first country invaded was Japan, where *D. kuriphilus* was accidentally introduced to Okayama prefecture in 1941 (Oho and Umeya 1975). It spread throughout Japan within 25 years, with a dramatic impact on chestnut production (Oho and Shimura 1970; Shiraga 1951). *D. kuriphilus* was then reported from Chaenchun, Chungchungpuk-do, Korea in 1958 (Cho and Lee 1963), and over 37 years spread across South Korea (Murakami et al. 1995). In 1974, *D. kuriphilus* colonized the American continent, and was first reported in Peach County, Georgia (Payne et al. 1976). By 1976, it had already spread to 3 adjacent counties (Houston, Crawford and Bibb). Despite a severe impact on the chestnut industry, no further information on its impact in the USA was found in the literature. *D. kuriphilus* was first recorded from Europe in 2002, in Piemonte, Italy (Brussino et al. 2002). Once recognised, the galls were recorded at 6 localities (Boves, Peveragno, Chiusa Pesio, Borgo San Dalmazzo, Roccavione and Robilante) over an area of 160 km^2. This leads us to believe that the actual introduction date is 1995–1996, when eight Chinese chestnut cultivars were introduced to the region. This is the only recorded event that could explain the invasion of *D. kuriphilus* in Italy. By 2005, *D. kuriphilus* had reached five new Italian regions (Lazio, Campania, Toscana, Abruzzo and Marche) covering a major part of the peninsula, and was recorded for the first time in neighboring regions of France (Val de Blore and Isola) (G. Bosio and J.-C. Malausa, personal communication, 2005). In 2004, a total of 1250 chestnut plants were imported from Piemonte into Slovenia. After a warning from the phytosanitary service of Piemonte, in 2005, Slovenian researchers were able to trace and inspect 47% of the imported material (450 chestnut plants in 18 plantations). Ten infested plants were found in 4 localities across the whole country, and destroyed.

9.3 The Natural Enemies of *D. kuriphilus* in China, and Initiation of Biological Control in Japan

Over the 65 years following its introduction to Japan (Murakami et al. 1980), scientists have achieved biological control of *D. kuriphilus* and left an extensive literature on its invasion biology. After its arrival around

1941 and rapid spread, initial control attempts using chemical pesticides were frustrated by protection of the immature stages within the gall (Murakami 1981; Torii 1959). Subsequent breeding of resistant chestnut varieties enabled chestnut growers to control *D. kuriphilus* for about 20 years, until the appearance of a strain able to attack resistant chestnut varieties (around 1960).

The finding that *D. kuriphilus* did not seem to have a serious impact on chestnut trees in its native China suggested that populations there were held at low densities by natural enemies (Murakami et al. 1980). A total of 11 species in 5 chalcid families (Torymidae, Ormyridae, Eurytomidae, Eulophidae and Eupelmidae) are known from the native range of *D. kuriphilus* (see Table 1). In 1975, a delegation of the Japanese Ministry of Agriculture and Forestry was sent to China to investigate the potential for biological control of *D. kuriphilus* in Japan. Sixty nine *D. kuriphilus* galls were collected in Hsi-an, Shensi and imported to Japan, where 8 parasitoid species were reared (Table 1) (Murakami 1980). In 1981, Murakami added *Eurytoma setigera* and *Megastigmus nipponicus*, and Luo and Huang (1993) added *Sycophila* (= *Decatoma*) *concinna*. A striking feature of the *D. kuriphilus* community, discussed further below, is the absence of cynipid inquilines—a major component of the communities associated with most oak cynipid galls (Stone et al. 2002).

Of these, only *T. sinensis* (named by Kamijo in 1982) showed high host specificity and a life cycle matching that of its host—traits required for a successful biological control agent. In 1979 and 1981, approximately 5000 *D. kuriphilus* galls were imported from China and in 1982, 260 mated *T.*

Table 1. The *D. kuriphilus* parasitoid community in its native range (China).

Parasitoid species	Family	References
Eupelmus urozonus *	Eupelmidae	1, 2
Eurytoma brunniventris *	Eurytomidae	1, 2
Eurytoma setigera	Eurytomidae	2
Megastigmus maculipennis *	Torymidae	1, 2
Megastigmus nipponicus	Torymidae	2
Ormyrus pomaceus (= *O. punctiger*) *	Ormyridae	1, 2
Sycophila concinna	Eurytomidae	3
Sycophila variegata *	Eurytomidae	1, 2
Tetrastichus sp. *	Eulophidae	1, 2
Torymus geranii *	Torymidae	1, 2
Torymus sinensis *	Torymidae	1, 2, 4

Species sampled from the original 69-gall rearing are marked with an asterisk (*). Key to references: 1, Murakami et al. (1980); 2, Murakami (1981); 3, Luo and Huang (1993); 4, Kamijo (1981) *sinensis* females were released onto Japanese chestnut trees at the Fruit Tree Research Station (FTRS) in Ibaraki prefecture (Ôtake et al. 1984). A study carried on in 1983 showed that *T. sinensis* was successfully attacking the local target host population (Moriya et al. 1990). There were some initial failures in establishment of *T. sinensis*. Research at Kumamoto Fruit Tree Experiment Station (Kumamoto Prefecture) using a third *D. kuriphilus* gall shipment from China showed that establishment of *T. sinensis* was slowed by both mortality inflicted by facultative hyperparasitoids and a low female sex ratio (Murakami and Gyoutoku, 1991, 1995; Murakami and Kiyota 1983; Murakami et al. 1985, 1989, 2001).

Nevertheless, the *T. sinensis* population at the FTRS grew by a factor of 25 times by 1989 (Moriya et al. 1990), becoming the commonest parasitoid reared from *D. kuriphilus* at FTRS. *T. sinensis* was reared from target galls more than 12 km from its release point, showing good dispersal ability. Importantly for the chestnut industry, the proportion of chestnut shoots infested with *D. kuriphilus* galls decreased rapidly from a maximum of 40% in 1983 to 3% in 1988 (Moriya et al. 1990), well below the estimated tolerable injury level of 30% (Gyoutoku and Uemura 1985).

9.4 Parasitoids Attacking *D. kuriphilus* in Japan

In Japan *D. kuriphilus* is now attacked by a rich parasitoid assemblage (Table 2) of 24 chalcid species in 7 families and 1 braconid (*Aspilota yasumatsui*). Its galls have also been colonized though at very low rates by a cynipid inquiline *Synergus* sp. A striking feature of this community is its high richness compared to data from the Chinese native range of *D. kuriphilus*, and the latter is almost certainly understudied. The 20 parasitoids identified to species in Japan can be divided into 10 shared with the Chinese community and 10 absent in China but recorded in Japan.

Ten of the 11 parasitoids attacking *D. kuriphilus* in China also attack it in Japan, raising the possibility that parasitoids, as well as the gall-inducer, were introduced from China. Of the 10 shared species, 8 are known to attack native Japanese oak gall wasps (Kamijo 1981; Yasumatsu and Kamijo 1979), and for these species both introduction and shifts from native hosts are possible. However, one parasitoid (*Tetrastichus* sp.) was never recorded in Japan prior to introduction of *D. kuriphilus* (Murakami 1981) and no alternative Japanese host is known. This species may have been imported with *D. kuriphilus* around 1940, or have been recruited from an unknown Japanese host. *T. sinensis* was recorded from Tsushima Island, the island nearest to Korea in Japan, by Murakami et al. (1993). However,

Table 2. Parasitoids attacking *D. kuriphilus* in Japan.

Parasitoid species	Family	References
Chalcids		
Amblymerus sp.	Pteromalidae	1
*Arthrolytus usubai**	Pteromalidae	2
*Caenacis peroni**	Pteromalidae	2
*Cecidostiba fushica**	Pteromalidae	2
*Cecidostiba semifascia**	Pteromalidae	2
*Cynipencyrtus flavus**	Encyrtidae	3, 4, 5
Cynipencyrtus sp.	Encyrtidae	6
Eupelmus urozonus*	Eupelmidae	3, 4, 5, 6, 7, 8
Eupelmus sp.	Eupelmidae	7
Eurytoma brunniventris*	Eurytomidae	1, 3, 4, 5, 6, 7, 8, 9
*Eurytoma schaeferi**	Eurytomidae	5
Eurytoma setigera*	Eurytomidae	4, 5, 6, 7
Megastigmus maculipennis*	Torymidae	1, 3, 4, 5, 6
Megastigmus nipponicus*	Torymidae	3, 4, 5, 6, 8, 10
*Mesopolobus yasumatsui**	Pteromalidae	2, 3, 6
*Ormyrus flavitibialis**	Ormyridae	4, 5, 6, 7, 8
Ormyrus pomaceus* (= *O. punctiger*)	Ormyridae	1, 4, 5, 7, 8
*Pteromalus apantelophagus**	Pteromalidae	2, 7
Sycophila variegata*	Eurytomidae	1, 3, 4, 5, 6
Tetrastichus sp.	Eulophidae	3, 4, 6
*Torymus beneficus**	Torymidae	4, 5, 6, 7, 8, 10, 11
Torymus geranii*	Torymidae	1, 4, 5, 6, 7, 8
Torymus sinensis (native)	Torymidae	12
Torymus sp.	Torymidae	14
Non-chalcid parasitoids		
Aspilota yasumatsui	Braconidae	13
Cynipid inquiline		
Synergus sp.	Cynipidae	7

Species recorded in China are named in **bold**. *, Species associated with Japanese oak gall wasps. Key to references: 1, Yasumatsu (1955); 2, Kamijo (1981); 3, Alam (1994); 4, Murakami et al. (1994); 5, Yasumatsu and Kamijo (1979); 6, Murakami and Gyoutoku (1995); 7, Ôtake et al. (1982); 8, Ôtake (1989); 9, Murakami et al. (1977); 10, Kato and Hijii (1999); 11, Toda et al. (2000); 12, Murakami et al. (1993); 13, Watanabe (1957); 14, Ito and Hijii (2000)

this population differs in adult emergence phenology from wasps imported from China and is thought to be native to Japan. No host other than *D. kuriphilus* is known for this strain of *T. sinensis*. This situation may parallel that seen in *T. sinensis* in Korea (see below).

18 parasitoid species associated with Japanese oak gall wasps also attack *D. kuriphilus* (Table 2) (Kamijo 1981; Yasumatsu and Kamijo 1979), representing host shifts from native hosts to the invader. *Torymus beneficus* is particularly interesting because it exists as two phenologically distinct ecotypes in galls of *D. kuriphilus*. At least two strains, an "early-season strain" and a "late-season strain", can be differentiated based on the

period of adult emergence (Murakami 1988; Ôtake 1987). It is thought that these strains represent ecotypes of the same species attacking different native cynipid hosts prior to the arrival of *D. kuriphilus* (Murakami 1988). *T. beneficus* also illustrates another form of interaction between native cynipid communities and *D. kuriphilus*, through its formation of fertile hybrids with introduced *T. sinensis* (Moriya et al. 1992, 2003). Individuals morphologically intermediate between *T. sinensis* and *T. beneficus* appeared in the field (Moriya et al. 1992; Yara et al. 2000), and their hybrid origin has been shown using molecular markers (Izawa et al. 1992, 1995, 1996; Toda et al. 2000; Yara, 2004, 2006; Yara et al. 2000).

The abundance of literature records allows us to examine the recruitment history of parasitoids on *D. kuriphilus* in Japan. For data published between 1978 and 1993 there is no simple pattern of community enrichment over time, as might be expected. Most of the species in Table 2 were present in most of the samples collected over this period (Alam 1994; Murakami and Gyoutoku 1995; Murakami et al. 1994; Ôtake et al. 1982, Yasumatsu and Kamijo 1979). The only exceptions are five pteromalids (*Arthrolytus usubai, Caenacis peronni, Cecidostiba fushica, Cecidostiba semifascia* and *Pteromalus apanthelophagus*) that were only recorded in 1981, suggesting that these records represent opportunistic and eventually unsuccessful host shifts early in the invasion process. Overall, data from Japan suggest that 38 years after its introduction, the parasitoid community associated with *D. kuriphilus* has stabilized, after a rapid early recruitment unrecorded in the literature.

Japan is the only region in which any cynipid inquiline has been reared from galls of *D. kuriphilus*. Only a single female *Synergus* sp. was reared (Ôtake et al. 1982), emphasizing the extreme rarity of inquilines in this gall.

9.5 Parasitoids Attacking *D. kuriphilus* in Korea

After its arrival in Korea, *D. kuriphilus* recruited a rich parasitoid assemblage of 15 chalcid species (Table 3).

This community shows substantial overlap with those recorded in China and Japan. All but 6 (including 2 undescribed species) of the parasitoids recorded from Korea were recorded from China, and so may represent host shifts from native Korean hosts. Of this group of 4 species, all but one (*Torymus koreanus*) were among the parasitoids recruited in Japan. The lower richness in Korea compared to Japan is hard to interpret: it may be a

Table 3. Parasitoids attacking *D. kuriphilus* in Korea.

Parasitoid species	Family	References
Caenacis peroni	Pteromalidae	1
Eupelmus sp.	Eupelmidae	2, 3, 4, 5, 6
Eupelmus urozonus	Eupelmidae	3, 6, 7
Eurytoma brunniventris	Eurytomidae	3, 6, 7
Eurytoma setigera	Eurytomidae	3, 6
Megastigmus maculipennis	Torymidae	6, 7
Megastigmus nipponicus	Torymidae	6, 7
Mesopolobus yasumatsui	Pteromalidae	1
Ormyrus flavitibialis	Ormyridae	6, 7
Ormyrus pomaceus (= *O. punctiger*)	Ormyridae	3, 6, 7
Sycophila variegata	Eurytomidae	3, 6, 7, 8
Torymus geranii	Torymidae	3, 6, 7
Torymus koreanus	Torymidae	9
Torymus sinensis**	Torymidae	6, 10
Torymus sp.	Torymidae	6

Species recorded from *D. kuriphilus* in China are named in **bold**. Key to references: 1, Kamijo (1981); 2, Murakami et al. (1994); 3, Murakami et al. (1995); 4, Murakami and Gyoutoku (1995); 5, Ôtake (1989); 6, Kim (1998); 7, Yasumatsu and Kamijo (1979); 8, Ko (1971); 9, Kamijo (1982); 10, Murakami et al. (1985)
**T. sinensis* in Korea exists in two phenologically distinct strains (see main text).

product of the more recent arrival (by circa 20 years) of the invading host in Korea, or an artifact of lower sampling effort in Korea.

Although *T. sinensis* was never imported to Korea, this species was recorded by Murakami et al. (1995) throughout the country. Based on differences in adult emergence phenology, Korean populations were described as a different strain to those in China (Murakami et al. 1993). Furthermore, in a manner paralleling the observations for *T. beneficus* in Japan, Korean *T. sinensis* were further subdivided into two strains (the KA and the G-3 populations) on the basis of differences in adult emergence phenology and exploitation of different native oak cynipid hosts before the *D. kuriphilus* invasion (Murakami et al. 1995).

9.6 Parasitoids Attacking *D. kuriphilus* in the USA

Despite a dramatic impact of *D. kuriphilus* on the chestnut industry in the USA, very little has been published on its natural enemies. Only two torymid species, *Torymus advenus* and *Torymus tubicola* (Payne 1978), are recorded.

9.7 Parasitoids Attacking *D. kuriphilus* in Italy

Galls of *D. kuriphilus* were reared to census natural enemy recruitment immediately after its discovery in 2002, 2003, 2004 and 2005. Fifteen chalcid parasitoid species belonging to six families were recorded (see Table 4).

Of the parasitoids found in Italy, four are shared with the communities reared in China, Korea and Japan. While it is possible that these species were introduced with the Chinese *D. kuriphilus*, the cultivars introduced in Italy in 1995–1996 carried no dried galls, suggesting that *D. kuriphilus* was probably present in buds as eggs or first instar larvae. These hosts werc probably too young to have been attacked by these parasitoids. In fact, all of the species in Table 4 are well known as parasitoids of oak cynipid hosts in Europe, and many of them (see Table 4) are widespread in the Western Palaearctic. The remainder are characteristic of oak cynipid communities in southern and Mediterranean Europe. All of these parasitoid records almost certainly represent recruitment from local cynipid communities.

Table 4. Parasitoids attacking *D. kuriphilus* in Italy, with year of first record (Year), geographic distribution in the Western Palaearctic (Dist: WP, western palaearctic; M, Southern and mediterranean Europe), and the proportion of parasitism (%) made up by each species in Italian rearings in 2003, 2004 and 2005.

Parasitoid species	Family	Year	Dist	2003	2004	2005
Sycophila iracemae	Eurytomidae	2004	M		4	
Sycophila variegata	Eurytomidae	2003	WP	2		
Sycophila biguttata	Eurytomidae	2005	WP			3
Eurytoma pistacina	Eurytomidae	2004	M		1	8
Eurytoma brunniventris	Eurytomidae	2002	WP		2	
Mesopolobus mediterraneus	Pteromalidae	2004	M		1	1
Mesopolobus sericeus	Pteromalidae	2003	WP	5		
Mesopolobus tarsatus	Pteromalidae	2005	M			1
Torymus auratus	Torymidae	2005	WP			1
Torymus flavipes	Torymidae	2003	WP	2	5	1
Torymus scutellaris	Torymidae	2005	WP			1
Megastigmus dorsalis	Torymidae	2002	WP	14		7
Eupelmus urozonus	Eupelmidae	2002	WP	77	87	77
*Baryscapus pallidae**	Eulophidae	2005	M			
Ormyrus pomaceus*	Ormyridae	2002	WP			

Abundance estimates were unavailable (*) for *B. pallidae* (first recorded in 2005) and *O. pomaceus* (first recorded in 2002). Species also attacking *D. kuriphilus* in China, Korea and Japan are named in **bold**

Early recording of parasitoid communities in Italy captured the rapid recruitment process that also occurred in Japan. Community richness rose from 4 species in 2002 to 7 in 2003, 10 in 2004 and 14 in 2005 (Table 4). Despite the increase in community richness, gall attack rates (estimated per inhabitant) have remained very low: from 1.6% in 2003 (n = 1900 galls), to 0.8% in 2004 (n = 2500 galls) and 0.5% in 2005 (n = 6713 galls). Of the parasitoids attacking *D. kuriphilus* in Italy, the most significant and consistent is *Eupelmus urozonus* (Table 4).

9.8 Initiation of Biological Control of *D. kuriphilus* in Italy

Following successful use of *T. sinensis* to control *D. kuriphilus* in Japan, in 2003 four of us (A. Alma, G. Bosio, A. Quacchia and L. Picciau) conducted preliminary tests to investigate the efficacy of this parasitoid in Italy. For these trials, 80 pairs of Japan-sourced *T. sinensis* were released on young infested chestnut trees in outdoor net cages. Because of a discrepancy between adult parasitoid emergence and gall development, the females were not able to successfully attack *D. kuriphilus*. In 2004, while further experiments were conducted in outdoor net cages, 55 pairs of *T. sinensis* were sleeved over shoots infested with *D. kuriphilus* in 4 localities. Again, phenological mismatch largely prevented successful attack and only 2 adult parasitoids emerged from the sleeved experiments. In 2005, the development of *D. kuriphilus* galls shipped from Japan was delayed by cooling, allowing artificial synchronization of *T. sinensis* emergence with gall development. Ninety mated females were released in the field in three localities, in ongoing trials.

9.9 Common Patterns in the Development of *D. kuriphilus* Communities

Communities associated with *D. kuriphilus* show consistent patterns worldwide—with the exception of North America, where the community is clearly under-studied.

1. The parasitoids attacking *D. kuriphilus* show strong overlap with those attacking oak cynipid gall wasps locally.
2. The parasitoids shared with oak cynipids are dominated by species with a broad host range—a feature epitomized by the pan-Palaearctic species *Eurytoma brunniventris*, *E. urozonus*, *Ormyrus pomaceus* and *Sycophila variegata*. These are among the most cosmopolitan of all parasitoids found

in cynipid galls (Csóka et al 2005; Stone et al. 2002), and *E. urozonus* even attacks inquiline cynipids in galls induced by cecidosid moths on *Rhus* in South Africa (van Noort et al. 2006). This pattern is extended in local parasitoid faunas: in Italy, the species attacking *D. kuriphilus* include almost all of the parasitoids known to attack cynipid hosts on at least two of oaks, sycamores and roses (*Baryscapus pallidae*, *Eurytoma pistacina*, *Mesopolobus sericeus*, *Sycophila biguttata*, *Sycophila iracemae* and *Torymus flavipes*) (Csóka et al 2005; Schönrogge and Askew 2006; Stone et al. 2002). This suggests that ability to exploit a diversity of plant environments is associated with rapid detection and exploitation of *D. kuriphilus* in its invaded range.

3. The interaction between parasitoid communities centered on native and invading cynipid hosts covers a broad parasitoid taxonomic spectrum. In Italy, for example, *D. kuriphilus* is already attacked by members of all 6 of the parasitoid families attacking native oak cynipids (Csóka et al. 2005; Stone et al. 2002).

4. The time lag between arrival of the invader and development of a community is very short. This implies that the novel gall morphology of *D. kuriphilus* and its development on a novel plant host (this species is the only Cynipid to develop on *Castanea* worldwide) are not significant barriers to its detection and exploitation by parasitoids normally resident on oaks, *Quercus*.

5. Even endoparasitoids characteristic of oak cynipids are able rapidly to exploit *D. kuriphilus*. *S. biguttata* is known as an endoparasitoid (Stone et al. 2002), and the same may be true for *S. iracemae* and *S. variegate*, though the latter has been recorded as an ectoparasitoid (Ito and Hijii 2004). This highlights the fact that though *D. kuriphilus* has a unique ecology among the Cynipini, it is phylogenetically nested within them (Liljeblad 2002) and so, once discovered, is probably physiologically exploitable by many oak cynipid parasitoids.

6. As elsewhere, inquiline cynipids are extremely rare in *D. kuriphilus* galls in Europe. As phytophages, these inquilines may be more sensitive than entomophagous parasitoids to the divergent host exploited by *D. kuriphilus*.

9.10 Future Development of the Community Associated with *D. kuriphilus* in Europe.

There are two chestnut species in Europe: *Castanea sativa* is native to Turkey, Greece and the southern Balkans, perhaps also native in Italy, and planted throughout Europe as far north as Britain. *Castanea crenata* is native to Japan and planted in Italy, Spain and Portugal. The combined native and planted ranges of these species cover much of the temperate Western Palaearctic, and define the potential scope of range expansion by *D. kuriphilus*. Europe has seen a wave of gall wasp invasions associated with human dispersal of Turkey oak, *Quercus cerris*, in which multiple species have escaped southern glacial refugia and expanded their distributions north as far as Scotland (Stone et al. 2002). In principle, *D. kuriphilus* could invade, interact with native parasitoids and, via them, native cynipids over a similar range. The same applies to *T. sinensis*, if it becomes established in Italy.

From an ecological perspective, *D. kuriphilus* represents an experiment (albeit unwanted) in the recruitment of natural enemies, and in the changing mortality these inflict on the host. In comparison with the invading oak cynipids, *D. kuriphilus* is recruiting natural enemies rather rapidly (see the chapter by K. Schönrogge et al., this volume; Stone et al. 2002). For example, the asexual generation galls of the invader *Andricus quercuscalicis* in Britain were free of parasitoids and inquilines for about 20 years (Schönrogge et al. 2000). The high rate of recruitment in Italy may reflect in part the high diversity of parasitoids (60+ species) attacking oak cynipids in southern central Europe (Schönrogge and Askew 2006). The extensive information available on the host ranges of European oak cynipids makes it possible to identify the following Europe-native parasitoids as possible future recruits to the community centered on *D. kuriphilus*. All are polyphagous and recorded from large, woody oak cynipid galls (Schönrogge and Askew 2006): *Torymus geranii* (Torymidae, already known to attack *D. kuriphilus* in China, Japan and Korea), *Ormyrus nitidulus* (Ormyridae), *Caenacis lauta, C. semifascia* (already known to attack *D. kuriphilus* in Japan), *Cecidostiba fungosa, Mesopolobus amaenus, Cyrtoptyx robustus* (all Pteromalidae), *Eupelmus annulatus* (Eupelmidae) and *Baryscapus berhidanus* (Eulophidae).

A key issue in development of the *D. kuriphilus* community is whether or not cynipid inquilines are recruited. These represent additional hosts for parasitoids and have a significant positive impact on parasitoid community richness (Schönrogge et al. 1994, 1996, this volume). However, it remains to be seen whether native European cynipid inquilines will be able to exploit galls induced on *Castanea* spp. The unique position of *D. kuriphilus* as a chestnut galler among cynipid gall inducers suggests that this host shift is not an easy one.

Invading European *Andricus* gall wasps also experienced an increase over time in the mortality inflicted by natural enemies, reaching more than 40% in their sexual generation galls (e.g. Schönrogge et al. 1994, 1996,

2000, this volume). A qualitatively similar response is to be expected for *D. kuriphilus*, particularly if this host becomes abundant relative to oak cynipids. Female lineages within parasitoid species able to locate the novel host and with appropriate phenology are then expected to achieve high reproductive success, and so spread through parasitoid populations. This in turn may result in the development of genetically discrete *D. kuriphilus* ecotypes within parasitoid species (Stone and Schönrogge 2003). A potential candidate for this pattern already reared from *D. kuriphilus* is *Torymus scutellaris*, a member of the *Torymus erucarum* species group, which, like *T. sinensis* and *T. beneficus*, are typically univoltine and show close phenological matching to their host (see also the chapter by K. Schönrogge et al., this volume). Increasing attack rates are also expected to reduce the high between-year variation in community composition evident in Table 4. The basis for predictions of this type would be much improved by further work on the biodiversity and population genetic structure of parasitoid communities of oak and chestnut cynipids throughout the native and introduced range of *D. kuriphilus*.

9.11 Impacts of *D. kuriphilus* on native European cynipid communities

The high local abundance achieved by *D. kuriphilus* in Italy suggests that, once discovered, this host may represent a significant resource for parasitoids. By exploiting galling sites on an otherwise unexploited plant host, *D. kuriphilus* avoids possible competition for oviposition sites with native cynipids (Stone et al. 2002) and has the potential to massively elevate local host (and hence parasitoid) populations. The rapid recruitment of generalist parasitoids shared with oak cynipids suggests that *D. kuriphilus* may have a negative impact on native cynipids through apparent competition (Holt 1977; see also the chapter by K. Schönrogge et al., this volume). Whether this effect materializes depends on the extent to which parasitoids show a positive density dependent response to increasing *D. kuriphilus* density and distribution, and on levels of parasitoid population flow between chestnuts and oaks. These key issues require detailed quantitative monitoring of parasitoid populations on native hosts and on *D. kuriphilus*, ideally using quantified web approaches (e.g. Schönrogge and Crawley 2000), and the deployment of genetic approaches to analyze parasitoid population structure (Hayward and Stone 2006; Yara et al. 2000; Yara 2004, 2006). Finally, release of *T. sinensis* could have a range of potential impacts. If this parasitoid can successfully track the phenology of *D. ku-*

riphilus in Europe, it has the potential to act as a control agent (see the chapter by K. Schönrogge et al, this volume). Its high specificity to *D. kuriphilus* to date suggests that it has limited potential for exploitation of native hosts, but enforced changes in phenology may mean that extrapolation from experience elsewhere is unreliable. As *T. sinensis* populations evolve, so probably will their host range. Finally, the possibility remains of interaction with native torymids through hybridization. These possibilities again call for detailed monitoring of the communities associated with both native cynipids and *D. kuriphilus*.

9.12 Acknowledgments

This work was supported by a SNSF postdoctoral fellowship (PBNEA-106766) and a Joint JSPS/SNSF individual short visit grant (RC 20530002/PIJSA-111328) to A. Aebi, by NERC grant numbers NER/B/S/2003/00856 to G. N. Stone, NE/B504406 to G. N. Stone and K. Schönrogge and by a British–Hungarian Intergovernmental collaborative research grant (2002–2004) to G. N. Stone and G. Melika.

9.13 References

Alam MZ (1994) Within-gall distribution of chestnut gall wasp *Dryocosmus kuriphilus* Yasumatsu (Hymenoptera: Cynipidae) and its parasitoids in wild and cultivated chestnut trees. Thai Journal of Agricultural Science 27:169-178

Brussino G, Bosio G, Baudino M, Giordano R, Ramello F, Melika G (2002) Pericoloso insetto esotico per il castagno europeo. Informatore Agrario 58:59-61

Cho DY, Lee SO (1963) Ecological studies on the chestnut gall wasp, *Dryocosmus kuriphilus* Yasumatsu, and observation on the damages of the chestnut trees by its insect (in Korean). Korean Journal of Plant Protection 2:47-54

Csóka G, Stone GN, Melika G (2005) The biology, ecology and evolution of gall wasps. In: Raman A, Schaeffer CW, Withers TM (eds) Biology, ecology and evolution of gall-inducing arthropods. Science Publishers, Enfield, New Hampshire, pp 569-636

Gyoutoku Y, Uemura M (1985) Ecology and biological control of the chestnut gall wasp, *Dryocosmus kuriphilus* Yasumatsu (Hym.: Cynipidae). 1. Damage and parasitization in Kumamoto Prefecture (in Japanese). Proceedings of the Association for Plant Protection of Kyushu 31:213-215

Hayward A, Stone GN (2006) Comparative phylogeography across two trophic levels: the oak gall wasp *Andricus kollari* and its chalcid parasitoid *Megastigmus stigmatizans*. Molecular Ecology (in press)

Holt RD (1977) Predation, apparent competition, and the structure of prey communities. Theoretical Population Biology 12:197-229

Ito M, Hijii N (2000) Life-history traits in the parasitoid complex associated with cynipid galls on three species of Fagaceae. Entomological Science 3:471-479

Ito M, Hijii N (2004) Relationship among abundance of galls, survivorship, and mortality factors in a cynipid wasp, *Andricus moriokae* (Hymenoptera: Cynipidae). Journal of Forest Research 9:355-359

Izawa H, Osakabe M, Moriya S (1992) Isozyme discrimination between an imported parasitoid wasp, *Torymus sinensis* Kamijo and its sibling species *T. beneficus* Yasumatsu et Kamijo (Hymenoptera, Torymidae) attacking *Dryocosmus kuriphilus* Yasumatsu (Hymenoptera, Cynipidae) (in Japanese with English summary). Japanese Journal of Applied Entomology and Zoology 36:58-60

Izawa H, Osakabe M, Moriya S (1995) Relation between banding-patterns of malic enzyme by electrophoresis and a morphological character in exotic and native *Torymus* species. Applied Entomology and Zoology 30:37-41

Izawa H, Osakabe M, Moriya S, Toda S (1996) Use of malic enzyme to detect hybrids between *Torymus sinensis* and *T. beneficus* (Hymenoptera: Torymidae) attacking *Dryocosmus kuriphilus* (Hymenoptera: Cynipidae) and possibility of natural hybridization (in Japanese with English summary). Japanese Journal of Applied Entomology and Zoology 40:205-208

Kamijo K (1981) Pteromalid wasps (Hymenoptera) reared from cynipid galls on oak and chestnut in Japan, with descriptions of four new species. Kontyû 49:272-282

Kamijo K (1982) Two new species of *Torymus* (Hymenoptera, Torymidae) reared from *Dryocosmus kuriphilus* (Hymenoptera, Cynipidae) in China and Korea. Kontyû 50:505-510

Kato K, Hijii N (1997) Effects of gall formation by *Dryocosmus kuriphilus* Yasumatsu (Hym, Cynipidae) on the growth of chestnut trees. Journal of Applied Entomology 121:9-15

Kato K, Hijii N (1999) Mortality factors of the chestnut gall-wasp, *Dryocosmus kuriphilus* Yasumatsu (Hymenoptera: Cynipidae) after gall formation. Entomological Science 2:483-491

Kim JK (1998) Studies on the parasitoids of chestnut gall wasp, *Dryocosmus kuriphilus* Yasumatsu (Hymenoptera: Cynipidae) in Korea. Journal of Korean Forestry Society 87:475-482

Ko JH (1971) Notes on *Eudecatoma variegata* Curtis (Hymenoptera: Eurytomidae) as a parasite of the gall wasps (Cynipidae) in Korea. The Korean Journal of Entomology 1:25-26

Liljeblad J (2002) Phylogeny and evolution of gall wasps (Hymenoptera, Cynipidae). PhD thesis, University of Stockholm, Stockholm

Luo YQ, Huang JF (1993) A preliminary morphological study on immature stage of natural enemies of *Dryocosmus kuriphilus* Yasumatsu (in Chinese). Scientia Silvae Sinicae (Linye Kexue) 29:33-39

Moriya S, Inoue K, Mabuchi M (1990) The use of *Torymus sinensis* (Hymenoptera, Torymidae) for controlling the chestnut gall wasp, *Dryocosmus kuriphi-*

lus (Hymenoptera, Cynipidae), in Japan. FFTC-NARC International Seminar on 'The use of parasitoids and predators to control agricultural pests', p 21

Moriya S, Inoue K, Shiga M, Mabuchi M (1992) Interspecific relationship between an introduced parasitoid, *Torymus sinensis* Kamijo, as a biological control agent of the chestnut gall wasp, *Dryocosmus kuriphilus* Yasumatsu, and an endemic parasitoid, *T. beneficus* Yasumatsu et Kamijo. Acta Phytopathologica et Entomologica Hungarica 27:479-483

Moriya S, Shiga M, Adachi I (2003) Classical biological control of the chestnut gall wasp in Japan. In: VanDriesche RG (ed) Proceedings of the 1st international symposium on biological control of arthropods. USDA Forest Service, Washington, pp 407-415

Murakami Y (1980) Recent topics on the chestnut gall wasp, with special reference to a report from China (in Japanese). Nogyô oyobi Engei (Agriculture and Horticulture) 55:249-253

Murakami Y (1981) Comparison of the adult emergence periods between *Torymus (Syntomaspis) beneficus* a native parasitoid of the chestnut gall wasp and a congeneric parasitoid imported from China (Hymenoptera: Torymidae) (in Japanese). Proceedings of the Association for Plant Protection of Kyushu 27:156-158

Murakami Y (1988) Ecotypes of *Torymus (Syntomaspis) beneficus* Yasumatsu et Kamijo (Hymenoptera: Torymidae) with different seasonal prevalence of adult emergence. Applied Entomology and Zoology 23:81-87

Murakami Y, Gyoutoku Y (1991) Colonization of the imported *Torymus* (Syntomaspis) *sinensis* Kamijo (Hymenoptera: Torymidae) parasitic on the chestnut gall wasp (Hymenoptera: Cynipidae). (5) Mortality of *Torymus* spp. by native facultative hyperparasitoids (in Japanese). Proceedings of the Association for Plant Protection of Kyushu 37:194-197

Murakami Y, Gyoutoku Y (1995) A delayed increase in the population of an imported parasitoid, *Torymus (Syntomaspis) sinensis* (Hymenoptera: Torymidae) in Kumamoto, Southwestern Japan. Applied Entomology and Zoology 30:215-224

Murakami Y, Kiyota Y (1983) Colonization of the imported *Torymus* (Syntomaspis) *sinensis* Kamijo (Hymenoptera: Torymidae) parasitic of the chestnut gall wasp (Hymenoptera: Cynipidae). (1) A release test in Kumamoto Prefecture (in Japanese). Proceedings of the Association for Plant Protection of Kyushu 29:155-1577

Murakami Y, Umeya K, Oho N (1977) A preliminary introduction and release of a parasitoid (Chalcidoidea, Torymidae) of the chestnut gall wasp, *Dryocosmus kuriphilus* Yasumatsu (Cynipidae) from China (in Japanese with English summary). Japanese Journal of Applied Entomology and Zoology 21:197-203

Murakami Y, Ao HB, Chang CH (1980) Natural enemies of the chestnut gall wasp in Hopei Province, China (Hymenoptera: Chalcidoidea). Applied Entomology and Zoology 15:184-186

Murakami Y, Uemura, M, Gyoutoku Y (1985) Colonization of imported *Torymus (Syntomaspis) sinensis* Kamijo (Hymenoptera: Torymidae) parasitic on the chestnut gall wasp (Hymenoptera: Cynipidae). (2) Recovery in Kumamoto

Prefecture (in Japanese). Proceedings of the Association for Plant Protection of Kyushu 31:216-219

Murakami Y, Uemura M, Gyoutoku Y, Kiyota Y (1989) Colonization of imported *Torymus (Syntomaspis) sinensis* Kamijo (Hymenoptera: Torymidae) parasitic on the chestnut gall wasp (Hymenoptera: Cynipidae). (4) Trends in host densities and parasitization during six years following release (in Japanese). Proceedings of the Association for Plant Protection of Kyushu 35:134-137

Murakami Y, Ohkubo N, Gyoutoku Y (1993) Origin of *Torymus* (*Syntomaspis*) *sinensis* native to Tsushima Islands (in Japanese). Proceedings of the Association for Plant Protection of Kyushu 39:124-126

Murakami Y, Hiramatsu T, Maeda M (1994) Parasitoid complexes of the chestnut gall wasp (Hymenoptera: Cynipidae) in two localities before introduction of *Torymus* (*Syntomaspis*) *sinensis* (Hymenoptera: Torymidae) with special reference to prediction of results after release of thc parasitoid (in Japanese with English summary). Japanese Journal of Applied Entomology and Zoology 38:29-41

Murakami Y, Ohkubo N, Moriya S, Gyoutoku Y, Kim CH, Kim JK (1995) Parasitoids of *Dryocosmus kuriphilus* (Hymenoptera: Cynipidae) in South Korea with particular reference to Ecologically different types of *Torymus* (*Sntomaspis*) *sinensis* (Hymenoptera: Torymidae). Applied Entomology and Zoology 30:277-284

Murakami Y, Toda S, Gyoutoku Y (2001) Colonization by imported *Torymus* (*Syntomaspis*) *sinensis* Kamijo (Hymenoptera: Torymidae) of the chestnut gall wasp (Hymenoptera: Cynipidae). (7) Success in the eighteenth year after release in Kumamoto (in Japanese). Kyushu Plant Protection Research 47:132-134

Nohara K (1956) Considerations on the reproductive capacity of *Dryocosmus kuriphilus* Yasumatsu (Hymenoptera: Cynipidae) (in Japanese with English summary). Science Bulletin of the Faculty of Agriculture, Kyushu University 15:441-446

Oho N, Shimura I (1970) Process of study on *Dryocosmus kuriphilus* Yasumatsu (Hymenoptera: Cynipidae) and several problems about recent infestation (in Japanese). Shokubutsu Boeki (Plant Protection) 24:421-427

Oho N, Umeya K (1975) Chestnut gall wasp is found in the People's Republic of China (in Japanese). Shokubutsu Boeki (Plant Protection) 29:463-464

Ôtake A (1987) Comparison of some morphological characters among two strains of *Torymus beneficus* Yasumatsu et Kamijo and *T. sinensis* Kamijo (Hymenoptera: Torymidae). Applied Entomology and Zoology 22:600-609

Ôtake A (1989) Chestnut gall wasp *Dryocosmus kuriphilus* Yasumatsu (Hymenoptera: Cynipidae) analysis of records of cell contents inside galls and on emergence of wasps and parasitoids outside galls. Applied Entomology and Zoology 24:193-201

Ôtake A, Shiga M, Moriya S (1982) A study on parasitism of the chestnut gall wasp, *Dryocosmus kuriphilus* Yasumatsu (Hymenoptera: Cynipidae) by parasitoids indigenous to Japan. Bulletin of the Fruit Tree Research Station A9:177-192

Ôtake A, Moriya S, Shiga M (1984) Colonization of *Torymus sinensis* Kamijo (Hymenoptera: Torymidae), a parasitoid of the chestnut gall wasp, *Dryocosmus kuriphilus* Yasumatsu (Hymenoptera: Cynipidae), introduced from China. Applied Entomology and Zoology 19:111-114

Payne JA (1978) Oriental chestnut gall wasp: new nut pest in North America. In: Macdonald WL, Cech FC, Luchok J, Smith C (eds) Proceedings of the American Chestnut Symposium. West Virginia University Press, Morgantown, pp 86-88

Payne JA, Green RA, Lester CD (1976) New nut pest: an oriental chestnut gall wasp in North America. Annual Report of the Northern Nut Growers Association 67:83-86

Payne JA, Jaynes RA, Kays SJ (1983) Chinese chestnut production in the United States: practice, problems and possible solutions. Economic Botany 37:187-200

Schönrogge K, Askew RR (2006) Oak gall communities. In: Stone GN, Melika G, Csóka G (eds) The oak gall wasps of the Western Palaearctic: ecology, evolution and systematics. The Royal Society, London (in press)

Schönrogge K, Crawley MJ (2000) Quantitative webs as a means of assessing the impact of alien insects. Journal of Animal Ecology 69:841-868

Schönrogge K, Stone GN, Crawley MJ (1994) Spatial and temporal variation in guild structure: parasitoids and inquilines of *Andricus quercuscalicis* (Hymenoptera: Cynipidae) in its native and alien ranges. Oikos 2:51-60

Schönrogge K, Stone GN, Crawley MJ (1996) Alien herbivores and native parasitoids: rapid development of guild structure in an invading gall wasp, *Andricus quercuscalicis* (Hymenoptera: Cynipidae). Ecological Entomology 21:71-80

Schönrogge K, Walker P, Crawley MJ (2000) Parasitoid and inquiline attack in the galls of four alien, cynipid gall wasps: host switch and the effect on parasitoid sex ratios. Ecological Entomology 25:208-219

Shiraga T (1951) Chestnut gall wasps and the control (in Japanese). Nogyo oyobi Engei (Agriculture and Horticulture) 26:167-170

Stone GN, Schönrogge K (2003) The adaptive significance of insect gall morphology. Trends in Ecology and Evolution 18:512-522

Stone GN, Schönrogge K, Atkinson RJ, Bellido D, Pujade-Villar J (2002). The population biology of oak gall wasps (Hymenoptera: Cynipidae). Annual Review of Entomology 47:633-668

Toda S, Miyazaki M, Osakabe M, Komazaki S (2000) Occurrence and hybridization of two parasitoid wasps, *Torymus sinensis* Kamijo and *T. beneficus* Yasumatsu et Kamijo (Hymenoptera: Torymidae) in the Oki islands. Applied Entomology and Zoology 35:151-154

Torii T (1959) Studies on the biological control of the chestnut gall wasp, *Dryocosmus kuriphilus* Yasumatsu (Hym.: Cynipidae), with particular reference to the utilization of its indigenous natural enemies. Journal of the Faculty of Agriculture, Shinshu University 2:71-149

van Noort S, Stone GN, Whitehead VB, Nieves-Aldrey J-L (2006) Biology and redescription of *Rhoophilus loewi* (Cynipidae: Cynipoidea: Hymenoptera), with evolutionary implications on the inquilinism in gall wasps. Biological Journal of the Linnean Society (in press)

Watanabe C (1957) A new species of *Aspilota* Förster parasitic on the chestnut gall wasp, *Dryocosmus kuriphilus* Yasumatsu (Hymenoptera: Braconidae). Mushi 30:35-36

Yara K (2004) Relationship between the introduced and indigenous parasitoids *Torymus sinensis* and *T. beneficus* (Hymenoptera: Torymidae) as inferred from mt-DNA (COI) sequences. Applied Entomology and Zoology 39:427-433

Yara K (2006) Identification of *Torymus sinensis* and *T. beneficus* (Hymenoptera: Torymidae), introduced and indigenous parasitoids of the chestnut gall wasp *Dryocosmus kuriphilus* (Hymenoptera: Cynipidae), using the ribosomal ITS2 region. Biological Control 36:15-21

Yara K, Yano E, Sasawaki T, Shiga M (2000) Detection of hybrids between introduced *Torymus sinensis* and native *T. beneficus* (Hymenoptera: Torymidae) in central Japan, using malic enzyme. Applied Entomology and Zoology 35:201-206

Yasumatsu K (1955) Investigations on the parasites of the chestnut gall wasp (in Japanese). Shinrin Boeki News (Forest Pests) 4:100-102

Yasumatsu K, Kamijo K (1979) Chalcidoid parasites of *Dryocosmus kuriphilus* Yasumatsu (Cynipidae) in Japan, with descriptions of five new species (Hymenoptera). Esakia 14:93-111

10 Cynipid Gall Wasps in Declining Black Oak in New York: Relationships with Prior Tree History and Crown Dieback

Carolyn C. Pike[1], Daniel J. Robison[2], and Lawrence P. Abrahamson[3]

[1]University of Minnesota, Cloquet Forestry Center, 175 University Rd, Cloquet, MN 55720, USA
[2]North Carolina State University, Box 8008, Jordan Hall Room 3118, Raleigh, NC 27695-8008, USA
[3]State University of New York's College of Environmental Science and Forestry, 241 Illick Hall, Syracuse, NY 13210, USA

Summary. Decline of black oak (*Quercus velutina* Lamarck) on New York's Long Island in the early 1990's was associated with a cynipid gall wasp on large urban trees. Symptoms included swollen twigs, gnarled branches, fungal cankers, and crown dieback. A study conducted from 1994 to 1996 showed that symptoms occurred on trees in urban and forested areas. Trees with greater incremental diameter growth before the gall wasp outbreak suffered less dieback than those that grew more slowly prior to the outbreak. A series of periodic environmental stress factors, such as drought and defoliation, played an important role in triggering susceptibility and decline in affected trees.

Key words. Oak decline, Black oak, Crown dieback, *Bassettia ceropteroides*

10.1 Introduction

In the mid-1990s, black oak trees (*Quercus velutina* Lamarck) across New York's Long Island (ca. latitude 40°50'N, longitude 72°40'W) exhibited severe crown dieback. Swollen gnarled twigs and fungal twig cankers were associated with dead/dying branches. A *Botryosphaeria* spp. fungus was identified in association with the twig cankers. Because *Botryosphaeria* is generally infective as an opportunistic secondary organism, its presence indicated the likelihood of prior stress in this decline syndrome. The swollen twigs were also associated with a cynipid gall wasp, first mis-identified as a *Callirhytis crypta* and later confirmed as *Bassettia ceropteroides* Bassett (R. Lyon, personal communication, 1996). Only female *B. ceropter-*

oides have been found in black oak branches (Bossart and Raupp 1995) and its life cycle remains poorly understood.

Black oak grows naturally on the sandy soils of the outwash plains of Long Island. It grows best on moist, well-draining soils, and on suboptimal sites black oak seldom lives more than 200 years (Sander 1990). Oaks are susceptible to a myriad of pests and abiotic conditions that, together or in succession, can result in decline. The most commonly cited predisposing and inciting factors to decline include drought and defoliation, respectively (Nichols 1968; Staley 1965). Due to the positive association of cynipid activity with tree health, gall wasps are generally not associated with classic three-step declines as outlined by Sinclair (1965). On the contrary, many cynipids are cited as true parasites, whose survival is often favored on vigorous host material (Caouette and Price 1989; Price 1991; Washburn and Cornell 1981).

In 1994, a study was undertaken to determine the extent and severity of the symptoms on Long Island. This survey revealed that symptoms (crown dieback, swollen twigs) were more severe on trees in urban areas (residential areas) than those in nearby forested, or undeveloped, areas (Pike 1998). In 1995, the association between historical tree growth and cynipid population in urban and forested areas was investigated and the findings are reported here.

10.2 Methods

Two locations in Suffolk County on New York's Long Island were chosen for intensive study ("north shore" and "south shore" areas). Both sites were found to be highly symptomatic from crown dieback during the 1994 survey and occurred on different soil types. Forty trees were chosen at each of two locations, distributed between each of two "ecotypes," defined as "urban" areas and "forested" areas. Trees in urban areas were open-grown and found in the yards of private residences. Forested areas were woodlots at least two hectares in size lacking urban development, and selected trees had crowns which were co-dominant in the stand. In each ecotype, 10 trees were chosen from each of two crown classes: "low dieback," those exhibiting no obvious or limited symptoms of decline (<20% crown dieback), and "high dieback" or trees that were highly symptomatic, with >50% crown dieback. Trees were selected that had average heights and diameters relative to others in that area.

To assess historical populations of cynipid gall wasps, branches were dissected and all dead insects were removed and tallied. To do this, one live terminal branch at least 1 m long was removed from each tree with pole-pruners from the lower-south portion of the canopy, or the nearest accessible side. Each branch was placed in a plastic bag and refrigerated (4°C) until processing. Using pruning shears, each branch was dissected by making cross-sectional cuts at approximately 1 mm intervals from the terminal bud down to the base of the branch sample. Only living tissue was dissected. The growth ring (year) in which each gall wasp (or empty gall) was found was recorded. Only insects residing in the most recent five years of growth were counted. Insects were stored in alcohol until their identities could be verified by a cynipid expert (R. Lyon, personal communication, 1996).

To assess overall sample tree vigor, one increment core was taken to the pith at breast height (1.37 m above the ground) from each tree, and also from 10 co-dominant white oak (*Q. alba*) trees from the forested plots at each site to provide a comparative control. White oak has similar site index (productivity) to black oak (Carmean 1965), is found in close association in this region with black oak, and is not a known host for *B. ceropteroides*. Ring widths were measured for all but current year growth using a Unislide Tree Ring Measuring System equipped with the Acu-Rite sliding scale and linear encoder interfaced with the quick-Check QC-1000 digital measuring device (Velmex, Inc., East Bloomfield NY). Measurements were taken at a precision of 0.001 mm. Basal area growth increment was calculated for each growth ring. Samples from the same site and ecotype were cross-dated graphically by comparing basal area fluctuations over the past 25 years (Fritts 1976). Mean annual (basal area) increment was calculated for five-year intervals (1970–1974, 1975–1979, 1980–1984, 1985–1989, 1990–1994) spanning 25 years of growth. Tree rings were also counted to age each tree. Rainfall history was obtained from weather stations near the north and south shore locations (Northeastern Regional Climate Center, Ithaca NY).

Analysis of variance was used to compare mean number of gall wasps and mean five-year basal area increment in urban vs forested areas, and between the two crown classes (low dieback and high dieback). All statistical significance was evaluated at $P < 0.05$ using SAS (1999). Fisher's Least-Significant Difference (LSD) test was used to compare means.

10.3 Results

Identifiable insects extracted from the dissected twigs were adult *B. ceropteroides*. The total number of gall wasps excised from branches was statistically similar between the two locations (north and south shore), (ANOVA, d.f. = 1, $F = 1.00$, $P = 0.32$) and therefore data from the two sites was pooled. The number of gall wasps exhumed was significantly greater in urban areas than in forested areas, averaging across crown category and twig year (Table 1). In urban areas, gall wasps were significantly more numerous on trees with high dieback than those with low dieback. In forested areas no significant differences between crown categories was found.

In urban areas, trees with high dieback had significantly greater number of insects during the period 1990 to 1992 than trees with low crown dieback (Fig. 1a). In addition, gall wasp populations peaked in 1991 and 1992 and crashed in 1994–1995. In forested areas their numbers gradually decreased from 1990 to 1995 and no significant differences between crown classes were evident (Fig. 1b).

Tree diameters ranged from 45 to 60 cm in urban areas and 22 to 36 cm in forested areas. Of the trees sampled, 94% exceeded 40 years in age. Mean basal area increment in residential areas was up to four times greater than the growth of trees in forested areas. In both urban and forested areas trees with low dieback had significantly greater annual growth than trees with high dieback (Table 2).

In urban areas, mean basal area increment was statistically similar between crown classes in 1970 to 1974 period, but significantly different for all other periods (Fig. 2a). In forested areas, annual increment differed significantly between the crown classes for all five periods (Fig. 2b).

In the white oak comparative control trees, an obvious decrease in radial growth in 1980–1981 was strongly associated with region-wide severe defoliation of oaks by the gypsy moth caterpillar (*Lymantria dispar*) (Fig. 3). A smaller, but obvious reduction in radial growth during a severe regional drought in 1987–1988 was also evident at both sites. Average basal area increment of white oak trees generally increased during the period from 1990 to 1995, when *B. ceropteroides* was present on black oak.

Table 1. Mean number of gall wasps (SE) found in black oak twigs in urban and forested areas from trees with low or high crown dieback. Significant differences between crown dieback levels are indicated with different letters using Fisher's LSD. Significant differences between ecotype, across the bottom row, are indicated similarly. N = 40 for each ecotype

	Urban areas	Forested areas
Low dieback	3.8 (0.9) [b]	3.4 (0.7) [d]
High dieback	8.3 (1.2) [a]	4.5 (0.9) [d]
Ecotype means	5.9 (0.8) [g]	3.9 (0.6) [h]

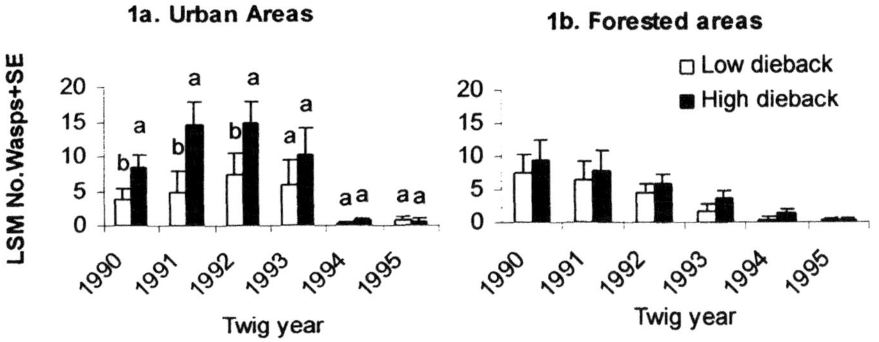

Fig. 1a, b. Least-squared mean *(LSM)* number of gall wasps exhumed from black oak twigs from **(a)** urban and **(b)** forested areas. Significant differences between crown categories for each year are indicated by *different letters* using Fisher's LSD ($P < 0.05$). No significant differences between crown categories occurred at any year in forested areas.

10.4 Discussion

The swollen twigs associated with crown decline revealed large numbers of adult cynipid wasps upon dissection. No live gall wasps were exhumed; however branches were collected in the summer reducing the likelihood of finding live insects. Identification of each insect removed from gall chambers was not possible, many having been entombed in the twig galls for several years. The vast majority of intact insects extracted were adult *B. ceropteroides*, or strongly putatively so. No other cynipid species were extracted. On current year's growth, several empty gall chambers with a corresponding exit hole were found, validating the hypothesis that *B. ceropteroides* completes part of its life cycle early in the spring on current year's growth. As described in Felt (1965), the gall chambers, positioned within the pith of new growth, were outwardly inconspicuous. The swollen twigs associated with the dieback were probably not visible until several years after the insects' departure. Subsequent twig diameter growth (addition of annual rings) slowly exaggerated the impact of the gall on twig shape. Relatively few insects and gall chambers were found on current year's growth indicating that the cynipid population was greatly diminished by 1994. By completion of the study in 1996, many trees appeared to be in a state of recovery (Pike 1998).

Table 2. Mean annual basal area growth increment (SE) of black oak trees in urban and forested areas with low and high crown dieback. Significant differences between crown dieback levels are indicated with different letters using Fisher's LSD. Significant differences between ecotype, across the bottom row, are indicated similarly. N = 40 for each ecotype

	Urban areas	Forested areas
Low dieback	36.8 (1.0) [a]	10.7 (0.5) [c]
High dieback	25.3 (0.8) [b]	5.9 (0.2) [d]
Ecotype means	30.8 (0.7) [e]	8.3 (0.3) [f]

Cynipids are notoriously poor fliers and yet are adept at re-colonizing host trees following disturbance (Washburn and Cornell 1981), and can propagate readily on isolated host trees. Many cynipid species have alternate sexual and asexual generations, existing as a morphologically distinct type until an environmental change triggers the alternate form. For several branch-galling cynipids a sexual leaf-galling generation has been identified (Eliason and Potter 2000a; Lyon, 1969). It is possible that *B. ceropteroides* also produces a leaf-galling sexual generation that has been overlooked. Leaf galls were present on black oak trees, but only inquilines, and not the cynipid gall-formers themselves, could be reared from leaves that were collected for observation (R. Lyon, personal communication). The presence of an endemic alternate sexual generation might explain the proficiency of this cynipid at simultaneously colonizing isolated trees across a large geographic area.

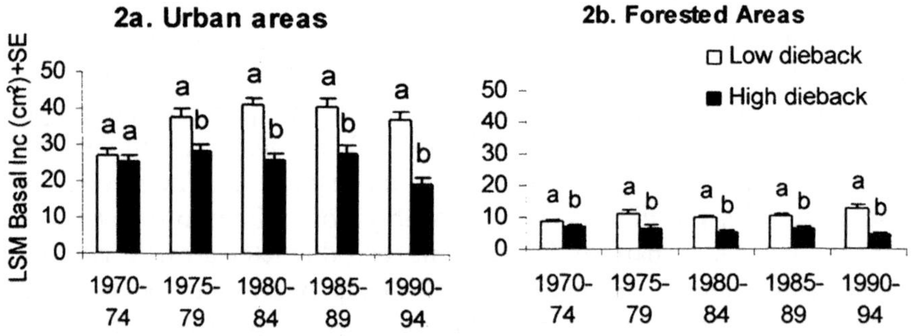

Fig.2a, b. Least-squared mean *(LSM)* periodic basal increment for black oak trees in **(a)** urban and **(b)** forested areas. Significant differences between crown classes within each time period are indicated by *different letters* using Fisher's LSD ($P < 0.05$).

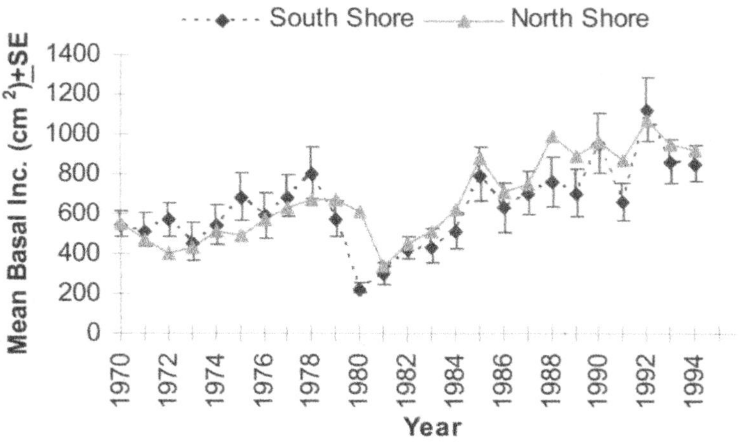

Fig. 3. Mean annual basal area growth increment for asymptomatic white oaks at north and south shore study locations. N = 10 for each site.

In the early 1990s, the apparent impact of *B. ceropteroides* differed dramatically in urban vs forested areas. Crown symptoms were much greater in urban than in forested areas, and on trees with high dieback vs low dieback. Some important predators of gall wasps, mainly chalcids and inquilines, overwinter in leaf litter and are deterred by the sanitary conditions found in urban areas (Frankie et al. 1992). There is likely an assemblage of insects associated with gall wasps in forested settings, including those that parasitize the developing larvae or that inhabit the gall after the cynipids' departure. The isolation and sanitary conditions of urban areas may disrupt such assemblages, considerably reducing predation of gall wasps. Unlike the urban areas, trees in forested areas fostered similar numbers of gall wasps regardless of crown condition. The black oak observed in urban areas had full crowns, while trees in forested areas were devoid of lower branches and co-dominant to nearly suppressed. It may be that urban trees had more surface area available for gall wasps to inhabit. As suggested by Eliason and Potter (2000b), a local wasp population might become epidemic on a tree with adequate host material through reinfection.

All black oaks that were afflicted with this syndrome were mature and moderate to large in stature. Other twig-galling cynipids have afflicted large urban trees (Bailey and Stange 1966; Dixon 1992; Frankie et al. 1992; Taft and Bissing 1988). Eliason and Potter (2000b) suggested that differences in host vigor might explain the patchy dispersal of *C. cornigera* among pin oak trees, but this association was not investigated further. It is impossible to deduce whether *B. ceropteroides* selectively oviposited on declining black oak trees or if these trees provided better host material than vigorous trees resulting in high local populations. Dieback in the absence of decline likely occurred, but in general, trees with the most crown dieback fostered the largest numbers of gall wasps. Vigorous trees (not declining) either sustained fewer attacks or were somehow ill-suited as host material for the insect.

On Long Island, natural regeneration of black oaks is hindered in forested areas because of dense crown cover. In urban areas, black oak has been replaced by other species, and remaining trees represent an aging cohort. Mueller-Dombois (1992) presents a decline theory based on synchronous cohort senescence that could be applied to the black oak in this study. Several juvenile trees were found in unmanaged, disturbed areas, but none were observed declining and no signs of the cynipid were present. A paucity of young black oak trees for study prohibited further data collection.

The urban black oak trees were growing largely in favorable micro environments, being open-grown on lawns that received intensive management (mowing, irrigation, and fertilization). In spite of these favorable conditions, drought was likely still a predisposing factor due to the well-draining soils in this region. In addition, gypsy moth had been prevalent in 1980–1981 and most trees were subjected to intense defoliation. The fact that the dieback was associated with reduced incremental growth prior to 1990 suggests that underlying poor tree health contributed to susceptibility to decline. The dieback observed was likely a combination of disrupted vascular flow from epidemic numbers of gall wasps and fungal cankers, and prior stresses (drought and defoliation). The absence of decline on neighboring white oak trees further supports the hypothesis that gall wasps were a major factor in the decline of black oak. Increment cores from adjacent white oak in the current study showed that they were also subjected to prior drought and defoliation but did not sustain crown dieback or other symptoms characteristic of black oak.

Previous studies have shown that gall wasps may be the benefactors of a host plant's nutritive gains, and may target vigorous hosts as described in Price's (1991) Plant Vigor Hypothesis. However, this hypothesis has been refuted by Williams and Cronin (2004) who present evidence that cynipids perform best on plant tissue that is less vigorous (as measured by nitrogen deficiency). Likewise, *B. ceropteroides* appeared to target mature, dominant trees in an advanced state of decline. The unique combination of underlying stress by drought and defoliation, coupled with some unidentified set of factors favoring *B. ceropteroides,* resulted in a decline in this circumstance induced by a reticent cynipid that was never before associated with widespread damage.

10.5 References

Bailey SF, Stange LA (1966) The twig wasp of cork oak – its biology and control. Journal of Economic Entomology 59:663-668

Bossart JL, Raupp MJ (1995) Identification of cynipid gall wasp associated with black oak decline on Long Island. Internal Report, Department of Entomology, University of Maryland

Caouette MR, Price PW (1989) Growth of Arizona rose and attack and establishment of gall wasps *Diplolepsis fusiformans* and *D. spinosa* (Hymenoptera: Cynipidae). Environmental Entomology 18:822-828

Carmean WH (1965) Black oak site quality in relation to soil and topography in southeastern Ohio. Soil Science Society of America Proceedings 29:308-312

Dixon WN (1992) Spined or horned galls on oaks in Florida induced by galls wasps, *Callirhytis quercusclaviger* (Ashmead) and *Callirhytis cornigera* (Osten Sacken) (Hymenoptera: Cynipidae). Entomology Circular No. 355, Florida Department of Agriculture and Consumer Services, Gainesville

Eliason EA, Potter DA (2000a) Biology of *Callirhytis cornigera* (Hymenoptera: Cynipidae) and the arthropod community inhabiting its galls. Environmental Entomology 29:551-559

Eliason EA, Potter DA (2000b) Budburst phenology, plant vigor, and host genotype effects on the leaf-galling generation of *Callirhytis cornigera* (Hymenoptera: Cynipidae) on pin oak. Environmental Entomology 29:1199-1207

Felt HP (1965) Plant galls and gall makers. Comstock Publishing Company, Ithaca

Frankie GW, Morgan DL, Grissell EE (1992) Effects of urbanization on the distribution and abundance of the cynipid gall wasp *Disholcaspis cinerosa*, on ornamental live oak in Texas, USA. In: Shorthouse JD, Rohfritsch O (eds) Biology of insect-induced galls. Oxford University Press, New York

Fritts HC (1976) Tree rings and climate. Academic Press, New York

Lyon RJ (1969) The alternate generation of *Callirhytis quercussuttonii* (Bassett) (Hymenoptera: Cynipoidea). Proceedings of the Entomological Society of Washington 71:1-65

Mueller-Dombois D (1992) A natural dieback theory, cohort senescence as an alternative to the decline disease theory. In: Manion PD, Lachance D (eds) Forest decline concepts. The American Pathological Society, St Paul, pp 26-37

Nichols JO (1968) Oak mortality in Pennsylvania. Journal of Forestry 66:681-694

Pike CC (1998) Characterization of black oak decline and associated cynipid gall wasp on New York's Long Island. MS Thesis, SUNY ESF, Syracuse

Price PW (1991) The plant vigor hypothesis and herbivore attack. Oikos 62:244-251

Sander IL (1990) *Quercus velutina* Lamarack. In: Burns RM, Honkala BH (eds) Silvics of North America. USDA Forest Service, Washington DC, pp 744-750

SAS Institute (1999) Statistical Analysis System, version 9. SAS Institute, Cary

Sinclair WA (1965) Comparison of recent declines of white ash, oaks, and sugar maple in northeastern woodlands. Cornell Plantations 20:62-67

Staley JM (1965) Decline and mortality of red and scarlet oaks. Forest Science 11:2-17

Taft JB, Bissing DR (1988) Developmental anatomy of the horned oak gall induced by *Callirhytis cornigera* on *Quercus palustris* (pin oak). American Journal of Botany 75:26-36

Washburn JO, Cornell HV (1981) Parasitoids, patches, and phenology: their possible role in the local extinction of a cynipid gall wasp population. Ecology 62:1597-1607

Williams MA, Cronin JT (2004) Response of a gall-forming guild (Hymenoptera: Cynipidae) to stressed and vigorous prairie roses. Environmental Entomology 33:1052-1061

11 Gall-forming Cecidomyiidae from Acacias: Can New Parasitoid Assemblages be Predicted?

Robin J. Adair[1] and Ottilie C. Neser[2]

[1]Department of Primary Industries, Primary Industries Research Victoria, PO Box 48, Frankston 3199, Australia
[2]Plant Protection Research Institute, PB X134, Pretoria 0121, South Africa

Summary. The Australian trees *Acacia mearnsii* and *A. cyclops* are invasive in South Africa and are targets for biological control. Gall-forming cecidomyiids are under consideration as biocontrol agents for these plants, but are parasitised by a diverse range of hymenoptera. As high parasitism levels can disadvantage biological control agents, general criteria are developed to determine if parasitoid composition can be predicted prior to introduction. Agents with low risk of parasitoid attack should receive higher priority than those identified at high risk. The potential parasitoids of *Dasineura rubiformis* are identified. Parasitoid predictions for *D. dielsi* are compared with parasitoids reared from galls after the insect established in South Africa. Parasitoid predictions failed to match parasitoids recorded from *D. dielsi*. Parasitism of *D dielsi* in South Africa has not been detrimental to this insect as a biological control agent; therefore expected parasitism should not preclude release of similar species.

Key words. *Acacia*, Wattle, Parasitoids, *Dasineura*, Gall

11.1 Introduction

The Australian trees *Acacia mearnsii* and *A. cyclops* are invasive in South Africa and cause substantial environmental harm (Versfeld et al. 1998). However, both species are utilised commercially and contribute to both national and local economies (de Wit et al. 2001; Theron et al. 2004). Classical biological control of *A. mearnsii* and *A. cyclops* is implemented in South Africa, but because of the conflict of interest with forestry and wood-harvesting interests, the program is restricted to organisms that reduce seed production without negatively affecting vegetative growth (Adair 2004).

A diverse gall-forming cecidomyiid fauna occurs on acacias in Australia, with many species having highly specialised feeding niches on host inflorescences. Some are under evaluation as potential biological control agents in South Africa (Adair 2004; Kolesik et al. 2005). Gall-forming insects are susceptible to natural enemies (Fernandes and Price 1991) and attack by parasitoids has compromised the effectiveness of cecidomyiids released for weed suppression (Harris and Shorthouse 1996). However, cecidomyiids are not always prone to debilitating attack by parasitoids and numerous introduced and indigenous species are important pests of plants of economic importance. As susceptibility to parasitism varies across feeding guilds of galling insects (Askew and Shaw 1986), estimating the susceptibility of new agents to parasitoid attack before their introduction could help in priority ranking of species under consideration for release (McFadyen and Spafford-Jacob 2004).

In this paper, we predict the composition of South African parasitoids that may affect *Dasineura rubiformis*, a potential agent for *A. mearnsii*, and *D. dielsi*, an agent released for suppression of *A. cyclops*. Parasitoid predictions for *D. dielsi* are compared with recoveries two years after the insect established in South Africa.

11.2 Biology of *Dasineura rubiformis* and *D. dielsi*

Dasineura rubiformis is stenophagous and restricted to acacias in the section Botrycephalae from south-eastern Australia, principally to *A. mearnsii*. In Western Australia the insect and its host are introduced (Adair 2004). Eggs of *D. rubiformis* are laid in open flowers and galls develop by evagination of expanded ovaries. Oviposition occurs in spring to early summer and larvae develop until July when late instars drop from the galls to pupate in the soil. Adults emerge from September to November when the host tree is in flower. *Dasineura rubiformis* is univoltine.

In contrast, *D. dielsi* develops on acacias in the sections Plurinerves and the Juliflorae with *A. cyclops* as its main host. Galls develop from ovaries, but larval development and pupation are completed within gall chambers. *Dasineura dielsi* is multivoltine with up to 5 generations per year, but some larvae diapause over several seasons (Adair 2005).

11.3 *Parasitism* of *D. rubiformis* and *D. dielsi* in Australia

General host utilisation patterns of parasitoids from gall-forming cecidomyiids of acacias in Australia may be important in predicting parasitoid patterns in the country of introduction, in this case, South Africa.

In an extensive survey of parasitoids of cecidomyiid galls from Australian acacias, six families and 22 genera of hymenoptera were reared from 18 cecidomyiid species feeding on 45 species of host acacia (Adair 2004). Two distinct host utilisation patterns were present: a small group of parasitoid species that were confined to either *Dasineura* or *Asphondylia* and associated with particular gall morphologies; and a larger group of parasitoid species that were polyphagous and utilised both *Dasineura* and *Asphondylia* across a range of gall morphology categories (Adair 2004).

Endoparasitic Platygastridae were confined to species of *Dasineura* and predominantly to those with fluted woody galls. A *?Synopeas* species was the most abundant parasitoid of *D. rubiformis* in both eastern and Western Australia, and was also a common parasitoid of *D. dielsi*. In Western Australia, parasitoid diversity of *D. rubiformis* (2 endoparasitoid species) was less than that in eastern Australia (2 endoparasitoid species, 4 ectoparasitoid species) (Table 1). Similarly, total parasitism of *D. rubiformis* in Western Australia was lower (44.4%) than in eastern Australia (61.4%), where the insect is indigenous (Adair 2004). Attack rates by endoparasitoids were similar between eastern and Western Australia, but differences in ectoparasitoid attack and mortality from other causes, possibly from the predatory mite *Pyemotes* sp. (Pyemotidae), accounted for most of the difference in mortality between the two regions.

Reduced parasitoid-predator pressure associated with the establishment of disjunct populations of *D. rubiformis* across a geographical barrier (e.g. Nullarbor desert) appears associated with high galling loads in populations outside the insect's natural range and indicates favourable prospects for biological control of *A. mearnsii* in South Africa. However, the potential impact of native South African parasitoids on Australian cecidomyiids is uncertain. Pre-release parasitoid impact predictions could be assisted by examining parasitoid utilisation patterns within gall-forming Cecidomyiidae in South Africa. Parasitoid prediction criteria are then needed to determine likely assemblages of native South African parasitoids that could exploit Australian cecidomyiids introduced as biological control agents.

Table 1. Parasitoids reared from *Dasineura rubiformis* and *D. dielsi* in Australia

Cecidomyiid	Location	Ectoparasitoids	Endoparasitoids
Dasineura rubiformis	Western Australia	-	?*Synopeas* sp. A (Platygastridae) ?*Conostigmus* sp. (Ceraphronidae)
	Eastern Australia	*Gastrancistrus* sp. (Pteromalidae) *Systasis* sp. A (Pteromalidae) Genus indet. (Pteromalidae) *Aprostocetus* sp. (Eulophidae)	?*Synopeas* sp. A (Platygastridae) Genus indet. (Platygastridae)
Dasineura dielsi	Southern Australia	*Gastrancistrus* sp. (Pteromalidae) *Systasis* sp. A (Pteromalidae) *Systasis* sp. B (Pteromalidae) *Torymoides* sp. (Torymidae) *Megastigmus* sp. (Torymidae) *Aprostocetus* sp. (Eulophidae) *Euderus* sp.* (Eulophidae) *Elasmus* sp*. (Elasmidae)	?*Synopeas* sp. A (Platygastridae) ?*Inostemma* sp. (Platygastridae)

*Feeding mode uncertain.

11.4 Concepts for Predicting new Parasitoid Assemblages

Parasitoids are more likely to attack introduced biocontrol agents when a close taxonomic, ecological and phenological match occurs between a parasitoids normal host range and the new introduction. Understanding the composition and host utilisation patterns of parasitoids from South African cecidomyiids then forms the basis for parasitoid predictions for introduced Australian species. We propose that eight factors may influence this process. Parasitoids of introduced biological control agents are more likely to come from: (1) taxonomically similar insect hosts in the country of introduction, (2) indigenous galls with similar morphological features to those of the introduced agent, particularly gall size and wall thickness (3) indigenous host plants taxonomically close to the host plant of the introduced agent, (4) species with similar bioclimatic requirements to the introduced agent, (5) species with a close phenological match with introduced agent, (6) species that are polyphagous, (7) species belonging to the same or closely allied genera utilising the agent in its country of origin, and (8) species that parasitise previously introduced agents related to those proposed for release. While the relative importance of each factor may vary, new parasitoid associations are more likely to occur in proportion to the number of factors that are satisfied. Biological control candidates with a large number of prospective parasitoids are more likely to be at risk and their impact on the targeted host to be limited than those predicted to acquire fewer parasitoids (Hawkins 1994; McFadyen and Spafford-Jacob 2004). This could then influence the evaluation and introduction of prospective agents for biological control. Candidates with a lower risk of parasitoid attack should be given higher priority than those with a higher risk of attack, providing their impact efficacies are similar.

To evaluate the validity of the conceptual approach described above, the parasitoid fauna of South African gall-forming cecidomyiids was surveyed, with an emphasis on African acacias and other Mimosaceae.

11.5 Parasitioids in South Africa: Surveys and Predictions

Between 1999 and 2003, cecidomyiid galls were collected from 15 families and 38 species of host plants in South Africa. Emerging cecidomyiids and parasitoids were collected, mounted and lodged with the South African National Collection of Insects in Pretoria. In all cases, determination to species level was not possible as all were undescribed taxa; therefore specimens were sorted into morphospecies using similarity of external morphological characters (Table 2). The results are discussed following the prediction criteria outlined above.

Criterion 1. Nine families of hymenopteran parasitoids from 29 genera and 56 morphospecies were reared from African cecidomyiid galls. Parasitoids utilised a broad range of cecidomyiid hosts with both 'generalist,' and possible 'specialist' feeding, patterns, particularly in *Eurytoma*, present. The genera *Eurytoma* and *Eupelmus* had the highest diversity of cecidomyiid hosts.

Australian and South African cecidomyiids of acacias have few taxonomic similarities. The known Australian cecidomyiid fauna consists of species from *Dasineura* and *Asphondylia*, while those known from South African acacias are species of *Acacidiplosis*, *Aposchizomyia*, *Contarinia* and *Asphondylia*. In South Africa, the supertribe Cecidomyiidi predominates, while in Australia the Lasiopteriidi are clearly the most diverse and abundant group. The genus *Dasineura* appears to be poorly represented in South Africa, with only a small number of taxa recorded galling hosts in the Asteraceae, therefore its parasitoids were not commonly encountered.

Criterion 2. Gall morphology influences parasitoid composition and abundance (Dixon et al. 1998) where increasing gall hardness or gall diameter may impede attack by parasitoids (eg. Craig et al. 1990; Mangoni and Hoffmann 1995). *Dasineura* from Australian acacias generally form lignified, thick-walled galls. In South Africa, woody cecidomyiid galls occur on the African acacias *A. luederitzii*, *A. mellifera* and *A. nigrescens*. The galls induced by a *Contarinia* sp. on the ovaries of *A. mellifera* most closely resembled the Australian fluted and flower gall morphology groups (Kolesik et al. 2005) induced by *Dasineura*. Thick-walled and woody galls were also found on species of *Chrysanthemoides*, *Tarchonanthus*, *Rhynchosia* and *Polygala*. Parasitoids reared from African galls with close morphological similarities to Australian *Dasineura* scored against this criterion (Table 2).

Criterion 3. As parasitoids can utilise host plant cues to locate suitable insect hosts (Tumlinson et al. 1993), foraging patterns associated with African acacias could be transposed to allied Australian acacias, possibly influencing parasitoid attack patterns.

A small proportion of parasitoids appeared restricted to cecidomyiids from African Mimosaceae. Nineteen cecidomyiid species forming galls on African Mimosaceae were parasitised by 31 parasitoid taxa from seven families. The parasitoids *Entedon* sp., *Neanastatus* sp., nr *Omphale* sp., ?*Baryscapus* sp., three Tetrastichinae spp., *Aprostocetus* sp. 1, two *Torymoides* spp., *Gastrancistrus* sp., three *Eurytoma* spp., two *Elasmus* spp. and three undetermined platygastrid species were only reared from galls on African acacias. Twenty-six parasitoid taxa were reared from galls occurring on the inflorescences of African acacias with 18 species associated with galled ovaries or receptacles induced by the cecidomyiids *Acacidiplosis*, *Aposchizomyia* and *Contarinia* (Table 2).

Criterion 4. Acacia mearnsii occurs in eastern and southern South Africa where mean annual rainfall exceeds 500mm, while *A. cyclops* is mostly confined to coastal or near-coastal localities (Henderson 2001). Cecidomyiids from African acacias are mostly restricted to the summer-rainfall, lowveld regions, reflecting the distribution of their hosts, which are most diverse in that region. Widespread acacias, such as *A. karroo*, support cecidomyiids over a broad geographic region. Although gall-forming cecidomyiids mirror host distribution patterns, their parasitoids appear to be less restricted geographically. The majority of parasitoids reared from South African cecidomyiids appear to overlap with the current distribution of *A. mearnsii* and *A. cyclops*.

Criterion 5. Galling insects show temporal variation in their susceptibility to parasitoids, which exploit "windows of vulnerability" in their hosts' biological patterns (Craig et al. 1990). Phenological compatibility is therefore required between susceptible stages of the host insect and reproductive activity of the parasitoid. This is likely to be the case for Australian *Dasineura* in South Africa. While some parasitiods will be synchronised with susceptible periods, many are likely to be excluded, particularly those that emerge after the gall wall has lignified and thickened. Insufficient details on the life cycles of South African parasitoids prevent us from making accurate predictions, but it seems most parasitoids collected have adult emergence periods that overlap with the early gall development stages of *D. rubiformis* and *D. dielsi.*

Criterion 6. Generalist parasitoids are more likely to attack introduced biological control agents than specialist parasitoids that are synchronised with the biology of indigenous hosts.

In this survey, species in the families Pteromalidae, Eurytomidae, Elasmidae and Platygastridae appear to be host-specific, perhaps confined to the supertribe Cecidomyiidi, indicating that *Dasineura* may not be susceptible to many of these parasitoids. Although *Dasineura* occur in southern Africa, relatively few records are known and these seem confined to woody Asteraceae. A low diversity of southern African *Dasineura* suggests that there may also be a poor *Dasineura*-specific parasitoid fauna, thus favouring low parasitism attack on Australian *Dasineura* introduced into South Africa. Evidence for the existence of genus-restricted parasitoids within the Cecidomyiidae occurs in Australia, where parasitoids restricted to *Dasineura* and *Asphondylia* are known (Adair 2004).

Criterion 7. The Australian cecidomyiids *Dasineura rubiformis* and *D. dielsi* are attacked in their natural range by 12 species of parasitoids, but predominantly by the endoparasitoid ?*Synopeas* sp. (Platygastridae). A similar tri-trophic relationship occurs in South Africa from galls induced by *Contarinia* on *A. mellifera,* which is parasitised by an undetermined platygastrid, a *Gastrancistrus* sp. and a *Systasis* sp.. *Gastrancistrus* and *Systasis* are important parasitoid genera of *D. rubiformis* and *D. dielsi* in Australia, suggesting that these genera may yield species that are capable of utilising the Australian cecidomyiids in South Africa.

Criterion 8. Two introduced Cecidomyiidae are utilised as biological control agents in South Africa: an Australian *Dasineura* sp. galling buds of *Leptospermum laevigatum* (Myrtaceae), and *Zeuxidiplosis giardi,* a native of Europe, released for suppression of *Hypericum perforatum* (Clusiaceae). In addition, the galling pteromalid *Trichilogaster acaciaelongifoliae* is well established as a biological control agent of *A. longifolia* (Olckers and Hill 1999). All three galling insects are parasitised by South African hymenoptera. Although the gall morphology and biology of these galling insects differ from those of *Dasineura* on acacia, their parasitoids could be candidates as parasitoids of *D. rubiformis* and *D. dielsi.*

Candidate parasitoids we predict to utilise *D. rubiformis* and *D. dielsi* in South Africa score five or more selection criteria and these species are shown in bold in Table 2.

Table 2. Parasitoids of African gall-forming Cecidomyiidae and their potential to attack Australian *Dasineura* from *Acacia*

Parasitoids[A]	1 Utilises host allied to *Dasineura*[B]	2 Similarity in gall morphology	3 Recorded from *Acacia* or other Mimosaceae	4 Bioclimatic overlap	5 Phenological match[C]	6 Polyphagous	7 Genus recorded from *Dasineura* in Australia	8 Genus recorded from introduced galling insects[E]	Host cecidomyiids (genera only)	Host plants (genera only except for *Acacia*)
Eulophidae										
***Pediobius* sp.**		*	*	*	*	*			nr *Bruggmanniella*, *Asphondylia*	*Dichrostachys*, *Psoralea*
Entedon sp. †			*	*					*Asphondylia*	*Acacia karroo*
Entedoninae indet.					*				Lasiopteridi indet.[D]	*Eriocephalus*
nr *Omphale* †			*		*	*			*Acacidiplosis*, *Aposchizomyia*, indet.	*A. tortilis*, *A. nilotica*, *A. gerradii*
?*Baryscapus* sp. †			*	*					*Asphondylia*	*A. karroo*
Tetrastichinae sp. 1†			*		*				?*Acacidiplosis*	*A. nigrescens*
Tetrastichinae sp. 2	*			*					Lasiopteridi indet.	*Eriocephalus*
Tetrastichinae spp. †		*	*		*				*Aposchizomyia*	*A. nilotica*
***Aprostocetus* sp. 1**	*			*	*		*	*	*Lasioptera*	*Carpobrotus*
***Aprostocetus* sp. 2**	*			*	*		*	*	*Lasioptera*	*Salvia*
?*Aprostocetus* sp. 1 †		*	*		*				*Aposchizomyia*	*A. nilotica*
?*Aprostocetus* sp. 2	*	*		*	*	*			*Schizomyia*, nr *Dasineura*	*Rhynchosia*, *Convolvulus*, *Psoralea*, *Zygophyllum*
?*Aprostocetus* sp. 3				*	*	*			*Asphondylia*, indet.	*Helichrysum*, *Senecio*
***Sigmophora* sp.** †		*	*	*	*	*	*		*Schizomyia*, *Contarinia*, *Asphondylia*	*A. karroo*, *Dipogon*, *Achyranthes*, *Centella*
Eupelmidae										
***Eupelmus* sp. 1**	*	*	*	*	*	*		*	Indet., *Lasioptera*, *Schizomyia*, *Aposchizomyia*	*Erica*, *Chrysanthemoides*, *Helichrysum*, *Eriocephalus*, *Convolvulus*, *A. nilotica*

Table 2. Continued

Eupelmidae continued										
***Eupelmus* sp. 2** †	*	*	*	*	*	*		*	*Afrodiplosis*, *Lasioptera*, indet.,	*Tarchonanthus*, *Chrysanthemoides*,
									Acacidiplosis, *Contarinia*	*A. tortilis*, *A. karroo*, *A. mellifera*
***Eupelmus* sp. 3** †	*	*	*	*	*	*		*	*Aposchizomyia*, *Lasioptera*,	*A. nilotica*, *A. mellifera*, *Conyza*,
									Asphondylia, *Contarinia*	*Diospyros*
Eupelmus sp. 4				*				*	*Asphondylia*	*Diospyros*
Neanastatus sp. †		*	*	*		*			*Acacidiplosis*, *Asphondylia*	*A. tortilis*, *A. karroo*, *Diospyros*
Torymidae										
***Pseudotorymus* sp. 1** †			*	*	*	*		*	*Acacidiplosis*, ?*Cecidomyia*,	*A tortilis*, *A. nigrescens*, *A. nilotica*,
									?*Acacidiplosis*	*Salvia*, *Chrysanthemoides*
Pseudotorymus sp. 2	*			*					Lasiopteridi indet.	*Eriocephalus*
?*Pseudotorymus* sp. †		*	*			*			?*Acacidiplosis*, ?*Cecidomyia*	*A. nigrescens*, *Chrysanthemoides*
***Torymoides* spp.**		*	*	*	*	*	*		nr *Bruggmanniella*, indet.	*A. tortilis*, *Dichrostachys*, *Erica*
?*Torymoides* spp. †	*	*	*	*	*	*			*Aposchizomyia*	*A. nilotica*, *A. exuvialis*
Torymus sp. 1				*	*			*	Indet.	*Helichrysum*
Torymus sp. 2				*				*	?*Cecidomyia*	*Chrysanthemoides*
Pteromalidae										
***Systasis* sp.** †	*	*	*	*	*	*	*		Indet., ?*Lobesia*, *Acacidiplosis*,	*A. tortilis*, *A. karroo*, *A. mellifera*,
									nr *Dasineura*, *Contarinia*	*Zygophyllum*
Gastrancistrus sp. †		*	*		*		*		*Contarinia*	*A. mellifera*
nr *Spathopus*				*	*				*Contarinia*	*Asparagus*
Mesopolobus sp.				*	*			*	Indet.	*Elytropappus*
nr Pseudocatolaccus †			*	*	*				?*Asphondylia*, *Asphondylia*	*A. karroo*, *Psoralea*
Macroglenes sp.		*		*	*		*		*Schizomyia*	*Erica*, *Centella*
Indetermined sp.				*					*Asphondylia*	*Diospyros*
Eurytomidae										
Eurytoma sp. 1				*	*		*		*Afrodiplosis*	*Tarchonanthus*
***Eurytoma* sp. 2** †		*	*		*	?	*		*Acacidiplosis*, indet.	*A. nilotica*, *A. tortilis*
***Eurytoma* sp. 3** †		*	*		*	*	*		?*Asphondylia*, indet.	*A. luederitzii*, *A. nilotica*
Eurytoma sp. 4	*			*	*		*		*Lasioptera*	*Salvia*
***Eurytoma* sp. 5**	*	*	*	*	*		*		*Lasioptera*	*Chrysanthemoides*
Eurytoma sp. 6				*	*		*		*Asphondylia*	*Psoralea*
Eurytoma sp. 7	*	*		*			*		*Lasioptera*	*Convolvulus*

Table 2. Continued

Eurytomidae continued										
Eurytoma sp. 8		*	*		*		*		nr *Bruggmanniella*	*Dichrostachys*
Eurytoma sp. 9 †			*	*	*		*		*Aposchizomyia*	*A. karroo*
Ormyridae										
Ormyrus spp. †		*	*		*	*			*Asphondylia*, *Acacidiplosis*, nr *Bruggmanniella*	*A. luederitzii*, *A. tortilis*, *Dichrostachys*
Encyrtidae										
Indetermined sp.	*			*	*		?		*Dasineura*	*Stoebe*
Elasmidae										
***Elasmus* sp. 1**		*	*	*	*		*	*	*Acacidiplosis*	*A. karroo*
***Elasmus* sp. 2** †		*	*	*			*	*	*Asphondylia*	*A. luederitzii*
Platygastridae										
?*Synopeas* sp. 1 †	*	*	*	*	*	*	*		*Contarinia*, nr *Dasineura*	*A. erioloba*, *Zygophyllum*
?*Synopeas* sp. 2		*		*	*		*		?*Cecidomyia*	*Chrysanthemoides*
Indetermined sp. 1 †		*	*		*				*Acacidiplosis*	*A. tortilis*, *A. karroo*
Indetermined sp. 2	*			*	*				?*Rhopalomyia*	*Cliffortia*
Indetermined sp. 3 †			*		*				*Aposchizomyia*	*A. nilotica*
Indetermined sp. 4				*	*				Indet.	*Berzelia*
Indetermined sp. 5	*			*	*				*Lasioptera*	*Carpobrotus*
Indetermined sp. 6	*			*	*				*Lasioptera*	*Salvia*
Indetermined sp. 7 †		*	*		*				*Contarinia*	*A. mellifera*
Indetermined sp. 8 †	*	*	*	*	*	*			*Lasioptera*, *Acacidiplosis*, Lasiopteridi indet.	*A. tortilis*, *Chrysanthemoides*, *Eriocephalus*

†Species reared from galled acacia inflorescences.

[A]All listed genera are known to parasitise Cecidomyiidae (Noyes 2002); parasitoids that scored 5 or more selection criteria are indicated in bold type.

[B]Defined as genera occurring in the supertribe Lasiopteridi.

[C]Adult parasitoids emerging during the main flowering period of *A. mearnsii*/*A. cyclops* (September–January) and the early stages of gall development of *D. rubiformis* and *D. dielsi* (October–February).

[D]Indetermined taxa.

[E]Genus recorded from the introduced biological control agents: *Dasineura*, *Zeuxidiplosis*, *Trichilogaster*.

11.6 Parasitism of *D. dielsi* in South Africa

Dasineura dielsi was approved for release in South Africa as a biological control agent, first established at Stellenbosch, Western Province in 2001, dispersed rapidly and now occurs over most of the range of *A. cyclops* (Adair 2005).

Two years following the establishment of *D. dielsi*, parasitoids were recovered from galls at levels ranging from 0.7–14.6% of the emergent fauna. The parasitoid richness in South Africa was considerably lower than in Australia, with only four species (cf. 12 species) recorded (Adair 2005). The most abundant parasitoids of *D. dielsi* in South Africa were *?Synopeas* sp. 3 and an undetermined platygastrid, but these were only recorded from one of three sites sampled. Smaller numbers of *Torymus* sp. 3 and *Mesopolobus* sp. 3 were present at all three sites (Adair 2005). At several sites in Western Province where *D. dielsi* had recently arrived, no parasitoids were recovered from gall collections (J. Moore, Agricultural Research Council, personal communication, 2003).

The composition and impact of parasitoids of *D. dielsi* in South Africa may be influenced by temporal and spatial factors. Parasitoids generally-find new hosts soon after introduction, with significant parasitoids often detected relatively early in the establishment phase of the new host (Dhileepan et al. 2005). However, parasitoid faunas can accumulate with time and may take many decades to achieve maximum diversity (Cornell and Hawkins 1993), therefore, new parasitoid accessions for *D. dielsi* are likely. Monitoring parasitoid composition over time would provide a convenient case study for developing species' accumulation models. As the geographic range of *D. dielsi* expands in South Africa, additional parasitoids are likely to appear. However, low parasitoid impact on *D. dielsi* in South Africa indicates this insect has considerable potential as a biological control agent. Whether high gall densities and low parasitism rates can be sustained remains to be determined.

11.7 Actual versus Predicted Parasitoids of *D. dielsi*

South African parasitoids that met five or more prediction criteria were considered by us to be strong candidates as parasitoids of Australian *Dasineura* from the fluted and flower galler complex of acacias. Nineteen species of parasitoid from seven families satisfied these criteria (Table 2). Of the four parasitoid species recovered from *D. dielsi* in South Africa two years after establishment, none matched those predicted. However, *?Syn-*

opeas sp. 3 (Adair 2005) was allied to two *?Synopeas* that were recovered from cecidomyiids surveyed in this study. Clearly, the parasitoid predictions have failed to match the actual range of parasitoids recovered from *D. dielsi* in South Africa. Several factors may have contributed to this: (1) *D. dielsi* has not accumulated its full complement of parasitoids because of limited temporal and spatial exposure to prospective parasitoid candidates, (2) generalist parasitoids from other insect hosts may be an important source of new parasitoids, (3) surveys undertaken in South Africa inadequately sampled the fauna associated with Cecidomyiidae, (4) South African parasitoids are insufficiently known, especially the Platygastridae, and (5) stochastic interference masks natural patterns making predictions inaccurate regardless of sampling intensity. However, increased sampling effort focussing on *D. dielsi* parasitism in South Africa and the parasitoids associated with other galling insects may improve the prediction success using the general criteria outlined in this paper. While we believe that sorting parasitoid samples into morphospecies provided sufficient sensitivity for this study, a more thorough understanding of the taxonomy and ecology of the organisms involved may have improved prediction accuracies. Retrospective analysis could determine the relative importance of these criteria, the potential for 'weighting' key criteria, or the need to change them.

Dasineura dielsi has established readily and significantly contributes to the biological control of *A. cyclops* in South Africa, despite being attacked by parasitoids. The prediction criteria were unable to enumerate the prospective parasitoid fauna of *D. dielsi* and the same is expected to apply to *D. rubiformis,* if released. The success to date of *D. dielsi* in South Africa indicates that the introduction of *D. rubiformis* for biological control of seeds of *A. mearnsii* should not be discounted on grounds that it may be excessively parasitised.

11.8 Acknowledgments

The Working for Water program in South Africa kindly provided funding for this study. We thank R Gagné and P Kolesik for identification of cecidomyiids, and GL Prinsloo, M Clark and J La Salle for assistance with identification of parasitoid collections. Stefan Neser and Rachel McFadyen provided valuable comment on an earlier draft of this paper.

11.9 References

Adair RJ (2004) Seed-reducing Cecidomyiidae as potential biological control agents for invasive Australian wattles in South Africa, particularly *Acacia mearnsii* and *A. cyclops*. PhD Thesis, University of Cape Town, Cape Town

Adair RJ (2005) The biology of *Dasineura dielsi* Rübsaamen (Diptera: Cecidomyiidae) in relation to the biological control of *Acacia cyclops* (Mimosaceae) in South Africa. Australian Journal of Entomology 44:446-456

Askew RR, Shaw MR (1986) Parasitoid communities: their size, structure and development. In: Wage J, Greathead D (eds) Insect parasitoids. Academic Press, London, pp 225-264

Cornell HV, Hawkins BA (1993) Accumulation of native parasitoid species on introduced herbivores: a comparison of hosts as natives and hosts as invaders. The American Naturalist 141:847-865

Craig TP, Itami JK, Price PW (1990) The window of vulnerability of a shoot-galling sawfly to attack by a parasitoid. Ecology 59:297-308

de Wit MP, Crookes DJ, van Wilgen BW (2001) Conflicts of interest in environmental management: estimating the costs and benefits of a tree invasion. Biological Invasions 3:167-178

Dhileepan K, Lockett CJ, McFadyen RE (2005) Larval parasitism by native insects on the introduced stem-galling moth *Epilema strenuana* Walker (Lepidoptera: Tortricidae) and its implications for biological control of *Parthenium hysterophorus* (Asteraceae). Australian Journal of Entomology 44:83-88

Dixon KA, Lerma RR, Craig TP, Hughes KA (1998) Gall morphology and community composition in *Asphondylia flocossa* (Cecidomyiidae) galls on *Atriplex polycarpa* (Chenopodiaceae). Environmental Entomology 27:592-599

Fernandes GW, Price P (1991) Comparison of tropical and temperate galling species richness: the roles of environmental harshness and plant nutrient status. In: Price PW, Lewinsohn TM, Fernandes GW, Benson WW (eds) Plant animal interactions: evolutionary ecology in tropical and temperate regions. Wiley, New York, pp 91-115

Harris P, Shorthouse JD (1996) Effectiveness of gall inducers in weed biological control. Canadian Entomologist 128:1021-1055

Hawkins BA (1994) Pattern and process in host-parasitoid interactions. Cambridge University Press, Cambridge

Henderson L (2001) Alien weeds and invasive plants: a complete guide to declared weeds and invaders in South Africa. Agricultural Research Council, Pretoria

Kolesik P, Adair RJ, Eick G (2005) Nine new species of *Dasineura* (Diptera: Cecidomyiidae) from flowers of Australian *Acacia* (Mimosaceae). Systematic Entomology 30:454-479

Manongi FS, Hoffmann JH (1995) The incidence of parasitism in *Trichilogaster acaciaelongifoliae* (Froggatt) (Hymenoptera: Pteromalidae), a gall-forming biological control agent of *Acacia longifolia* (Andr.) Willd. (Fabaceae) in South Africa. African Entomology 3:147-151

McFadyen R, Spafford-Jacob H (2004) Insects for the biocontrol of weeds: predicting parasitism levels in the new country. In: Briese JM, Kriticos DT, Lonsdale WM, Morin L, Scott JK (eds) Proceedings of the XI International Symposium on Biological Control of Weeds. CSIRO Entomology, Canberra, pp 135-140

Noyes JS (2002) Interactive Catalogue of World Chalcidoidea. 2nd Edition. The Natural History Museum

Olckers T, Hill M (1999) Biological control of weeds in South Africa (1990-1998). African Entomology Memoir No. 1:1-182

Theron JM, van Laar A, Kunneke A, Bredenkamp BV (2004) A preliminary assessment of utilizable biomass in invading *Acacia* stands on the Cape coastal plains. South African Journal of Science 100:123-125

Tumlinson JH, Lewis WJ, Vet LEM (1993) How parasitic wasps find their hosts. Scientific American 268:46-52

Versfeld DB, le Maitre DC, Chapman RA (1998) Alien invading plants and water resources in South Africa: A preliminary assessment. WRC Report No. TT99/98, CSIR No. ENV/S-C 97154

12 Recent Outbreaks of the Maize Orange Leafhopper *Cicadulina bipunctata* Inducing Gall-like Structures on Maize in Japan

Masaya Matsumura[1], Makoto Tokuda[2], and Nobuyuki Endo[1]

[1] National Agricultural Research Center for Kyushu Okinawa Region, 2421 Suya, Nishigoshi, Kumamoto 861-1192, Japan
[2] National Institute of Advanced Industrial Science and Technology, 1-1-1 Higashi, Tsukuba, Ibaraki 305-8566, Japan

Summary. The maize orange leafhopper *Cicadulina bipunctata* is distributed widely from Africa to Asia including Japan, and northern Australia. Some cereals infested by *C. bipunctata*, such as maize and rice, exhibit stunted growth and severe swelling of leaf veins, symptoms commonly referred to as 'wallaby ear disease'. Though previous studies attributed the symptoms to a leafhopper-transmitted virus, recent studies suggest that chemicals injected by *C. bipunctata* during feeding are important contributors to these symptoms. Therefore the damage is considered a sort of insect gall. Following the initiation of biyearly plantings of forage maize in Kyushu, Japan, *C. bipunctata* became recognized as a serious insect pest of forage maize. Since 2001 the total area of forage maize fields damaged by *C. bipunctata* has gradually increased, and outbreaks of *C. bipunctata* occurred in 2004. We speculate that relatively low winter mortality and early appearance of *C. bipunctata* are possible factors contributing to recent outbreaks. In central Kyushu, an increase of 1.3 generations per year following global warming is estimated for *C. bipunctata* by 2100, relative to 1990. Thus, *C. bipunctata* has the potential to become a serious insect pest of cereal crops other than forage maize in the future.

Key words. *Cicadulina bipunctata*, Forage maize, Global warming, Leaf galls, Maize wallaby ear disease

12.1 Introduction

A diverse group of insect orders such as Diptera, Hymenoptera, and Homoptera are known to have the ability to induce galls (e.g. Yukawa and Masuda 1996). In Homoptera, most gall inducers occur among aphids, psyllids, and coccids, while very few are leafhoppers (Cicadellidae) (Raman et al. 2005). The maize orange leafhopper *Cicadulina bipunctata* (Melichar) (Homoptera: Cicadellidae) induces stunted growth and severe swelling of leaf veins on some cereals, which have been commonly referred to as 'wallaby ear disease' (Agati and Calica 1949; Maramorosch et al. 1961).

Since the initiation of biyearly plantings of forage maize in central and southern Kyushu, Japan in the late 1980's, *C. bipunctata* became recognized as a serious pest causing wallaby ear disease in the second planting of forage maize (Ohata 1993). In this paper, we first describe the symptom and characteristics of maize wallaby ear disease, then summarize the biology of *C. bipunctata*, and lastly review the recent outbreaks of *C. bipunctata* and maize wallaby ear disease in Japan. In addition, we discuss the cause of recent outbreaks of *C. bipunctata* in relation to global warming.

12.2 Maize Wallaby Ear Disease Induced by *C. bipunctata*

Some cereals infested by *C. bipunctata*, such as maize, rice, and wheat, exhibit symptoms of wallaby ear disease (Fig. 1a) (Agati and Calica 1949; Kawano 1994). Among these cereals, severe damage by *C. bipunctata* was reported on maize in several countries, including Australia (Grylls 1975), the Philippines (Agati and Calica 1949; Maramorosch et al. 1961), China (Li and Liu 2004; Li et al. 2004c), Taiwan (Chen 1991), and Japan (Matsumura et al. 2005; Ohata 1993).

Previous studies attributed the symptoms to a leafhopper-transmitted virus (Agati and Calica 1949; Boccardo et al. 1980; Grylls 1975; Maramorosch et al. 1961; Reddy et al. 1976), but recent studies suggest

Fig. 1a, b. a Symptom of wallaby ear disease on forage maize. **b** Adult of *Cicadulina bipunctata*; the body length of the adult *C. bipunctata* is ca. 3 mm.

that chemicals injected by *C. bipunctata* during feeding are important contributors to these symptoms (e.g., Kawano 1994; Ofori and Francki 1983;Ohata 1993). So the damage is considered a sort of insect gall. However, unlike other gallers, *C. bipunctata* induces the symptom not on the feeding site but from a distance away (see below) and no progeny lives in the galled plant tissue. Therefore, the gall induction by *C. bipunctata* should be distinguished from others when considering its ecological significance. Thus we represent the symptoms as 'gall-like structures' in this paper. The mechanism inducing the gall-like structures and the effect of induction on the performance of *C. bipunctata* have not yet been clarified.

The symptom of wallaby ear disease occurs also on seedlings of maize. Matsumura and Tokuda (2004) established a simple method, using maize seedlings, for evaluating varietal resistance of maize to wallaby ear disease. Using this method, we clarified the induction process of wallaby ear disease on maize (Fig. 2) (Matsumura and Tokuda 2004; M. Matsumura and M. Tokuda, unpublished data). When adults of *C. bipunctata* were released on a six-day-old maize seedling for three days, stunted growth and severe swelling of leaf veins were evident six days after the removal of adults on a new leaf that extended from the seedling, but not on the leaf that had been fed on by the adults. In addition, the symptom did not appear when further new leaves were produced after the removal of adults (Matsumura and Tokuda 2004; M. Matsumura and M. Tokuda, unpublished data). These results suggest that chemicals injected by *C. bipunctata* during feeding possibly affect shoot apical meristem and induce the symptom on the leaves that are produced during the feeding, perhaps as a manner similar to that reported for a galling adelgid (Sopow et al. 2003).

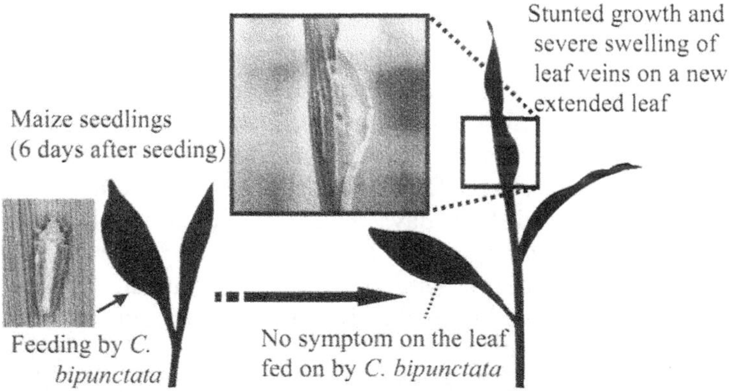

Fig. 2. The induction process of wallaby ear disease on maize seedlings by *Cicadulina bipunctata*.

12.3 General Information on *C. bipunctata*

The maize orange leafhopper *C. bipunctata* (Fig. 1b), which causes wallaby ear disease, is distributed widely from East and North Africa, across the Indian Ocean, southern Palaearctic and the Oriental Region to China, Taiwan, and Japan, and south to New Guinea and northern Australia (Chen 1991; Li and Liu 2004; Webb 1987; Wilson and Claridge 1991). In Japan, *C. bipunctata* was first collected at Kumamoto, central Kyushu (Matsumura 1914). Thereafter, *C. bipunctata* has been collected in central and southern Kyushu (Matsumura et al. 2005; Ohata 1993), the Ryukyu Islands (Hayashi 2002; Kawano 1994), and the Bonin Islands (Hayashi 2002). So *C. bipunctata* exists widely in southern Japan, while it had been recognized as an uncommon species until late 1990's (Matsumura et al. 2005). The northern limit of distribution of *C. bipunctata* in Japan has not been examined intensively, but it seems around Kumamoto even now.

Although native hosts of *C. bipunctata* are still unknown, Catindig et al. (1996) reported that *C. bipunctata* could complete development on 17 host plants of the Poaceae, including maize and rice. Among agricultural crops, *C. bipunctata* is usually found on maize in Australia (Grylls 1975), the Philippines (Agati and Calica 1949), Taiwan (Chen 1991), China (Li and Liu 2004), and Japan (Ohata 1993). According to Wilson and Claridge (1991), *C. bipunctata* is common on rice in Asia but not recorded as a serious pest. Shepard et al. (1995) mentioned that *C. bipunctata* is uncommon in rice in tropical Asia. In temperate Asia, Chen (1991) reported that *C. bipunctata* attacked rice in paddy fields in Taiwan. Though *C. bipunctata* has not been found in rice fields in Japan, there is a potential of *C. bipunctata* for becoming a serious pest of rice in Japan in the future as mentioned later.

Several studies have been conducted to compare development and reproduction of *C. bipunctata* on maize and rice (Catindig et al. 1996; Li et al. 2004a, b; Tokuda and Matsumura 2005). Catindig et al. (1996) reported higher nymphal survivorship, shorter nymphal developmental time, much higher fecundity, and longer female adult longevity on maize (94.3%, 11.1 days, 479.0 eggs, and 14.5 days at 27.6°C, respectively) than on rice (31.3%, 16.5 days, 24.9 eggs and 13.9 days at 27.6°C, respectively). Li et al. (2004a, b) also reported similar results. In contrast, Tokuda and Matsumura (2005) reported higher survivorship, faster nymphal development, higher fecundity, and longer female adult life span on rice (40.0 %, 12.8 days, 188.7 eggs, and 32.3 days at 28.4°C, respectively), than the results of Catindig et al. (1996) using rice. Such differences might be attributed to the rice varieties used for the experiments, because, in contrast with the experiment by Tokuda and Matsumura (2005) using a japonica rice variety 'Reiho', Catindig et al. (1996) employed an indica rice variety 'TN1'. The low survivorship, fecundity, and longevity of *C. bipunctata* on the indica

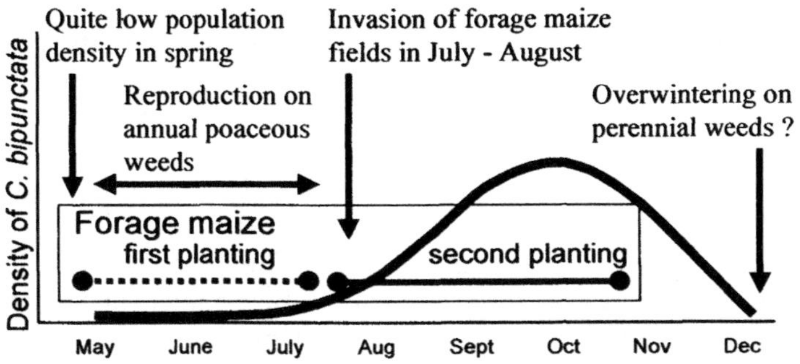

Fig. 3. Seasonal occurrence of *Cicadulina bipunctata* and cropping season of forage maize in central Kyushu, Japan (illustrated based on Matsumura et al. 2005).

variety could explain why *C. bipunctata* has not been recorded as a serious pest of rice in tropical Asian countries even though it is frequently common on rice there (e.g. Wilson and Claridge 1991). In contrast, the high survivorship, fecundity, and longer longevity on the japonica variety suggest that *C. bipunctata* has the potential to become a serious pest of rice in Japan in the future. Therefore, egg and nymphal survivorship of *C. bipunctata* needs to be examined on native plant hosts as well as on various rice and maize varieties.

Seasonal occurrence of *C. bipunctata* and cropping season of forage maize in central Kyushu are illustrated in Fig. 3, based on Matsumura et al. (2005). Population density of *C. bipunctata* is quite low in spring and *C. bipunctata* reproduces on annual poaceous weeds from May to July. The first planting of forage maize is present during this period but wallaby ear disease seldom occurs on the maize, presumably because of the low density of *C. bipunctata.* The density of *C. bipunctata* on annual poaceous weeds becomes high and leafhoppers invade the second planting of forage maize from late July to August, when they induce wallaby ear disease. Population density of *C. bipunctata* peaks around October. The detailed overwintering ecology is still unknown in Japan, but *C. bipunctata* seems to overwinter in the adult stage, probably on perennial weeds.

12.4 Recent Occurrence of Maize Wallaby Ear Disease and *C. bipunctata* in Japan

C. bipunctata only became recognized as an insect pest of forage maize following the initiation of biyearly plantings of forage maize (from late March to late July for the first planting and from mid August to October for the second planting) in central and southern Kyushu, Japan in the late 1980's (Ohata 1993). The distribution area of *C. bipunctata* and the fields damaged by wallaby ear disease were restricted to only two or three locations in Kikuchi, Kumamoto, central Kyushu until the late 1990's (Ohata 1993). Since 2001, however, the total area of forage maize fields damaged by *C. bipunctata* has gradually expanded, and outbreaks of *C. bipunctata* occurred in 2004 (Matsumura et al. 2005).

Matsumura et al. (2005) surveyed the distribution area, abundance, and seasonal occurrence of *C. bipunctata* in forage maize fields and surrounding areas in Kikuchi, Kumamoto in 2004. They showed that *C. bipunctata* inhabited almost all cultivated areas of forage maize in Kikuchi and that the density was high enough to cause severe damage to forage maize in many areas of Kikuchi. The peak density of *C. bipunctata* on poaceous weeds near maize fields occurred during October in 2004.

Since the late 1980's, when wallaby ear disease first occurred on forage maize in Kikuchi, a resistant variety of forage maize against wallaby ear disease (variety '30D44') was commoditized by Pioneer Hi-Bred Japan Co., Ltd. (Matsumura and Tokuda 2004). The extent of stunted growth and swelling of leaf veins is very small on this variety, even though *C. bipunctata* can feed and survive on it. However, when the population density was very high as in 2004, the damage of maize wallaby ear disease occurred even where the resistant variety was planted (Matsumura et al. 2005). Thus, the establishment of control measures against *C. bipunctata* using novel resistant varieties and/or insecticides is an urgent necessity.

To clarify whether the outbreaks of *C. bipunctata* are related with recent increase in the temperature or not, we examined average temperatures in Kumamoto and Kikuchi (Matsumura et al. 2005). As a result, average annual temperatures have increased about 1 to 1.5°C between 1960 and 2004 (Matsumura et al. 2005) and the trend was noticeable especially in winter (December–February; Fig. 4a) and in spring (April–June; Fig. 4b). In addi-

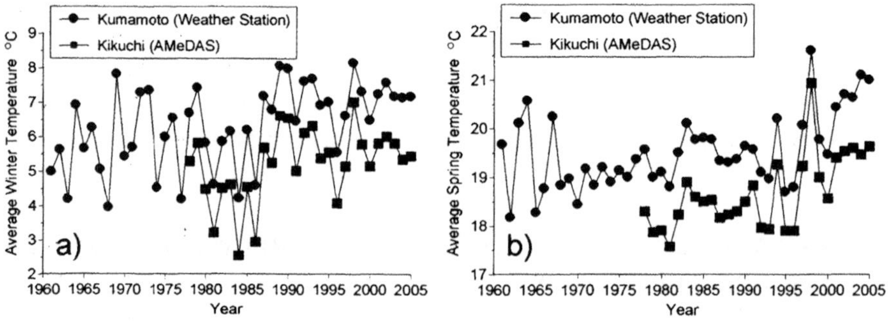

Fig. 4a, b. Long-term trend of average temperatures in **(a)** winter (December to February) and **(b)** spring (April to June) in Kumamoto and Kikuchi, Kumamoto, Japan (data from Japan Meteorological Agency web site: *http://www.jma.go.jp*). *AMeDAS*, Automated Meteorological Data Acquisition System.

tion, the average temperatures in these seasons have been continuously high and have not fluctuated much in recent years. These data suggest that there were no cold winters and cool springs in these years, related probably to the global warming and heat island effect. Thus, we speculate that the recent expansion of damaged area and outbreaks of *C. bipunctata* is due to relatively low mortality of *C. bipunctata* in winter and its early appearance in spring.

12.5 Potential of *C. bipunctata* for Becoming a Serious Pest of Various Cereals

At present, damage by *C. bipunctata* is conspicuous only on the second planting of forage maize, which grows in the summer season, in some restricted areas of central and southern Kyushu. However, the damage might also become serious on the first planting if further increase in the average temperature causes low winter mortality of *C. bipunctata* and promotes the earlier appearance in spring. Moreover, damage may also expand to other cereals such as rice and wheat, because the leafhopper has the ability to induce gall-like structures on a variety of cereals as mentioned earlier. Therefore, the effect of changes in thermal conditions on the development and life history traits of *C. bipunctata* needs to be clarified.

In laboratory experiments, the developmental periods of eggs and nymphs were not delayed even at high temperatures, such as 31 and 34°C (Tokuda and Matsumura 2005), suggesting that *C. bipunctata* has relatively high tolerance for high thermal conditions. In addition, adults of *C. bipunctata* also have some tolerance to temperatures as low as 5.0°C (M. Matsumura and M. Tokuda, unpublished data). Such thermal tolerance of *C. bipunctata* possibly explains the wide natural distribution range of *C. bipunctata* (Webb 1987; Wilson and Claridge 1991) and implies the possibility of a future expansion in the distributional area following global warming.

Adult longevity of *C. bipunctata* at 25°C, 50.8 days for females and 68.2 days for males, is much longer than that of other rice-associated leafhoppers (Tokuda and Matsumura 2005). The average fecundity of *C. bipunctata* at 25°C (291.5 eggs/female) is also higher than that of other rice-associated leafhoppers (Tokuda and Matsumura 2005). The long adult life span and high fecundity of *C. bipunctata* are of note, because these ecological traits can cause continuous and severe damage to cereal crops.

The intrinsic rate of natural increase (r_m) of *C. bipunctata* was highest at 31°C among the range of temperatures examined (16–34°C) (Tokuda and Matsumura 2005), suggesting that a high population growth rate of *C. bipunctata* would be achieved under high temperature conditions.

Based on the mid range scenario predicted by the Intergovernmental Panel on Climate Change (an increase of 2°C in average temperature by 2100 relative to 1990) (IPCC 1996), Tokuda and Matsumura (2005) estimated that the number of generations of *C. bipunctata* would increases by 1.3 generations per year in Kumamoto, central Kyushu, by 2100. Therefore, we need to address the possibility that *C. bipunctata* can cause severe damage in the future to cereal crops other than the second planting of forage maize due to an earlier appearance of *C. bipunctata* and higher overwintering survival, as well as an increase in the number of generations per year following global warming.

12.6 References

Agati JA, Calica C (1949) The leaf-gall disease of rice and corn in the Philippines. Philippine Journal of Agriculture 14:31-40

Boccardo G, Hatta T, Francki RIB, Grivell CJ (1980) Purification and some properties of reovirus-like particles from leafhoppers and their possible involvement in wallaby ear disease of maize. Virology 100:300-313

Catindig JLA, Barrion AT, Litsinger JA (1996) Plant host range and life history of the orange leafhopper *Cicadulina bipunctata* (Melichar) (Hemiptera: Cicadellidae). Philippine Entomologist 10:163-174

Chen CC (1991) The plant diseases transmitted by the leafhoppers and planthoppers in Taiwan (in Chinese). Chinese Journal of Entomology Special Publication 7:139-156

Grylls NE (1975) Leafhopper transmission of a virus causing maize wallaby ear disease. Annals of Applied Biology 79:283-296

Hayashi M (2002) Homoptera (in Japanese). In: Azuma S (ed) Check list of the insect of the Ryukyu islands, flora and fauna in Okinawa No.1, Second Edition. Biological Society of Okinawa, Okinawa, pp 97-112

IPCC (1996) Climate Change 1995: The Science of Climate Change. Cambridge University Press, Cambridge

Kawano S (1994) Occurrence of maize leaf gall disorder induced by the feeding of the leafhopper, *Cicadulina bipunctella* (Matsumura) (Hemiptera: Cicadellidae) in Okinawa, Japan (in Japanese with English summary). Bulletin of the Okinawa Agricultural Experiment Station 15:51-57

Li XY, Liu YH, Wei J, Qing L (2004a) Effect of temperature on development survival and reproduction of *Cicadulina bipunctella* (Homoptera: Cicadellidae) (in Chinese with English summary). Journal of Southwest Agricultural University (Natural Science) 26(1):35-39

Li XZ, Liu YH (2004) The biological characteristics and artificial rearing of *Cicadulina bipunctella* (Mots) (in Chinese with English summary). Journal of Southwest Agricultural University (Natural Science) 26(2):143-145

Li XZ, Liu YH, Tian Y (2004b) Effects of six host plants on the development and fecundity of *Cicadulina bipunctella* (in Chinese with English summary). Chinese Journal of Applied Ecology 15(8):1431-1434

Li XZ, Liu YH, Zhao ZM, Zhou LF (2004c) Temporal and spatial dynamics of *Cicadulina bipunctella* natural populations (in Chinese with English summary). Zoological Research 25(3): 221-226

Maramorosch K, Calica CA, Agati JA, Pableo G (1961) Further studies on the maize and rice leaf galls induced by *Cicadulina bipunctella.* Entomologia Experimentalis et Applicata 4:86-89

Matsumura M, Tokuda M (2004) A mass rearing method using rice seedlings for the maize orange leafhopper *Cicadulina bipunctata* (Melichar) (Homoptera: Cicadellidae) and a simple method for evaluating varietal resistance of maize to maize wallaby ear disease (in Japanese with English summary). Kyushu Plant Protection Research 50:35-39

Matsumura M, Tokuda M, Endo N, Ohata S, Kamitani S (2005) Distribution and abundance of the maize orange leafhopper *Cicadulina bipunctata* (Melichar) (Homoptera: Cicadellidae) in Kikuchi, Kumamoto, Japan, in 2004 (in Japanese with English summary). Kyushu Plant Protection Research, 51:36-40

Matsumura S (1914) Die Jassinen und einige neue Acocephalinen Japans. Journal of the College of Agriculture Tohoku Imperial University 5:165-240

Ofori FA, Francki RIB (1983) Evidence that maize wallaby ear disease is caused by an insect toxin. Annals of Applied Biology 103:185-189

Ohata S (1993) The occurrence of maize wallaby ear disease caused by the feeding of leafhopper, *Cicadulina bipunctata* Melichar (Homoptera, Cicadellidae) (in Japanese). Grassland Science 39:120-123

Raman A, Schaefer CW, Withers TM (2005) Galls and gall-inducing arthropods: An overview of their biology, ecology, and evolution. In: Raman A, Schaefer CW, Withers TM (eds) Biology, ecology, and evolution of gall-inducing arthropods Volume 1. Science Publishers, Enfield Plymouth, pp 1-33

Reddy DVR, Grylls NE, Black LM (1976) Electrophoretic separation of dsRNA genome segments from maize wallaby ear virus and its relationship to other phytoreoviruses. Virology 73:36-42

Shepard BM, Barrion AT, Litsinger JA (1995) Rice-feeding insects of tropical Asia. International Rice Research Institute, Los Baños

Sopow SL, Shorthouse JD, Strong W, Quiring DT (2003) Evidence for long-distance, chemical gall induction by an insect. Ecology Letters 6:102-105

Tokuda M, Matsumura M (2005) Effect of temperature on the development and reproduction of the maize orange leafhopper *Cicadulina bipunctata* (Melichar) (Homoptera: Cicadellidae). Applied Entomology and Zoology 40:213-220

Webb MD (1987) Species recognition in *Cicadulina* leafhoppers (Hemiptera: Cicadellidae), vectors of pathogens of Gramineae. Bulletin of Entomological Research 77:683-712

Wilson MR, Claridge MF (1991) Handbook for the identification of leafhoppers and planthoppers of rice. CAB International, Oxon
Yukawa J, Masuda H (1996) Insect and mite galls of Japan in colors (in Japanese with English explanations for color plates). Zenkoku Nôson Kyôiku Kyôkai, Tokyo

3. Galling Arthropods - Plant Interactions

13 Different Oviposition Strategies in Two Closely Related Gall Midges (Diptera: Cecidomyiidae): Aggregation versus Risk Spreading

Ken Tabuchi[1] and Hiroshi Amano[2]

[1] JSPS Research Fellow, Hokkaido Research Center, Forestry and Forest Products Research Institute, 7 Hitsujigaoka, Toyohira, Sapporo 062-8516, Japan
[2] Laboratory of Applied Entomology and Zoology, Faculty of Horticulture, Chiba University, 648 Matsudo, Chiba 271-8510, Japan

Summary. The oviposition strategies were compared between two closely related gall midges, *Asteralobia sasakii* and *A. soyogo* (Diptera: Cecidomyiidae). They induce spherical, multilocular galls on the axillary buds of *Ilex* species (Aquifoliaceae). *A. sasakii* is known to have a narrower host range and induce larger galls that contain more numerous larvae than *A. soyogo*. The large *A. sasakii* galls have been considered to act as physical barriers to protect larvae from ectoparasitoid attack. We observed that *A. sasakii* induced galls more frequently on the axillary bud at the tip of the shoot, indicating that the females concentrated their eggs at a single oviposition site. In contrast, *A. soyogo* induced several small galls in a shoot, scattering their progeny. The number of *A. sasakii* galls per shoot was significantly smaller than that of *A. soyogo*. Our current observation, together with the previous data, demonstrates that *A. sasakii* lays eggs in clusters to lessen the threat of ectoparasitoid attack (aggregation), whereas *A. soyogo* spread the risk to progeny by scattering eggs in a single shoot using several host plant species (risk spreading). The oviposition strategy of *A. soyogo* also seems to be supported by its diversified life cycles with a polymodal emergence pattern.

Key words. Aquifoliaceae, Egg allocation, *Ilex*, Resource use, Tritrophic interaction

13.1 Introduction

Herbivorous insects whose immatures spend all their time at a single site on a particular host plant cannot avoid predation by moving to other, less dangerous sites. Therefore, the females' choice of oviposition sites may be crucial for the survival of their progeny. The life-history strategies of insect herbivores, such as oviposition pattern, are largely dominated by the trophic levels above and below them.

To detect such adaptations in herbivorous insects, comparative study of the same tritrophic system can be used. We have studied two closely related gall midge species, *Asteralobia sasakii* (Monzen) and *Asteralobia soyogo* (Kikuti) (Diptera: Cecidomyiidae) on *Ilex* hosts (Aquifoliaceae) and the parasitoid communities that attack these gall midges (Tabuchi and Amano 2004). The clutch size (i.e., the number of larvae per gall) and gall size of *A. sasakii* is larger than those of *A. soyogo*; selection pressure by ectoparasitoids is stronger on the former species and the thick wall of the larger gall decreases the likelihood of successful attack.

In our previous work, the number of larvae in each gall was measured to examine the oviposition strategies (Tabuchi and Amano 2004). Based on observations of oviposition behavior, *A. sasakii* and *A. soyogo* females select one shoot as a unit for oviposition (K. Tabuchi, personal observation). Therefore, egg allocation occurs on a per-shoot basis. Moreover, the constraints of different host plants must be considered because gall distribution in shoots is determined by where available buds are distributed within the shoot.

To present an overview of the oviposition strategies of these gall midges, we compared the egg allocation pattern (i.e., gall distribution) in the shoots between *A. sasakii* on *Ilex crenata* Thunb. and *A. soyogo* on *Ilex integra* Thunb. Growth patterns of the axillary buds on shoots of each host plant were also examined. We discuss the oviposition strategies of the two gall midges with regard to their respective tritrophic systems.

13.2 Materials and Methods

13.2.1 Gall Midges Studied

A. sasakii and *A. soyogo* are widely distributed in Japan (Tokuda et al. 2002; Yukawa and Masuda 1996). Both gall midges form multilocular galls on the axillary buds of several species of *Ilex* tree (Aquifoliaceae) (Monzen 1937; Tokuda et al. 2002; Yukawa and Masuda 1996). The distributions of these gall midges overlap in many parts of Japan, but they segregate by using different host plants (Tokuda et al. 2002).

The life histories of *A. sasakii and A. soyogo* inhabiting *I. crenata* and *I. integra*, respectively, were described in Tabuchi and Amano (2003a) (These authors referred to *A. soyogo* as an *A. sasakii* population on *I. integra.*). All individuals of *A. sasakii* and many individuals of *A. soyogo* have a univoltine life cycle: adults emerge in late April to June after overwintering in galls on the shoots. After mating, females oviposit into the axillary buds of shoots. The galls mature by autumn, and the larvae complete their development by the end of October. The mature larvae overwinter in the galls and pupate the following March. Some individuals of *A. soyogo,* however, have 2- or 3-year life cycles, with first instars remaining in immature galls during this time.

Until recently, host ranges of both *A. sasakii* and *A. soyogo* were not determined because of morphological similarity between the two species and the necessity of reexamination of their host ranges was emphasized (Yukawa and Masuda 1996). Recently, Tokuda et al. (2004) clarified their host ranges based on DNA analysis and morphological characteristics. In addition, allochronic reproductive isolation caused by different adult emergence periods was noted between *A. sasakii* on *I. crenata* and *A. soyogo* on *I. integra* (Tabuchi and Amano 2003a, b).

13.2.2 Distribution of Galls and Larvae in Shoots

Field studies were conducted at the Matsudo campus (35°47′N, 139°54′E, about 26 m a.s.l.), Chiba University, central Honshu, Japan. Spatial distribution of galls and larvae in shoots were investigated to clarify which axillary bud positions adult females of *A. sasakii* and *A. soyogo* prefer for galling sites. Galls were sampled from November to May in 2000–2001 and 2001–2002. In 2001–2002, one *I. crenata* and one *I. integra* tree were selected for sampling galls. In 2000–2001, two *I. crenata* trees were used for sampling to ensure enough gall samples. In each field season, galls were sampled from different trees to avoid any influence from the preceding year's survey. In the census field, *I. crenata* and *I. integra* were distributed non-randomly; the sampling sites of *A. sasakii* on *I. crenata* and *A. soyogo* on *I. integra* were 470 m apart. Mature galls of *A. sasakii* on *I. crenata* (994 galls in 2000–2001 and 1006 galls in 2001–2002) and *A. soyogo* on *I. integra* (728 galls in 2000–2001 and 518 galls in 2001–2002) were randomly collected from the census trees. In the laboratory, the number of galls per shoot and galling position within the shoots were investigated. The relative galling position was calculated as the number of axillary buds from the most basal part of the shoot divided by the total number of axillary buds. Galls were dissected under a binocular microscope to count the larvae (i.e., chambers) in the gall. Because some galls had been eaten by unknown lepidopteran larvae, we could not count the gall midge larvae in those galls.

13.3 Results

The number of galls and larvae in the shoots differed significantly between *A. sasakii* and *A. soyogo* in 2000–2001 and 2001–2002 (Table 1). The number of *A. soyogo* galls per shoot was higher than that of *A. sasakii* in both survey years. The frequency distributions of the number of galls per shoot also differed between species ($\chi^2 = 50.8$, $P < 0.001$ in 2000–2001; χ^2

Table 1. The number of galls and larvae in the shoots of *Asteralobia sasakii* and *A. soyogo* in 2000–2001 and 2001–2002

	Year	Species		*t*
		A. sasakii	*A. soyogo*	
No. of galls / shoot ± SE (*n*)	2000–2001	1.4 ± 0.1 (140)	2.5 ± 0.1 (144)	6.73 *
	2001–2002	1.3 ± 0.1 (173)	1.6 ± 0.1 (167)	2.85 *
No. of larvae / shoot ± SE (*n*)	2000–2001	5.0 ± 0.4 (120)	6.2 ± 0.4 (135)	1.63
	2001–2002	6.9 ± 0.3 (162)	4.1 ± 0.3 (167)	6.30 *

* $p < 0.001$.

Table 2. Summary of the two-way ANOVA models testing the relative galling position in shoots

	SS	*F*-value	*P*
2000-2001			
Species	6.9	16.9	< 0.01
No. of galls / shoot	17.1	41.5	< 0.01
Interaction	8.9	21.7	< 0.01
2001-2002			
Species	0.9	2.6	0.11
No. of galls / shoot	23.0	65.2	< 0.01
Interaction	30.5	86.4	< 0.01

Data of relative galling position was arc-sign transformed.

= 16.8, $P < 0.01$ in 2001–2002). Shoots that bore a single gall accounted for 71.4% (2000–2001) and 76.3% (2001–2002) in the sample of *A. sasakii* on *I. crenata* and 34.7% and 62.3% in *A. soyogo* on *I. integra*. The number of larvae per shoot varied with the year; there was no significant difference between the species in 2000–2001, but the number of *A. sasakii* larvae per shoot of was higher in 2001–2002 (Table 1).

The relative gall position within the shoots differed significantly between the species in 2000–2001, with the galls of *A. sasakii* induced on the more terminal positions of shoots (Table 2). In 2001–2002, there was no significant difference between the species with regard to relative gall position because *A. sasakii* galls on the bottom halves of shoots were more numerous than the preceding year (Fig. 1). Galls of *A. sasakii* were more concentrated toward the tips of shoots than those of *A. soyogo*: for *A. sasakii*, 37.4% (2000–2001) and 42.5% (2001–2002) of galls were induced on the axillary bud at the tip of the shoot, whereas these values for *A. soyogo* were 11.8% (2000–2001) and 8.0% (2001–2002) ($\chi^2 = 65.7$, $P < 0.001$ in 2000–2001; $\chi^2 = 81.6$, $P < 0.001$ in 2001–2002). Growth patterns of galls and shoots also differed between the host plant and gall midge systems. Shoot growth of *I. crenata* was disturbed by *A. sasakii* galls, which were induced on axillary buds on the most terminal positions of shoots. In contrast, no *A. soyogo* gall was observed to have disturbed shoot growth on *I. integra*.

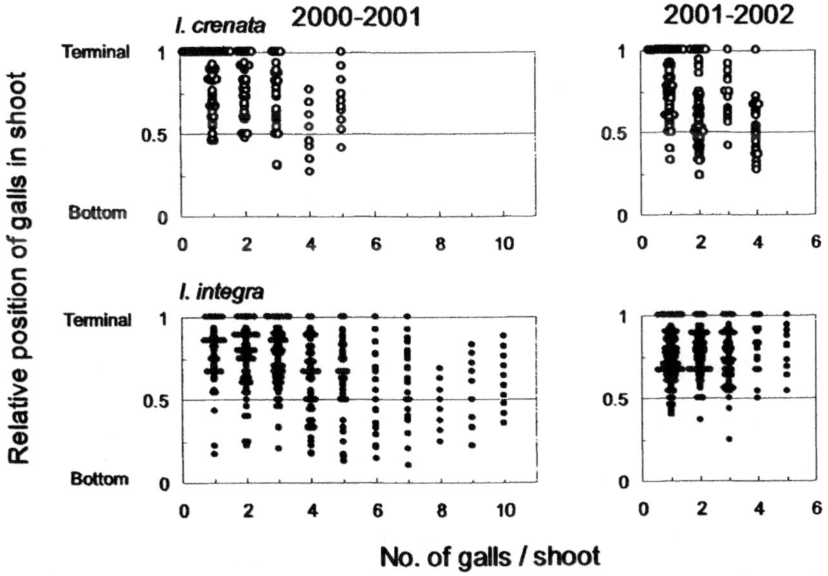

Fig. 1. Relative gall position in the shoot: *Asteralobia sasakii* on *Ilex crenata* and *A. soyogo* on *I. integra* in 2000–2001 and 2001–2002.

13.4 Discussion

The gall midges *A. sasakii* and *A. soyogo* are closely related species with similar life histories and galling habits (Tokuda et al. 2004; Yukawa and Masuda 1996). However, the oviposition patterns of these gall midges differ in relation to their plant hosts and parasitoid species.

Fig. 2 and Table 3 provide a summary of distributions of galls and larvae in shoots. The two gall midge species chose similar galling positions within the shoots. Most galls of *A. sasakii* and *A. soyogo* were formed on the terminal halves of shoots (Fig. 1). This trait would be adaptive for gall formation because terminal buds would have high plasticity for gall growth in the following season.

In *A. sasakii* on *I. crenata*, galls were typically observed on the tips of shoots, and these galls disturbed shoot growth. Such positioning and disturbance of shoot growth was not observed in *A. soyogo* on *I. integra*. Different gall distributions in shoots between these species may be due to the relative sizes of host plant organs: the axillary buds, leaves, and shoots of *I. crenata* are smaller than those of *I. integra* (Yamazaki 1989). The terminal buds of *I. crenata* become covered with overgrown cells of the gall and cease to grow. In addition, it is likely that *A. sasakii* females are inevitable to oviposit in the terminal region of the shoot because inflorescences are produced on the basal parts of current-year shoots in *I. crenata*, making these buds unsuitable for gall formation. In contrast, inflorescences of *I. integra* are produced on the shoot produced the preceding year. Thus, the position of available axillary buds in shoots differs between these plant species. Therefore, the growth patterns of the host plants offer constraints for oviposition site selection by each gall midge species.

Table 3. Distribution of *A. sasakii* and *A. soyogo* larvae and galls in shoots

	A. sasakii	*A. soyogo*
No. of larvae / shoot	NS or High	Low or NS
No. of larvae / gall [a]	High	Low
No. of galls / shoot	Low	High
Galling position in shoots	Concentrated on terminal region	Relatively Dispersed
Oviposition pattern	Concentrated	Dispersed

NS, not significantly different from another species.
[a] see Tabuchi and Amano (2004).

In conclusion, while the two *Asteralobia* gall midges are closely related, their oviposition strategies differ: *A. sasakii* females oviposit their eggs in clusters (aggregation), whereas *A. soyogo* females reduce the risk to their progeny by spreading oviposition both spatially and temporally (risk spreading). The host range of *A. soyogo* is wider (six *Ilex* species) than that of *A. sasakii* (two *Ilex* species), and *A. sasakii* can use only a single host plant in most regions of Japan (Tokuda et al. 2002). Moreover, polymodal emergence is reported in *A. soyogo* but not in *A. sasakii*, which is strictly univoltine. Some first instars of *A. soyogo* enter a prolonged diapause and adult emergence extends over 2 or 3 years. These first instars in immature galls could temporally avoid ectoparasitoid attack (Tabuchi and Amano 2003b), whereas all mature larvae of *A. sasakii* were exposed to the attack in the same season each year. Each female of *A. sasakii* lays her eggs into one oviposition site in clusters and larvae induce a single, large gall, in which the larvae are protected from ectoparasitoid attack by a thick barrier surrounding larval chambers. As compared to *A. soyogo* galls, *A. sasakii* galls were smaller in the number per shoot and more numerous in the number of larvae per gall. Thus, *A. sasakii* may reduce the risk of parasitoid attack by inhabiting the gall together in a crowd. In contrast, *A. soyogo* employs a risk-spreading strategy that provides spatial and temporal escape from parasitoids: scattering galls across several axillary buds in a shoot, using several host plant species, and entering prolonged diapause. Based on mitochondrial DNA analysis, divergence of the two gall midges is estimated to have occurred 3.2–4.3 million years ago (Tokuda et al. 2004), and their different oviposition strategies have possibly evolved in response to selective pressures within their respective tritrophic systems.

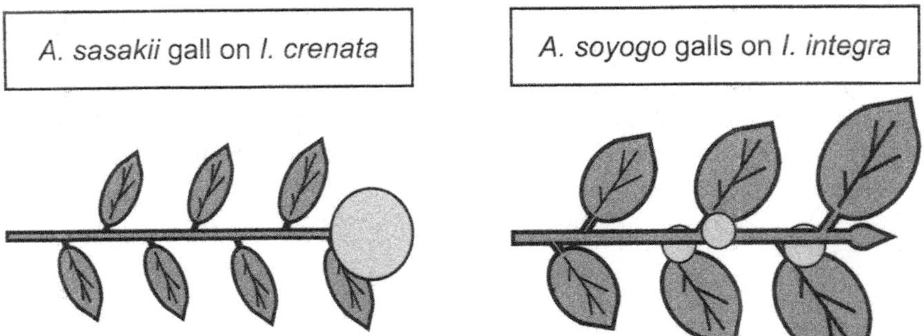

Fig. 2. Schematic representation of the galls of *A. sasakii* and *A. soyogo* in shoots of their respective host plants.

13.5 Acknowledgments

We express our thanks to H. Taki (Department of Environmental Biology, Guelph University) for his critical reading of an early draft. We are grateful to the members of the Laboratory of Applied Entomology and Zoology, Chiba University, for their help in field surveys.

13.6 References

Monzen K (1937) On some new gall midges (in Japanese). Kontyû 11: 180-194

Tabuchi K, Amano H (2003a) Polymodal emergence pattern and parasitoid composition of *Asteralobia sasakii* (Monzen) (Diptera: Cecidomyiidae) on *Ilex crenata* and *I. integra* (Aquifoliaceae). Applied Entomology and Zoology 38:493-500

Tabuchi K, Amano H (2003b) Host-associated differences in emergence pattern, reproductive behavior and life history of *Asteralobia sasakii* (Monzen) (Diptera: Cecidomyiidae) between populations on *Ilex crenata* and *I. integra* (Aquifoliaceae). Applied Entomology and Zoology 38:501-508

Tabuchi K, Amano H (2004) Impact of differential parasitoid attack on the number of chambers in multilocular galls of two closely related gall midges (Diptera: Cecidomyiidae). Evolutionary Ecology Research 6:695-707

Tokuda M, Tabuchi K, Yukawa J, Amano H (2004) Inter- and intraspecific comparisons between *Asteralobia* gall midges (Diptera: Cecidomyiidae) causing axillary bud galls on *Ilex* species (Aquifoliaceae): species identification, host range, and the mode of speciation. Annals of the Entomological Society of America 97:957-970

Tokuda M, Uechi N, Yukawa J (2002) Distribution of *Asteralobia* gall midges (Diptera: Cecidomyiidae) causing axillary bud galls on *Ilex* species (Aquifoliaceae) in Japan. Esakia 42:19-31

Yamazaki K (1989) Aquifoliaceae (in Japanese). In: Satake Y, Hara H, Watari S, Tominari T (eds) Wild flowers of Japan, Woody plants, vol. 2. Heibonsha, Tokyo, pp 26-32

Yukawa J, Masuda H (1996) Insect and mite galls of Japan in colors (in Japanese, with English explanation for color plates). Zenkoku Nôson Kyôiku Kyôkai, Tokyo

14 A Protective Mechanism in the Host Plant, *Aucuba*, against Oviposition by the Fruit Gall Midge, *Asphondylia aucubae* (Diptera: Cecidomyiidae)

Kensuke Imai

Laboratory of Insect Ecology, Graduate School of Agriculture, Kyoto University, Kitashirakawa-Oiwakecho, Sakyo, Kyoto 606-8502, Japan

Summary. The effect of mechanical traits of the dioecious evergreen shrub *Aucuba japonica* (Aucubaceae) on the efficiency of oviposition behavior of the monophagous fruit gall midge *Asphondylia aucubae* was examined in Japan. The hard endocarp of *A. japonica* provided effective mechanical protection of the young developing fruits from the gall midge, which deposits eggs by inserting its ovipositor into the fruit. The protection by the endocarp was retained only while the fruit was susceptible to gall induction by the gall midge, apparently alleviating constriction of fruit growth by the hard endocarp. The results showed that *A. aucubae* and *A. japonica* interact with each other intensely, even before gall induction.

Key words. *Aucuba*, Egg, Endocarp, Fruit, Gall midge

14.1 Introduction

Among insect herbivores, galling insects have particularly intimate interactions with their host plants, in that they are usually specialized to a particular organ of their host plant species (Yukawa and Masuda 1996). One important aspect of such an interaction takes place before gall induction, especially during oviposition. Physical (Ferrier and Price 2004) or chemical (Roininen et al. 1999; Tooker et al. 2005) traits of the host plants may affect the efficiency of the ovipositing behavior of galling insects. Oviposition efficiency should, in turn, determine the fecundity of galling insects, many species of which have very short adult life spans (Freese and Zwölfer 1996; Frey et al. 2004; Yukawa 2000) that are readily shortened further by predation or environmental stresses such as desiccation, rainfall, and wind. Such pre-galling interactions would play an important role in the ecology of galling insects.

To date, the effects of host plants on insect oviposition behavior have been frequently studied, primarily in the context of host choice or preference–performance linkages (e.g., Fritz et al. 2003; Horner and Abrahamson 1992; Larsson and Strong 1992; Pires and Price 2000; Prado and Vieira 1999; Price et al. 2004; Wilkens et al. 2005). In contrast, the host effects on oviposition efficiency have been addressed in relatively few studies: Ferrier and Price (2004) inferred that preference of the sawfly *Euura* sp. for vigorously developing shoots is to avoid tough shoots. Price and Hunter (2005) revealed that abortion of oviposition partially accounted for the variation in performance of *Euura lasiolepis* Smith (Hymenoptera: Tenthredinidae) across host clones. Similar abortion has also been observed in several sawflies (Naito 1996) and in a galling tephritid, *Eurosta solidaginis* Fitch (Diptera: Tephritidae) (Uhler 1950; see also review by Abrahamson and Weis 1997). Burkhardt and Zwölfer (2003) showed that the specialist tephritid galler *Urophora jaceana* Hering has more efficient oviposition behavior than other *Urophora* species that are less specialized to the host plant.

In this study, I focus on the mechanical traits affecting the oviposition efficiency of a galling insect. I provide an example of this type of interaction between an evergreen shrub, *Aucuba japonica* Thunberg (Aucubaceae), and the monophagous fruit gall midge *Asphondylia aucubae* Yukawa et Ohsaki (Diptera: Cecidomyiidae). I reveal a well-organized mechanism for the mechanical protection of young fruits of *A. japonica*, reviewing my two recent findings (Imai and Ohsaki 2004a, b), and presenting new histological data from scanning electron microscope observations.

14.2 Mechanical Protection by the Endocarp against Ovipositor Thrusting by *A. aucubae*

Recently, Imai and Ohsaki (2004a) reported that young fruits of *A. japonica* are protected from oviposition by the *Aucuba* fruit gall midge *A. aucubae*. In a manner similar to *E. solidaginis*, females of this gall midge repeatedly thrust their ovipositors into host fruits, leaving oviposition scars, and frequently aborting oviposition after leaving a scar. These scars were used to determine how each ovipositing thrust contributed to successful oviposition.

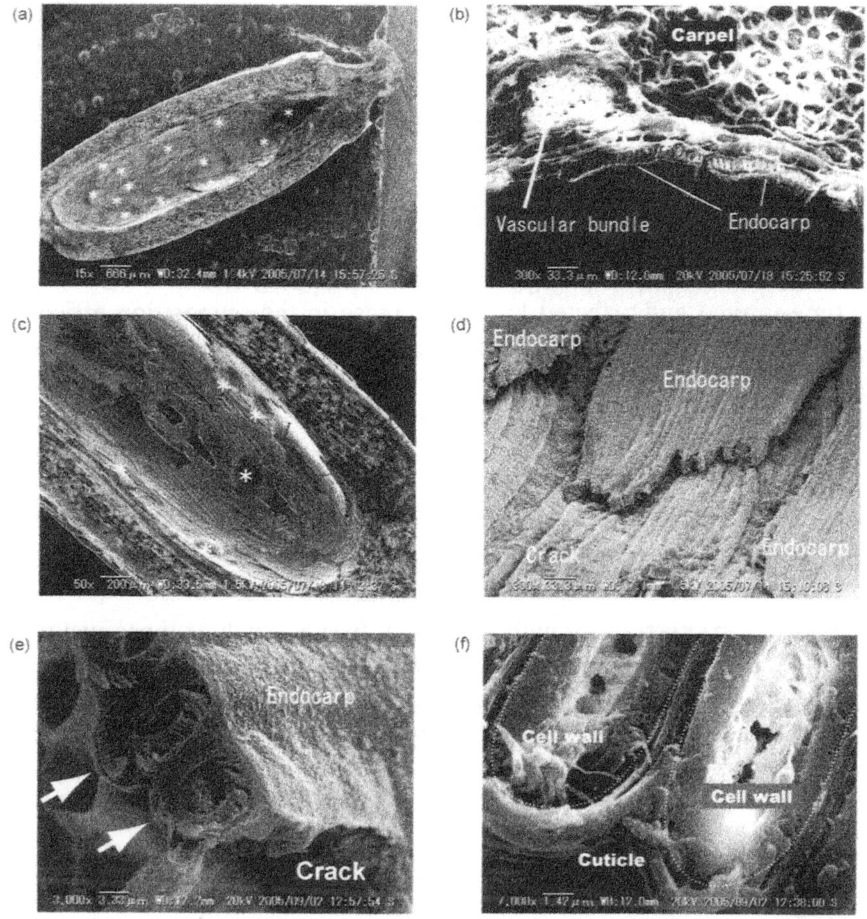

Fig. 1. Structure of the endocarp. **a** Cross section in which the ovule (integument and embryo sac) has been peeled from the center part. *Asterisks* indicate fragments of the endocarp. **b** Cross section in the horizontal direction. **c** Cross section with eggs *(asterisks)*. **d** Fine grooves along which cracks run. **e** Sectional microstructure of the endocarp cells *(arrows)*. **f** Microstructure of the endocarp cell junction. *Scale*, *magnitude*, *working distance*, *accelerating voltage*, and *date* of observation under scanning electron microscopy are listed under each photo.

The endocarp inside young host fruits (Fig. 1a) has a thicker cuticle and cell wall than other fruit tissues (Fig. 1b), and inhibits ovipositor thrusts by the gall midges. The gall midges succeed in oviposition only when the ovipositor is inserted through cracks in the endocarp (Fig. 1c); otherwise oviposition is almost completely unsuccessful (1845/1849, 99.8%). The endocarp, consequently, inhibits 33.2% (1211/3652) of observed thrusts (Imai and Ohsaki 2004a).

A. aucubae has a very short adult life span (1.5 days) (Yukawa and Ohsaki 1988b), during which it does not eat and suffer readily from predation (K. Imai and N. Ohsaki, unpublished data, 2004), desiccation (K. Imai and N. Ohsaki, unpublished data, 2004), and rainfall (Ohsaki and Yukawa 1990; Yukawa and Ohsaki 1988b). The number of *A. aucubae* ovarian eggs is 189 on an average, which is apparently fewer than 231–257 in other congeners (Yukawa and Ohsaki 1988b) that have wing length (an indicator of body size) similar to that of *A. aucubae* (Yukawa 1971). The function of the endocarp might reduce the fecundity of *A. aucubae*, which is time-limited and affected by the decrease in the oviposition efficiency. However, no data are available at the moment to compare among congeners the proportion of ovarian eggs that are realized during the short adult life span. Further study is needed to ascertain these speculations.

14.3 The Temporary Nature of the Hard Endocarp Enables Mechanical Protection of Growing Young Fruit

Mechanical protection through hard plant tissues, such as those just noted, has seldom been considered as a defense for young developing plant organs. One reason is for this that mechanical protection traits have been assumed to be impractical for such purposes due to ecological restrictions; hard plant tissues are thought to grow slower than the softer tissues under their protection, and thereby constrict the growth of the entire plant organ (Aide 1992; Stock et al. 1991).

However, in the interactions between *A. aucubae* and its host plant, this ecological restriction seems to have been alleviated; the protective function of the endocarp is retained only while the fruit is susceptible to gall midge attacks. The larval chamber, an essential structure of the gall, is constructed using the space and tissue of the integument (Imai and Ohsaki 2004b). As the fruit grows, the integument degenerates, making the fruit unavailable for gall induction (Imai and Ohsaki 2004a, b). The endocarp protects the fruit, mainly while the integument is retained, and once the integument begins to degenerate, the endocarp rapidly cracks and fragments (Imai and Ohsaki 2004a), which should ameliorate the detrimental effect on fruit growth. Thus, through the minimum provision of mechanical protection, the host shrub seems to have found a balance between the protection of the endocarp and the concomitant risk of constricting fruit growth.

Cracking is likely to be a programmed adaptive trait to avoid the constriction of the fruit growth. The cracks run along fine grooves on the surface of the endocarp (Fig. 1d), which correspond to the junction of endocarp cells (Fig. 1e, f). Thus, the cracks expand without breaking endocarp cells (Fig. 1d). Unlike other cells, the cell walls of the endocarp seem to be indirectly connected to each other by a cuticular layer (Fig. 1f), which may allow smooth cracking.

14.4 Well-organized Protective Mechanism against Oviposition by the Gall Midge

There are two possible explanations for the evolutionary interaction between *A. aucubae* and its host plant. One is that *A. japonica* started to develop endocarp after the arrival of *A. aucubae* on the plant. The endocarp protects young host fruit temporarily but efficiently, in well collaboration with the rapid degeneration of the integument. The hardness of the endocarp, the timing and speed of endocarp fragmentation and integument degeneration, their relative timing, and the cell junction structure allowing smooth endocarp cracking, are all essential and indispensable parts of this protective mechanism. From these observations, it might be natural to speculate that natural selection favors or maintains the well collaboration among these traits.

Another explanation is that *A. japonica* had already the trait of endocarp development before the arrival of *A. aucubae* and the endocarp has functioned for other purposes. *A. aucubae* has been well adapting to use *A. japonica* fruit as a host organ by concentrating their oviposition to the restricted time when the fruit is susceptible to the gall midge attacks.

At the moment, no data are available to determine whether or not *A. japonica* started to develop endocarp after *A. japonica* had arrived at the plant, although morphological and molecular phylogenetic studies indicate that the Japanese *Asphondylia* species are now spreading to various plant taxa (Yukawa et al. 2005).

In order to confirm the evolutionary interaction between the gall midge and its host plant, comparative morphological studies of fruit are needed between *A. japonica* populations attacked by the gall midge and those having been growing under gall midge free conditions. However, such a comparison is impossible because the geographic distribution of *A. aucubae* and *A. japonica* overlap almost everywhere in Japan (Yukawa and Ohsaki 1988a). In addition, *A. japonica* cannot be compared readily with its congeners that are not attacked by gall midges because in Japan it is the only representative of the genus, which was recently transferred from the family Cornaceae (Cornales) to the family Aucubaceae (Garryales) (The Angiosperm Phylogeny Group 2003). Therefore, comparative morphological studies of fruit among known host plants of *Asphondylia* (Yukawa and Masuda 1996) and non-host plants across various plant taxa would be more practical to obtain information on the evolutionary interaction.

In northern Japan, *A. aucubae* has been known to induce another type of fruit gall on a variety of the same host plant, *A. japonica* var. *borealis* Miyabe et Kudo (Yukawa and Masuda 1996). This gall has a mature seed and is larger in size with more larval chambers than the gall on *A. japonica* in southwestern Japan, suggesting that eggs might be oviposited in the fruit later in the season. According to Uechi and Yukawa (personal communication, 2005), gall midges on both *A. aucubae* and its variety have a common haplotype in the partial mtDNA COI region. The variation of gall morphology may be derived from variation in the interaction between the same gall midge species and its host plant before gall induction. Such variations could provide us with a good opportunity for a comparative study.

14.5 Acknowledgments

I would like to express our sincere appreciation to Dr. J. Yukawa (Kyushu University) and Dr. N. Ohsaki (Kyoto University) for their great efforts in revising my manuscript. I also express my appreciation to Drs. E. Kuno, K. Fujisaki, and T. Nishida (Kyoto University) for their useful comments. This research was supported in part by a Grant-in-Aid (No. 11308021 and No. 16054181) and the 21st Century COE program for Innovative Food and Environmental Studies Pioneered by Entomomimetic Sciences, from the Ministry of Education, Culture, Sports, Science, and Technology of Japan.

14.6 References

Abrahamson WG, Weis AE (1997) Evolutionary ecology across three trophic levels: goldenrod, gallmakers, and natural enemies. Princeton University Press, Princeton

Aide TM (1992) Dry season leaf production—an escape from herbivory. Biotropica 24:532-537

Burkhardt B, Zwölfer H (2002) Macro-evolutionary trade-offs in the tephritid genus *Urophora*: benefits and costs of an improved plant gall. Evolutionary Ecology Research 4:61-77

Ferrier SM, Price PW (2004) Oviposition preference and larval performance of a rare bud-galling sawfly (Hymenoptera: Tenthredinidae) on willow in northern Arizona. Environmental Entomology 33:700-708

Freese G, Zwölfer H (1996) The problem of optimal clutch size in a tritrophic system: the oviposition strategy of the thistle gallfly *Urophora cardui* (Diptera, Tephritidae). Oecologia 108:293-302

Frey JE, Frey B, Baur R (2004) Molecular identification of the swede midge (Diptera: Cecidomyiidae). Canadian Entomologist 136:771-780

Fritz RS, Crabb BA, Hochwender CG (2003) Preference and performance of a gall-inducing sawfly: plant vigor, sex, gall traits and phenology. Oikos 102:601-613

Horner JD, Abrahamson WG (1992) Influence of plant genotype and environment on oviposition preference and offspring survival in a gallmaking herbivore Oecologia 90:323-332

Imai K, Ohsaki N (2004a) Internal structure of developing aucuba fruit as a defense increasing oviposition costs of its gall midges *Asphondylia aucubae*. Ecological Entomology 29:420-428

Imai K, Ohsaki N (2004b) Oviposition site of and gall formation by the fruit gall midge *Asphondylia aucubae* (Diptera: Cecidomyiidae) in relation to internal fruit structure. Entomological Science 7:133-137

Larsson S, Strong DR (1992) Oviposition choice and larval survival of *Dasineura marginemtorquens* (Diptera, Cecidomyiidae) on resistant and susceptible *Salix viminalis*. Ecological Entomology 17:227-232

Naito T (1996) Tenthredinidae (in Japanese). In: Yukawa J, Masuda H (eds) Insect and mite galls of Japan in colors. Zenkoku Nôson Kyôiku Kyôkai, Tokyo, pp 374-378

Ohsaki N, Yukawa J (1990) Gradual increase in the number of *Asphondylia aucubae* (Diptera: Cecidomyiidae) due to inversely density-dependent mortality processes. Ecological Research 5:173-183

Pires CSS, Price PW (2000) Patterns of host plant growth and attack and establishment of gall-inducing wasp (Hymenoptera: Cynipidae). Environmental Entomology 29:49-54

Prado PIKL, Vieira EM (1999) The interplay between plant traits and herbivore attack: a study of a stem galling midge in the neotropics. Ecological Entomology 24:80-88

Price PW, Hunter MD (2005) Long-term population dynamics of a sawfly show strong bottom-up effects. Journal of Animal Ecology 74:917-925

Price PW, Ohgushi T, Roininen H, Ishihara M, Craig TP, Tahvanainen J, Ferrier SM (2004) Release of phylogenetic constraints through low resource heterogeneity: the case of gall-inducing sawflies. Ecological Entomology 29:467-481

Roininen H, Price PW, Julkunen-Tiitto R, Tahvanainen J, Ikonen A (1999) Oviposition stimulant for a gall-inducing sawfly, *Euura lasiolepis*, on willow is a phenolic glucoside. Journal of Chemical Ecology 25:943-953

Stock WD, Pate JS, Rasins E (1991) Seed developmental patterns in *Banksia attenuata* R. Br. and *B. laricina* C. Gardner in relation to mechanical defense costs. New Phytologist 86:138-146

The Angiosperm Phylogeny Group (2003) An update of the Angiosperm Phylogeny Group classification for the orders and families of flowering plants: APG II. Botanical Journal of the Linnean Society 141:399-436

Tooker JF, Crumrin AL, Hanks LM (2005) Plant volatiles are behavioral cues for adult females of the gall wasp *Antistrophus rufus*. Chemoecology 15:85-88

Uhler LD (1950) Biology and ecology of the goldenrod gall fly, *Eurosta solidaginis* (Fitch). Cornell University Agricultural Station Memoir 300:1-51

Wilkens RT, Vanderklein DW, Lemke RW (2005) Plant architecture and leaf damage in bear oak II: insect usage patterns. Northeastern Naturalist 12:153-168

Yukawa J (1971) A revision of the Japanese gall midges (Diptera: Cecidomyiidae). Memoirs of the Faculty of Agriculture, Kagoshima University 8:1-203

Yukawa J (2000) Synchronization of gallers with host plant phenology. Population Ecology 42:105-113

Yukawa J, Masuda H (1996) Insect and mite galls of Japan in colors (in Japanese, with English explanation for color plates). Zenkoku Nôson Kyôiku Kyôkai, Tokyo

Yukawa J, Ohsaki N (1988a) Separation of the aucuba fruit midge, *Asphondylia aucubae* sp. nov. from the ampelopsis fruit midge, *Asphondylia baca* Monzen (Diptera: Cecidomyiidae). Kontyû 56:365-376

Yukawa J, Ohsaki N (1988b) Adult behaviour of the aucuba fruit midge, *Asphondylia aucubae* Yukawa et Ohsaki (Diptera: Cecidomyiidae). Kontyû 56:645-652

Yukawa J, Uechi N, Tokuda M, Sato S (2005) Radiation of gall midges (Diptera: Cecidomyiidae) in Japan. Basic and Applied Ecology 6: 453-461

15 Genetic Variation in the Timing of Larval Mortality and Plant Tissue Responses Associated with Tree Resistance against Galling Adelgids

Kenichi Ozaki and Yasuaki Sakamoto

Hokkaido Research Center, Forestry and Forest Products Research Institute, Hitsujigaoka, Sapporo 062-8516, Japan

Summary. Large genetic variation in resistance to galling adelgids, *Adelges japonicus* exists in ezo-spruces, *Picea jezoensis*. To address genetic differences in the mechanisms of tree resistance against *A. japonicus*, we examined the timing of larval mortality and tissue responses in a number of susceptible and resistant clones of *P. jezoensis*. We placed a winged adult on a branch in each of 21 clones, and examined the survival of the next generation. Although the number of larvae that settled on buds in autumn did not differ between susceptible and resistant clones, galls were induced almost exclusively on susceptible clones, indicating that resistance was manifested as greater mortality of fundatrices when they were sedentary on buds. However, the mortality was greater only in spring on resistant clones from the Oketo population, while the mortality was greater in both winter and spring on resistant clones from the other populations than on susceptible clones. Brown lesions with necrotic cells were observed in attacked tissue in both resistant and susceptible clones. The proportion of buds with lesions was greater only in resistant clones from non-Oketo populations than in susceptible clones suggesting that hypersensitive responses may have been involved in these resistant clones. Larval size was smaller on resistant clones from non-Oketo populations than that on susceptible clones. These results suggest that resistance exhibited in different timings had different underlying mechanisms.

Key words. *Adelges japonicus*, Genetic variation, Hypersensitive response, Insect–plant interaction, Tree resistance

15.1 Introduction

Hypersensitive response, defined as the death of plant cells associated with plant resistance (Larsson 2002), has often been reported as a mechanism of plant resistance against galling arthropods (Fernandes 1990; Fernandes and Negreiros 2001). Hypersensitive response was originally used in plant pathology to refer to a rapid process that occurred within hours after attack, but when this term was introduced into entomology, it came to refer also to the process exhibited over much longer periods (Ollerstam et al. 2002). Thus, information about the timing of cell death after insect attack is important to address the mechanisms of plant resistance in various insect–plant interactions.

A galling adelgid, *Adelges abietis* is known to induce hypersensitive responses in bud tissue of their host trees (Rohfritsch 1981, 1988). Using its long stylet, *A. abietis* attacks bud tissue of susceptible and resistant trees from autumn to the next spring. However, attacked cells show necrosis with accumulation of phenolic substances in spring only in resistant trees. Because necrotic cells form a barrier to soluble proteins of the tree, Rohfritsch (1988) suggests that cell death is the primary mechanism of resistance. However, because only one resistant tree was examined, genetic variation in resistance traits is unknown. Large genetic variation in adelgid susceptibility has been reported from Norway spruce, *Picea abies* in Europe (Mattson et al. 1998) and from white spruce, *Picea glauca* in North America (Mattson et al. 1994).

The aim of this study is to address the genetic variation in the timing of resistant responses against galling adelgids. We examined the timing of larval mortality and tissue responses in a number of susceptible and resistant clones of ezo-spruces, *Picea jezoensis* (Sieb. et Zucc.) Carr. against galling adelgids, *Adelges japonicus* (Monzen) (Homoptera: Adelgidae).

A. japonicus is anholocyclic with two generations (fundatrix and gallicola), monoecious, and parthenogenetic. Larval fundatrices settle on the base of buds in autumn, and overwinter there. In spring, each fundatrix deposits an egg mass on the same place. Due to stimulation attributed to the fundatrix's long stylet, the infested bud does not develop into a normal shoot, but is transformed into a multichambered, pineapple-shaped gall. Larval gallicolae, which emerge from the egg mass, crawl into the larval chambers of the galls. Gallicolae develop in the galls during summer, and emerge from the galls to lay egg masses on leaves in autumn.

A transplant experiment of *P. jezoensis* demonstrated a large genetic variation in resistance to *A. japonicus* (Itahana and Ubukata 1995). In the experiment, seeds were collected from five populations in Hokkaido, Japan, and seedlings germinated from the seeds were planted in two plantations in Hokkaido. Gall infestation varies significantly among populations and is higher in two populations located in low elevation (<450 m) areas than in the other three populations located in high elevation (>700 m) areas. In addition, differences in gall infestation among families are consistent between years and between plantations, suggesting that tree resistance is consistent over time and space (Itahana et al. 2002).

15.2 Materials and Methods

15.2.1 Study Area and Study Clones

Our study was conducted in a young plantation in the Hokkaido Breeding Center, Forest Tree Breeding Institute, Ebetsu, Hokkaido, Japan (43°10'N, 141°30'E). In the plantation, trees from 54 clones were planted in 1997 with 1.4 m intervals in 12 blocks in a randomized complete block design. Fifty of the 54 clones were from the trees used in the transplant experiment mentioned above: 41 trees with little or no gall infestation and 9 trees with heavy gall infestation were selected. The other four clones were from trees in a breeder's stock garden in the Hokkaido Breeding Center. These 54 clones were generated by propagating them from shoots grafted onto non-clonal rootstocks. When adult gallicolae were placed on the trees in the plantation, there was a large variation among clones in the number of galls formed the next summer (Hoshi et al. 2002; Itahana et al. 2002). We selected the most heavily infested six clones (hereafter referred to susceptible clones) and 15 clones that had no galls (hereafter referred to resistant clones) for the following experiment. The susceptible clones originated from five populations, and the resistant clones originated from four populations in Hokkaido.

15.2.2 The Timing of Mortality on Susceptible and Resistant Clones

In autumn 2002, we haphazardly chose a branch with ca. 30 buds from each tree of the 21 clones in four blocks. We placed an adult gallicola on each branch, and after a few days, we confirmed their oviposition. We enclosed these branches with fine-mesh bags until late October to prevent predation. Then we examined the survival of fundatrices by recording the number of fundatrices at the base of all buds on the branches in late October (when fundatrices settled on buds), early the following April (when fundatrices had overwintered), and in mid May (when fundatrices oviposited). The number of galls formed was also counted.

15.2.3 Tissue Responses in Susceptible and Resistant Clones

In autumn 2003, we placed an adult gallicola on each branch of the 21 clones, and enclosed these branches with fine mesh bags as in 2002. Then ca. 10 infested buds were sampled from each clone in late October (when fundatrices settled on buds), early the following April (when fundatrices had overwintered), and in late April (when fundatrices matured). After we measured bud length and fundatrix body length under a stereo microscope,

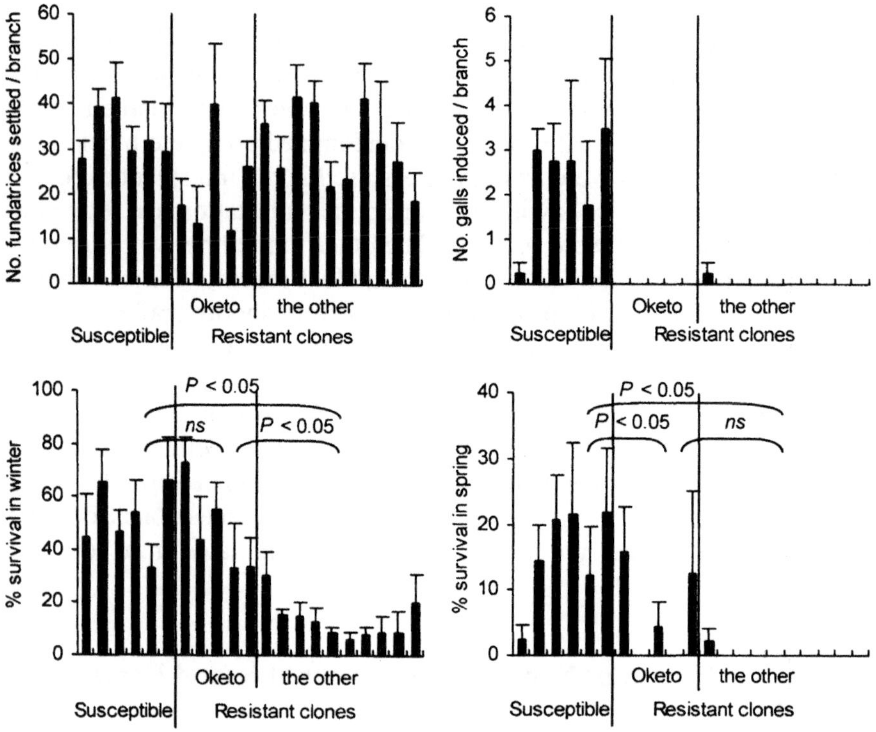

Fig. 1. The number of fundatrices that settled on buds, the number of galls induced, and fundatrix survival on each clone. Data on each branch to which an adult gallicola was placed were averaged for each clone and their values + SE are shown. *P* values indicate the results of Tukey-Kramer multiple comparisons between groups of clones. Resistant clones from the Oketo population and from the other populations were analyzed separately.

we cut the buds longitudinally into two pieces, from the point where fundatrices inserted their stylets, to observe the attacked tissue. In some buds, we observed brown lesions in cortical tissue near where fundatrices had inserted their stylets. In these buds, the area of lesions was estimated by multiplying the maximum length and the length at right angles to the maximum length of the lesions.

To examine the lesions at the cellular level, infested buds were sampled from susceptible and resistant clones in early November 2002. Bud samples were cut to examine the occurrence of lesions, and were fixed with FAA (formalin: acetic acid: 50% ethanol = 5:5:90) for five months. After washing under running water for 3–4 hours, the samples were dehydrated in an ethanol series, and embedded in paraffin. Tangential 14 μm sections were cut on a sliding microtome. They were stained with safranin O (1% solution in 50% ethanol) and fast green (0.1% solution in 95% ethanol), and were then observed under a light microscope.

15.2.4 Data Analyses

We analyzed data using nested ANOVA with clones nested within resistance. Data were pooled for each branch in the analysis of survival. Dependent variables were Box-Cox transformed prior to the analyses. Then we performed Tukey-Kramer multiple tests for post-hoc comparisons between groups of clones.

15.3 Results

15.3.1 The Timing of Mortality on Susceptible and Resistant Clones

The number of fundatrices that settled on buds in autumn did not differ between susceptible and resistant clones ($F = 1.64$, $P = 0.22$) (Fig. 1). However, these fundatrices induced galls almost exclusively on susceptible clones: fundatrices induced galls on all susceptible clones, but induced a total of only one gall on resistant clones. These findings showed that the resistance was manifested as lower survival of fundatrices when they were sedentary on buds. Thus we further analyzed fundatrix survival in this stage.

Fundatrix survival in winter was lower on resistant clones than on susceptible clones ($F = 8.90$, $P = 0.008$). However, winter survival on resistant clones varied greatly due to the greater survival on five resistant clon-

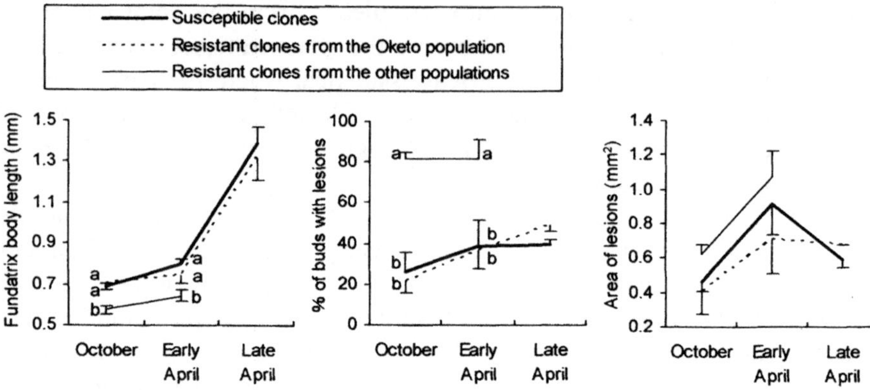

Fig. 2. Fundatrix body length, percentage of buds with lesions, and area of lesions in susceptible and resistant clones. Infested buds were sampled in late October (when fundatrices settled on buds), early the following April (when fundatrices had overwintered), and late April (when fundatrices matured). Mean and SE for clones are shown. Data on resistant clones from the other populations were not recorded in late April because all fundatrices died in this period. Different letters show significant differences between groups of clones in Tukey-Kramer multiple comparison.

es from the Oketo population (Fig. 1). When we analyzed resistant clones from the Oketo population and from the other three populations separately, winter survival on the clones from the Oketo population (average 47.7%) did not differ from that on susceptible clones (51.5%), but winter survival on the other resistant clones was 38.0% lower than that on susceptible clones.

Fundatrix survival in spring was also lower on resistant clones than on susceptible clones ($F = 26.5$, $P < 0.001$). In this season, both the survival on resistant clones from the Oketo population (6.6%) and from the other populations (0.2%) were lower than survival on susceptible clones (15.5%). In total, 15 fundatrices matured and oviposited on resistant clones from the Oketo population, but no gall was induced by these fundatrices. In contrast, only one fundatrix oviposited on resistant clones from the other populations, and this fundatrix induced a gall.

15.3.2 Tissue Responses in Susceptible and Resistant Clones

Because of the different survival between resistant clones from the Oketo population and those from the other populations, we analyzed the two groups separately in this experiment. The length of infested buds did not

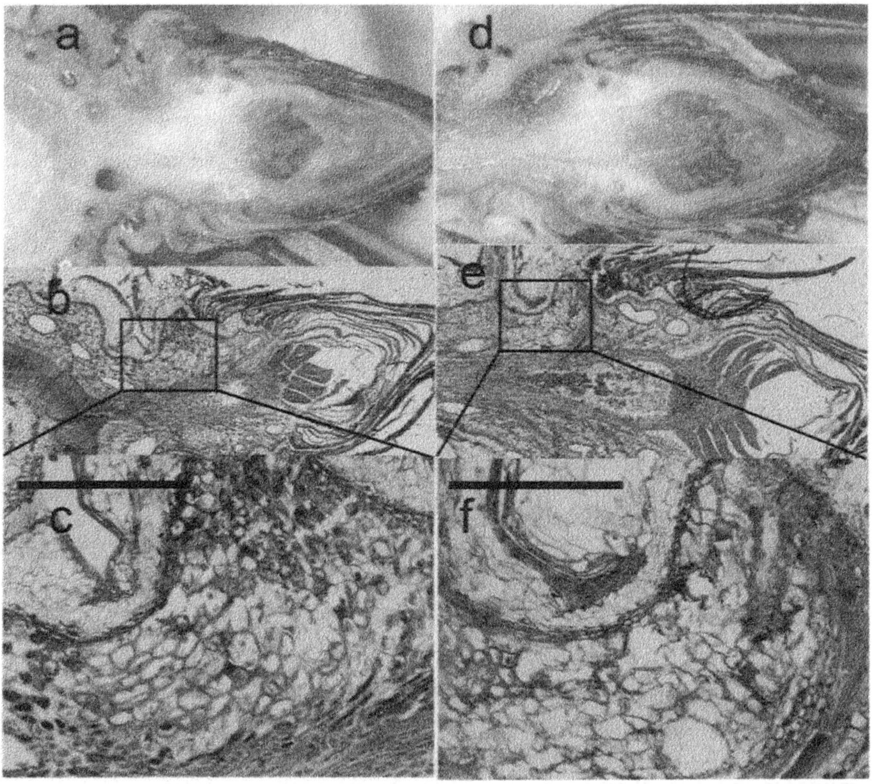

Fig. 3. Tangential views of infested buds on a susceptible (**a–c**) and a resistant (**d–f**) clone. Brown lesions can be seen on the left side of the bud bases where fundatrices inserted their stylets. Scale bar = 500 μm.

differ significantly among the susceptible and two groups of resistant clones in all periods. Fundatrix body length did not differ between resistant clones from the Oketo population and susceptible clones in all periods (Fig. 2). However, resistant clones from the other populations had 20% smaller fundatrix body length than susceptible clones in October and in early April.

The proportion of buds with lesions also did not differ between resistant clones from the Oketo population and susceptible clones in all periods. On these clones, lesions were observed in 24% of the buds in October, and this percentage increased to 38% in late April when fundatrices matured. In contrast, lesions were observed in 80% of the buds in resistant clones from the other populations in October and in early April, and this proportion was significantly greater than that in susceptible clones (Fig. 2). The area of lesions varied largely among buds, and did not differ among the susceptible and two groups of resistant clones in any period.

The color of lesions varied from light to dark brown. In microscopic observation of the lesions, cells near attacked sites showed necrosis with other cells that became hypertrophied or collapsed (Fig. 3). More cells collapsed in resistant clones than in susceptible clones. In contrast, no cell necrosis, hyperplasia, or collapse was observed in cortical tissue of buds without lesions.

15.4 Discussion

In galling adelgids, adult preference does not differ between susceptible and resistant trees (Björkman 2000; K. Ozaki, unpublished data, 1994). Thus the resistance should be caused by differences in larval performance. This study revealed that the resistance was primarily caused by lower survival of larval fundatrices when they were sedentary on buds. Different survival in the overwintering period was also found in a previous study that used *P. jezoensis* seedlings that differed in gall density (Ozaki 1991).

Our study also revealed that the timing of resistant responses differed between *P. jezoensis* populations: on resistant clones from the Oketo population, fundatrix survival was lower than on susceptible clones only in spring, whereas on resistant clones from the other populations, fundatrix survival was lower than on susceptible clones both in winter and in spring. In addition, no gall was induced on resistant clones from the Oketo population even if the fundatrices stimulated bud tissue until they matured and oviposited. In contrast, fundatrix body length was smaller and the proportion of buds with lesions was greater in resistant clones from the other populations than in susceptible clones. These findings suggest that resistance responses exhibited in different timings had different underlying mechanisms.

The proportion of buds with lesions was greater in resistant clones from non-Oketo populations than in susceptible clones, suggesting that the induction of lesions is related to the resistant mechanism in these clones. In *P. abies* trees infested by *A. abietis*, attacked tissue in bud bases is transformed into primary nutritive tissue, where cells become hypertrophied and enriched in lipids and starch, in autumn (Rohfritsch 1988). Then in spring, some cells in primary nutritive tissue show necrosis and accumulation of phenolic substances in a resistant spruce. Because necrotic cells form a barrier to soluble proteins of the tree, Rohfritsch (1988) suggests cell death as the primary mechanism of resistance. Because the primary nutritive tissue was induced in the same part of the buds with similar cell modification as the lesions, hypersensitive responses may have also been involved in tissue responses in the resistant clones from non-Oketo populations. However, our study differed from the *A. abietis*/*P. abies*-system in that necrosis was already induced in autumn, and was not restricted to resistant clones. In addition, 20% of the buds did not induce lesions on the resistant clones, suggesting that cell death was not the only cause of mortality. At present, we do not know the cause of resistance on the buds without lesions.

In resistant clones from the Oketo population, the proportion of buds with lesions was similar to that in susceptible clones. Furthermore, the timing of mortality was different from the timing of lesion induction, suggesting that lesions are not related to the resistant mechanism. In the *A. abietis*/*P. abies*-system, main nutritive tissue is induced in the cortical tissue of the first and second leaf bases by extensive probing in spring (Rohfritsch and Anthony 1992). Then in the resistant spruce, a white-yellowish nodule that contains necrotic cells is observed in this tissue (Rohfritsch 1988). In our study, lesions were observed only in bud bases, and no modification was observed in main nutritive tissue at the macroscopic level. However, no gall was formed in resistant clones from the Oketo population even if the fundatrices matured and oviposited. Because main nutritive tissue is later transformed into gall tissue, resistance responses in this tissue are likely to be exhibited not only as greater mortality of fundatrices but also as failure of gall formation. Further studies should examine the cell modification in main nutritive tissue at the microscopic level.

In plant pathology, hypersensitive response is considered to be a rapid process occurring within hours after attack, and is interpreted as plant-programmed cell death that evolved as a defense against pathogens (Heath 1998). However, in entomology the occurrence of necrosis has been characterized as a hypersensitive response without examining the timing of cell death (Fernandes and Negreiros 2001). One exception is in *Salix vimilis* trees infested by the gall midge *Dasineura marginemtorquens* (Ollerstam et al. 2002). In this system, a rapid (within hours after a larval attack) hypersensitive response was observed exclusively on resistant tree genotypes, which has features commonly found in plant resistance against pathogens. However, in galling adelgids, resistance occurs as a process that lasts for more than a few months, suggesting that the resistant mechanism is not equivalent to that in plant–pathogen interactions. Because adelgids can attack plant tissue with their long stylets, it is possible that necrosis is not a plant-programmed resistant mechanism against insects, but is simply caused by mechanical damage by the fundatrices. In spite of the aggressive probing activity of fundatrices, cell necrosis usually does not occur along the track of stylets because fundatrices have numerous enzymes that help their stylets to progress through tightly packed cells in buds (Rohfritsch and Anthony 1992). Thus it is important to examine the biochemical processes associated with the probing activity that may induce cell death in resistant clones.

15.5 References

Björkman C (2000) Interactive effects of host resistance and drought stress on the performance of a gall-making aphid living on Norway spruce. Oecologia 123:223-231

Fernandes GW (1990) Hypersensitivity: a neglected plant resistance mechanism against insect herbivores. Environmental Entomology 19:1173-1182

Fernandes GW, Negreiros D (2001) The occurrence and effectiveness of hypersensitive reaction against galling herbivores across host taxa. Ecological Entomology 26:46-55

Heath MC (1998) Apoptosis, programmed cell death and the hypersensitive response. European Journal of Plant Pathology 104:117-124

Hoshi H, Itahana N, Ozaki K (2002) Resistance breeding against *Adelges japonicus* (in Japanese). Tree Breeding 204:20-22

Itahana N, Hoshi H, Ozaki K (2002) Evaluation of ezo-spruce resistance against *Adelges japonicus* (in Japanese). Tree Breeding in Hokkaido 45:10-13

Itahana N and Ubukata M (1995) Differences in the resistance of ezo-spruce provenances to *Adelges japonicus* (in Japanese). Transactions of the 106th Annual Meeting of the Japanese Forestry Society, p 501

Larsson S (2002) Resistance in trees to insects – an overview of mechanisms and interactions. In: Wagner MR et al (eds) Mechanisms and deployment of resistance in trees to insects. Kluwer Academic Publishers, Dordrech Boston London, pp 1-29

Mattson WJ, Birr BA, Lawrence RK (1994) Variation in the susceptibility of North American white spruce populations to the gall-forming adelgid, *Adelges abietis* (Homoptera: Adelgidae). In: Price PW et al (eds) The ecology and evolution of gall-forming insects. General Technical Report NC-174, U. S. Department of Agriculture, Forest Service, North Central Forest Experimental Station, St. Paul, pp 135-147

Mattson WJ, Levieux J, Piou D (1998) Genetic and environmental contributions to variation in the resistance of *Picea abies* to the gall-forming adelgid, *Adelges abietis* (Homoptera: Adelgidae). In: Csóka G et al (eds) The biology of gall-inducing arthropods. General Technical Report NC-199, U. S. Department of Agriculture, Forest Service, North Central Research Station, St. Paul, pp 304-314

Ollerstam O, Rohfritsch O, Höglund S, Larsson S (2002) A rapid hypersensitive response associated with resistance in the willow *Salix viminalis* against the gall midge *Dasineura marginemtorquens*. Entomologia Experimentalis et Applicata 102:153-162

Ozaki K (1991) Resistance of young ezo-spruce, *Picea jezoensis*, to the fundatrix of the yezo-spruce gall aphid, *Adelges japonicus* Monzen (Homoptera, Adelgidae) (in Japanese with English abstract). Journal of Japanese Forestry Society 73:431-433

Rohfritsch O (1981) A "defense" mechanism of *Picea excelsa* L. against the gall former *Chermes abietis* L. (Homoptera, Adelgidae). Zeitschrift fur Angewandte Entomologie 92:18-26

Rohfritsch O (1988) A resistance response of *Picea excelsa* to the aphid, *Adelges abietis* (Homoptera: Aphidoidea). In: Mattson WJ et al (eds) Mechanisms of woody plant defenses against insects, Search for pattern. Springer-Verlag, New York, pp 253-266

Rohfritsch O, Anthony M (1992) Strategies in gall induction by two groups of homopterans. In: Shorthouse JD, Rohfritsch O (eds) Biology of insect-induced galls. Oxford University Press, New York, pp 102-117

16 Variable Effects of Plant Module Size on Abundance and Performance of Galling Insects

Dan Quiring, Leah Flaherty, Rob Johns, and Andrew Morrison

Population Ecology Group, Faculty of Forestry and Environmental Management, University of New Brunswick, Fredericton, New Brunswick E3B 6C2, Canada

Summary. We conducted a review of published literature reporting relationships between the size of plant modules and the abundance or performance of gall insects. Insects in the family Tenthredinidae were recently reviewed and thus omitted from this review. The abundance or performance of approximately half of 53 species examined was positively related to plant module size. However, negative and parabolic relationships were found for all major insect families with sample sizes ≥ 5 (i.e., Adelgidae/Aphididae, Cynipidae and Cecidomyiidae for abundance and Adelgidae/Aphididae for performance). This suggests that relationships between plant module size and the abundance or performance of non-tenthredinid gallers, although often positive, are best characterized as variable.

Key words. Gall, Module size, Abundance, Performance

16.1 Introduction

Much of the literature evaluating the relationship between plant growth rate or "vigor," and herbivore abundance or performance has been carried out with gall insects. Whereas all endoparasites must be extremely fine-tuned to the internal environment of their host, gallers must also "take over" the development of a plant part and induce it to make a foreign structure, the gall. The growth rate or size of the plant module could influence both gall induction success (Björkman 1998; McKinnon et al. 1999; Sopow et al. 2003) as well as the fitness of juveniles in the developing gall (Craig et al. 1989). The size of any particular plant part, such as a leaf, bud or shoot, varies not only among plants but also within them. For example, in open-grown conifers the longest shoot is usually the leader (i.e., most apical shoot), and shoot sizes usually decrease continuously as you move down the crown. Thus the influence of plant module size on galler abundance (commonly used to infer preference) and performance has been evaluated both within and among host plants.

Thirty-seven of 43 galling species of tenthredinids attack the longest shoots on host plants (reviewed in Price 2003; Price et al. 2004), supporting the plant vigor hypothesis (Price 1991), but a similar review has not been carried out for other insect groups. The plant vigor hypothesis states that galling insects do best on large, fast growing plant modules because cells of small plant modules are often too small or divide too slowly: i) to produce a suitable gall, leading to galler death before gall induction; or ii) to produce a large gall, where survival is high and the size of survivors is large. Although support for this hypothesis is strong, there are at least 8 tenthredinid species that sometimes display negative or insignificant relationships between the size of the module galled and galler abundance or performance (Fritz et al. 2003; Kokkonen 2000; Price et al. 1999, 2004). Similarly, studies with other gall insects have also reported variations in the relationships between the size of plant modules galled and the abundance or performance of gallers. Positive, negative, and parabolic relationships between plant module size and galler abundance or performance have resulted in numerous hypotheses (e.g., Björkman 1998; Larsson 1989; Larson and Whitham 1997; McKinnon et al. 1999; Price 1991; Price et al. 2004; Sopow et al. 2003) that will not be discussed here.

Here we present a review of the literature reporting relationships between plant module size, commonly used as a correlate for plant vigor or growth rate, and galler abundance or performance in non-tenthredinids.

16.2 Methods

The review was carried out using available reference sources, including Biological Abstracts (1989–May 2005), Agricola (1970–2005), General Science Index (May 1989–June 2005), Forest Science (1939–2005) and Google Scholar (dates unknown). Unfortunately the review concentrated almost exclusively on papers written in or translated into English, a constraint that probably resulted in the omission of some papers. We classified the relationships between plant module size and galler abundance or performance as positive, negative, parabolic or non-significant (Table 1). In four instances we inferred a parabolic distribution based on published graphs: Fig. 1 in Cornelissen et al. (1997), an inference that has been substantiated in three subsequent years of sampling (Fernandes, personal communication); Fig. 2 in Craig et al. (1999); Fig. 4 in Vieira et al. (1996); and Fig. 2 in Ito and Hijii (2001). We categorized the latter three as positive (the authors' conclusion) as well as parabolic, because the data presented in the figures appear to be parabolic. We plotted frequency histograms to illustrate the number of species for which there are published reports of positive, negative, parabolic or non-significant, relationships between the size of plant modules that are galled and the abundance or performance of gall-forming insects. Thus up to four different relationships could be reported for each species. In contrast, multiple studies reporting the same relationship for the same species were only counted as representing one relationship.

16.3 Results and Discussion

The relationship between plant module size and galler abundance or performance was reported for 53 non-tenthredinid species. Most species were cecidomyiids (29), with ten cynipids, eight adelgids/aphids and one species in each of six other families (Table 1). In 47% ($N = 53$) of all reported relationships, galler abundance was positively related to plant module size. This positive relationship was the most common relationship reported for all three major families (Fig. 1). However, it is noteworthy that 53% of reported relationships were not positive, and that some negative, parabolic and non-significant relationships were present for every major family (Fig. 1). Furthermore, there were several occasions, indicated by two or more relationships separated by a slash in Table 1, where the type of relationship for the same species changed depending on the sampling date, location or parameter used to estimate abundance.

The relationship between plant module size and galler performance was only reported for 24 non-tenthredinid species, including nine cecidomyiids and 8 adelgids/aphids. Forty-eight percent ($N = 27$) of these relationships were positive, but again 52% of reported relationships between plant module size and galler performance were negative, parabolic or non-significant (Fig. 1).

A negative relationship between galler performance and plant module size was reported for all major families except Cynipidae, and a parabolic relationship was reported for adelgids, but not for cynipids or cecidomyiids, where few studies have been reported (Fig. 1).

This review illustrates the paucity of data evaluating the influence of plant module size on the performance of non-tenthredinid gallers. Nevertheless, reports of both positive and negative, as well as parabolic, relationships for all families with sample sizes ≥ 5 suggests that the relationship between plant module size and galler abundance is variable within

Table 1. Summary of literature reporting positive (P), negative (N), parabolic (Par) or non-significant (NS) relationships between the size of the plant module galled (MG) and the abundance or performance of insects in different families (IF) forming galls in different plant genera (PG). Note that some papers evaluated more than one insect species and that reports for the same insect species sometimes occur in different papers

Insect species	IF[a]	PG[b]	MG[c]	Abundance	Performance	References[d]
Adelges abietis	A	*Pi*	S	Par	Par	McKinnon 1999
Adelges cooleyi	A	*Pi*	S	N	P/N[e]	Fay 1990
Adelges japonicus	A	*Pi*	B	-	P	Ozaki 2000
Sacchiphantes abietis	A	*Pi*	B	N	P	Björkman 1998
Aploneura lentisci	Ap	*Pi*	S	-	P	Wool 1988
Hormaphis hamamelidis	Ap	*Ha*	L	NS	NS	Rehill 2001
Tetraneura sp.	Ap	*Ul*	L	-	NS	Akimoto 1994
Pemphigus betae	Ap	*Po*	L	P	P	Whitham 1978
Asphondylia sp. 1	C	*La*	L	NS	-	Waring 1990
Asphondylia sp. 2	C	*La*	L	P	-	Waring 1990
Asphondylia sp. 3	C	*La*	L	NS	-	Waring 1990
Asphondylia sp. 6	C	*La*	L	NS	-	Waring 1990
Asphondylia sp. 7	C	*La*	L	NS	-	Waring 1990
Asphondylia sp. 8	C	*La*	L	NS	-	Waring 1990
Asphondylia sp. 10	C	*La*	F	NS	-	Waring 1990
Asphondylia sp. 11	C	*La*	S	NS	-	Waring 1990
Cecidomyia avicenniae	C	*Av*	L	N	-	Goncalves 2001
Contarinia sp.	C	*Ba*	L	P	NS	Fernandes 1998
Contarinia sp.	C	*Pa*	L	P/Par[f]	-	Vieira 1996
Dasineura marginemtorquens	C	*Sa*	L	P	P	Glynn 1994
Giraudiella inclusa	C	*Ph*	S	P/N[g]	P/N[h]	Tscharntke 1988
Izeniola obesula	C	*Su*	S	P	P	Dorchin 2004
Neolasioptera sp.	C	*Er*	S	P	P	Prado 1999
Rabdophaga sp. 3	C	*Sa*	S	NS	NS	Kopelke 2003
Rabdophaga sp. 4	C	*Sa*	S	P	NS	Kopelke 2003
Rabdophaga sp. 5	C	*Sa*	S	P	N	Kopelke 2003
Rhopalmyia n. sp.	C	*Tr*	B	-	P	Hinz 2000
unknown sp. 1	C	*Bo*	L	P	-	Fernandes 2001
unknown sp. 2	C	*Te*	L	NS	-	Fernandes 2001
unknown sp. 3	C	*Va*	L	P	-	Fernandes 2001
unknown sp. 4	C	*Da*	L	P	-	Fernandes 2001
unknown sp. 5	C	*By*	L	P	-	Fernandes 2001

Table 1. Continued

Insect species	IF[a]	PG[b]	MG[c]	Abundance	Performance	References[d]
unknown sp. 6	C	*Ou*	L	P	-	Fernandes 2001
unknown sp. 7	C	*St*	L	NS	-	Fernandes 2001
unknown sp. 8	C	*Se*	L	NS	-	Fernandes 2001
unknown sp. A	C	*Ba*	L	P	-	Cornelissen 1997
unknown sp. B	C	*Ba*	L	N/Par[f]	-	Cornelissen 1997
Lipara lucens	Ch	*Ph*	S	N	N	De Bruyn 1995
Andicus sp.	Cy	*Qu*	S	P	P	Pires 2000
Andricus moriokae	Cy	*Qu*	B	P/Par[f]	-	Ito 2001
Aphelonyx glanduliferae	Cy	*Qu*	B	P	-	Ito 2001
Callirhytis cornigera	Cy	*Qu*	L	NS	-	Eliason 2000
Diplolepis fusiformans	Cy	*Ro*	S	P	P	Caouette 1989
Diplolepsis ignota	Cy	*Ro*	L	N	-	Williams 2004
Diplolepsis nodulosa	Cy	*Ro*	S	N	-	Williams 2004
Diplolepsis rosaefolii	Cy	*Ro*	L	N	-	Williams 2004
Diplolepis spinosa	Cy	*Ro*	S	P	P	Caouette 1989
Dryocosmus kuriphilus	Cy	*Ca*	B	P	-	Kato 2001
Amblypalpis olivierella	G	*Ta*	S	P	-	Price 2004
Daktulosphaira vitifoliae	P	*Vi*	L	P	-	Kimberling 1990
Daktulosphaira vitifoliae	P	*Vi*	L	-	NS	Kimberling 1996
Neopelma baccharidis	Ps	*Ba*	L	NS	NS	Faria 2001
Eurosta solidaginis	T	*So*	S	P/Par[f]	Par	Craig 1999
Eurosta solidaginis	T	*So*	S	P	NS	Craig 2000
Epiblema strenuana	Tor	*Pa*	S	P	-	Dhileepan 2004

[a]A, Adelgidae; Ap, Aphididae; C, Cecidomyiidae; Ch, Chloropidae; Cy, Cynipidae; G, Gelechiidae; P, Phylloxeridae; Ps, Psyllidae; T, Tephritidae; Tor, Tortricidae.
[b]First two letters of genus of galled plant.
[c]B, bud; L, leaf; S, stem or shoot; F, flower.
[d]Only first author of reference is listed in table.
[e]Fundatrix survival decreases, and fundatrix and gallicolae body weight and number of gallicolae per gall increase, with shoot length.
[f]Potentially parabolic relationship inferred from data.
[g]Egg and gall abundance were positively and negatively related to shoot basal diameter, respectively.
[h]Larval survival and weight were lowest and highest, respectively, on large shoots.

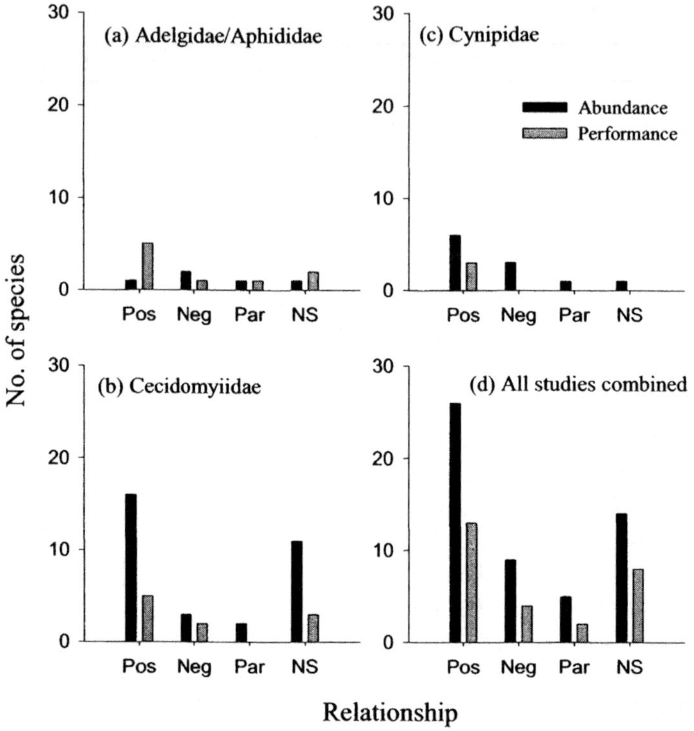

Fig. 1. Number of species in the families **(a)** Adelgidae and Aphididae, **(b)** Cecidomyiidae, **(c)** Cynipidae or **(d)** in all studies combined for which there are published reports of positive *(Pos)*, negative *(Neg)*, parabolic *(Par)* or non-significant *(NS)* relationships between the size of plant modules that are galled and the abundance *(dark bars)* or performance *(light bars)* of gall-forming insects. Note that totals from all studies combined **(d)** were obtained from studies on the gallers in the families Adelgidae and Aphididae **(a)**, Cecidomyiidae **(b)**, Cynipidae **(c)** and 6 other species (see Table 1).

non-tenthredinid families. Similarly, our review of the literature indicates that the effects of plant module size on galler performance is equally variable. Approximately half of the small number of studies evaluating the influence of plant module size on the performance of non-tenthredinid gallers reported a positive relationship, whereas the other half reported negative or non-significant relationships. Recent studies with two adelgids and one cecidomyiid in our lab support this conclusion, as the abundance of two was parabolically, and that of the other, positively related to plant module size (D. Quiring et al., unpublished data). Such results emphasize the need to establish theory that can explain positive as well as negative and parabolic relationships between plant module size and the abundance or performance of gall-forming insects.

16.4 References

Akimoto S, Yamaguchi Y (1994) Phenotypic selection on the process of gall formation of a *Tetraneura* aphid (Pemphigidae). Journal of Animal Ecology 63:727-738.

Björkman C (1998) Opposite, linear and non-linear effects of plant stress on a galling aphid. Scandinavian Journal of Forest Research 13:177-183

Caouette MR, Price PW (1989) Growth of Arizona rose and attack and establishment of gall wasps *Diploplepis fusiformans* and *D. spinosa* (Hymenoptera: Cynipidae). Environmental Entomology 18:822-828

Cornelissen TG, Madeira BG, Allain LR, Lara ACF, Araújo LM, Fernandes GW (1997) Multiple responses of insect herbivores to plant vigor. Ciência e Cultura 49:285-288

Craig TP, Itami JK, Price PW (1989) A strong relationship between oviposition preference and larval performance in a shoot-galling sawfly. Ecology 70:1691-1699

Craig TP, Abrahamson WG, Itami JK, Horner JD (1999) Oviposition preference and offspring performance of *Eurosta solidaginis* on genotypes of *Solidago altissima*. Oikos 86:119-128

Craig TP, Itami JK, Shantz C, Abrahamson WG, Horner JD, Craig JV (2000) The influence of host plant variation and intraspecific competition on oviposition preference and offspring performance in the host races of *Eurosta solidaginis*. Ecological Entomology 25:7-18

De Bruyn L (1995) Plant stress and larval performance of a dipterous gall former. Oecologia 101:461-466

Dhileepan K (2004) The applicability of the plant vigor and resource regulation hypotheses in explaining *Epiblema* gall moth–*Parthenium* weed interactions. Entomologia Experimentalis et Applicata 113:63-70

Dorchin N, Freidberg A (2004) Sex ratio in relation to season and host plant quality in a monogenous stem-galling midge (Diptera: Cecidomyiidae). Ecological Entomology 29:677-684

Eliason EA, Potter DA (2000) Budburst phenology, plant vigor, and host genotype effects on the leaf-galling generation of *Callirhytis cornigera* (Hymenoptera: Cynipidae) on pin oak. Environmental Entomology 29:1199-1207

Faria M, Fernandes GW (2001) Vigour of a dioecious shrub and attack by a galling herbivore. Ecological Entomology 26:37-45

Fay PA, Whitham TG (1990) Within-plant distribution of a galling adelgid (Homoptera: Adelgidae): the consequences of conflicting survivorship, growth, and reproduction. Ecological Entomology 15:245-254

Fernandes GW (1998) Hypersensitivity as a phenotypic basis of plant induced resistance against a galling insect (Diptera: Cecidomyiidae). Environmental Entomology 27:260-26

Fernandes GW, Negreiros D (2001) The occurrence and effectiveness of hypersensitive reaction against galling herbivores across host taxa. Ecological Entomology 26:46-55

Fritz RS, Crabb BA, Hochwender CG (2003) Preference and performance of a gall-inducing sawfly: plant vigor, sex, gall traits and phenology. Oikos 102:601-613

Glynn C, Larsson S (1994) Gall initiation success and fecundity of *Dasineura marginemtorquens* on variable *Salix viminalis* host plants. Entomologia Experimentalis et Applicata 73:11-17

Goncalves Alvim SJ, Vaz dos Santos MCF, Fernandes GW (2001) Leaf gall abundance on *Avicennia germinans* (Avicenniaceae) along an interstitial salinity gradient. Biotropica 33:69-77

Hinz HL, Müller-Schärer H (2000) Influence of host condition on the performance of *Rhopalomyia* n. sp. (Diptera: Cecidomyiidae), a biological control agent for scentless chamomile, *Tripleurospermum perforatum*. Biological Control 18:147-156

Ito M, Hijii N (2001) Effect of shoot size and phonological variation of host plants on the spatial patterns of cynipid galls. Journal of Forestry Research 6:147-151

Kato K, Hijii N (2001) Ovipositional traits of the chestnut gall wasp, *Dryocosmus kuriphilus* (Hymenoptera: Cynipidae). Entomological Science 4:295-299

Kimberling DN, Price PW (1996) Competition, leaf morphology, and host clone effects on leaf-galling grape phylloxera (Homoptera: Phylloxeridae). Environmental Entomology 25:1147-1153

Kimberling DN, Scott ER, Price PW (1990) Testing a new hypothesis: plant vigor and phylloxera distribution on wild grape in Arizona. Oecologia 84:1-8

Kokkonen K (2000) Mixed significance of plant vigor: two species of galling *Pontania* in a hybridizing willow complex. Oikos 90:97-106.

Kopelke JP, Amendt J, Schönrogge K (2003) Patterns of interspecific associations of stem gallers on willows. Diversity and Distributions 9:443-453

Larson KC, Whitham TG (1997) Competition between gall aphids and natural plant sinks: plant architecture affects resistance to galling. Oecologia 109:575-582

Larsson S (1989) Stressful times for the plant stress – insect performance hypothesis. Oikos 56:277-283

McKinnon ML, Quiring DT, Bauce E (1999) Influence of tree growth rate, shoot size and foliar chemistry on the abundance and performance of a galling adelgid. Functional Ecology 13:859-867

Ozaki K (2000) Insect-plant interactions among gall size determinants of adelgids. Ecological Entomology 25:452-459

Pires CSS, Price PW (2000) Patterns of host plant growth and attack and establishment of gall-inducing wasp (Hymenoptera: Cynipidae). Environmental Entomology 29:49-54

Prado P, Vieira EM (1999) The interplay between plant traits and herbivore attack: a study of a galling midge in the neotropics. Ecological Entomology 24:80-88

Price PW (1991) The plant vigor hypothesis and herbivore attack. Oikos 62:244-251

Price PW (2003) Macroevolutionary Theory on Macroecological Patterns. Cambridge University press, Cambridge

Price P, Gerling D (2004) Complex architecture of *Tamarix nilotica* and resource utilization by the spindle-gall moth *Amblypalpis olivierella* (Lepidoptera: Gelechiidae). Israel Journal of Entomology 34:1-17

Price P, Roininen H, Ohgushi T (1999) Comparative plant-herbivore interactions involving willows and three gall-inducing sawfly species in the genus *Pontania* (Hymenoptera: Tenthredinidae). Écoscience 6:41-50

Price PW, Ohgushi T, Roininen H, Ishihara M, Craig TP, Tahvanainen J, Ferrier SM (2004) Release of phylogenetic constraints through low resource heterogeneity: the case of gall-inducing sawflies. Ecological Entomology 29:467-481

Rehill BJ, Schultz, JC (2001) *Hormaphis hamamelidis* and gall size: a test of the plant vigor hypothesis. Oikos 95:94-104

Sopow SL, Shorthouse JD, Strong W, Quiring DT (2003) Evidence for long-distance, chemical gall induction by an insect. Ecology Letters 6:102-105

Tscharntke T (1988) Variability of the grass *Phragmites austalis* in relation to the behavior and mortality of the gall-inducing midge *Giraudiella inclusa* (Diptera, Cecidomyiidae). Oecologia 76:504-512

Vieira EM, Andrade I, Price PW (1996) Fire effects on a *Palicourea rigida* (Rubiaceae) gall midge: a test of the plant vigor hypothesis. Biotropica 28:210-217

Waring GL, Price PW (1990) Plant water stress and gall formation (Cecidomyiidae: *Asphondylia* spp.) on creosote bush. Ecological Entomology 15:87-95

Whitham TG (1978) Habitat selection by *Pemphigus* aphids in response to resource limitation and competition. Ecology 59:1164-1176

Williams MA, Cronin JT (2004) Response of a gall-forming guild (Hymenoptera: Cynipidae) to stressed and vigorous prairie roses. Environmental Entomology 33:1052-1061

Wool D, Manheim O (1988) The effects of host-plant properties on gall density, gall weight and clone size in the aphid *Aploneura lentisci* (Pass.) (Aphididae: Fordinae) in Israel. Researches on Population Ecology 30:227-234

17 Biology and Life History of the Bamboo Gall Maker, *Aiolomorphus rhopaloides* Walker (Hymenoptera: Eurytomidae)

Ei'ichi Shibata

Laboratory of Forest Protection, Graduate School of Bioagricultural Sciences, Nagoya University, Chikusa-ku, Nagoya 464-8601, Japan

Summary. *Aiolomorphus rhopaloides* (Hymenoptera: Eurytomidae) induces galls on two species of bamboo, *Phyllostachys heterocycla* and *Phyllostachys bambusoides* (Monocotyledoneae: Gramineae) in Japan. Our work showed eight hymenopteran species, including *A. rhopaloides*, emerge from bamboo galls. Larvae of *A. rhopaloides* appear in galls in July, with the percentage of larvae decreasing in September, before overwintering as pupae. The percentage of parasitoids in galls is low in July, but increases until winter. The percentages of inquiline *Diomorus aiolomorphi* emerging from the overwintering galls are relatively high, suggesting that inquilines might be a key mortality factor of bamboo galls. Females of *A. rhopaloides* with ca. 80 mature eggs in their ovaries start emerging from galls from mid-April to early May just after the bud burst of *P. heterocycla* and emergence continues during bud elongation. Phenological asynchrony between adult emergence and bud burst may have a large influence on the population dynamics of the gall maker *A. rhopaloides*; a slight advance or delay in emergence may reduce suitable oviposition sites, causing population fluctuations. Although more bamboo galls are found on longer branches, selection of longer branches for oviposition does not result in higher offspring survival.

Key words. *Aiolomorphus rhopaloides*, Bamboo, *Diomorus aiolomorphi*, Gall maker, Life-history traits

17.1 Introduction

The bamboo gall maker *Aiolomorphus rhopaloides* Walker (Hymenoptera: Eurytomidae) induces galls on two species of bamboo, *Phyllostachys heterocycla* Matsumura and *Phyllostachys bambusoides* Siebold and Zuc-

Fig.1. Bamboo gall induced by *Aiolomorphus rhopaloides*.

carini (Monocotyledoneae: Gramineae), in Japan (Kobayashi 1994; Takahashi and Mizuta 1971; Yukawa and Masuda 1996). Adult females of *A. rhopaloides* oviposit in the young buds of bamboo branchlets, which develop into unitary galls (length: 20–30 mm, width: 2–4 mm; Fig. 1) (Takahashi and Mizuta 1971). Outbreaks of this species occur in stands of *P. heterocycla* in western Japan (Kobayashi 1994).

Studies on the population dynamics of gall insects emphasize the importance of top-down forces (i.e., the effects of natural enemies) (Washburn and Cornell 1981; Weis and Abrahamson 1985). Some studies have reported many species of inquilines and parasitoids of gall makers (Askew 1961, 1980). To clarify the population dynamics of *A. rhopaloides*, information is needed on its life-history traits and related insects as mortality factors within bamboo galls. Price and Martinsen (1994) argued that bottom-up effects (i.e., plant quality) might be particularly important in host plants to which many herbivores respond positively compared with top-down forces. Variation in plant phenology may also be a major determinant of the population dynamics of herbivores on their host plant (Fox et al. 1997). Furthermore, information on female fecundity is essential for understanding population dynamics. This paper reviews recent literature that describes: (i) species composition of insects within bamboo galls; (ii) life-history traits of *A. rhopaloides*; (iii) synchronization of shoot elongation and emergence of *A. rhopaloides* as a bottom-up effect; (iv) fecundity of *A. rhopaloides*; and (v) oviposition site preference of adult females.

17.2 Insect Inhabitants

Eight hymenopteran species emerged from galls in bamboo stands in Nagoya City, central Japan (Shibata and Ito 2005): *Leptacis* sp. (Platygastridae), *A. rhopaloides* (Eurytomidae), *Eurytoma* sp. (Eurytomidae), *Sycophila* sp. (Eurytomidae), *Diomorus aiolomorphi* Kamijo (Torymidae), *Homoporus japonicus* Ashmead (Pteromalidae), *Norbanus* sp. (Pteromalidae), and *Eupelmus* sp. (Eupelmidae). One dipteran species also emerged from the galls: Cecidomyiidae sp. The *Leptacis* sp., *A. rhopaloides*, *Eurytoma* sp., *Sycophila* sp., *D. aiolomorphi*, *Norbanus* sp., *Eupelmus* sp. and Cecidomyiidae sp. are solitary with respect to galls while *H. japonicus* is gregarious (Shibata and Ito 2005).

Takahashi and Mizuta (1971) observed one inquiline and two parasitoids on bamboo galls induced by *A. rhopaloides* in Kyoto, western Japan. *D. aiolomorphi* is the known inquiline (Takahashi and Mizuta 1971), that is, a phytophagous insect that cannot make its own galls but feeds on gall tissues induced by the gall maker (Askew 1961). Takahashi and Mizuta (1971) also observed that *H. japonicus* and *Eupelmus* sp. parasitize the larvae of *A. rhopaloides* but not of *D. aiolomorphi*, which has a harder larval body surface than *A. rhopaloides*. Thus, both are thought to be primary parasitoids of the gall maker. *Eurytoma* sp., *Sycophila* sp. and *Norbanus* sp. may attack not only larvae of the gall maker *A. rhopaloides*, but also

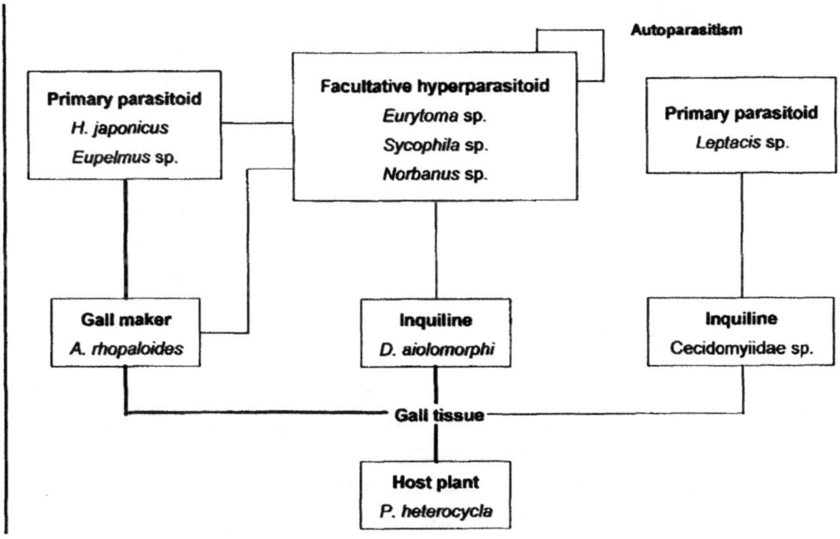

Fig. 2. Possible food web of insects inhabiting bamboo galls based on Takahashi and Mizuta (1971) and Shibata and Ito (2005) (modified from Shibata and Ito [2005]). *Thick lines* show known relationships. *Thin lines* show possible relationships.

larvae of the inquiline *D. aiolomorphi* as well as other parasitoids (Shibata and Ito 2005). In addition, these three species may parasitize each other and themselves (autoparasitism) as reported for cynipid gall wasps (Askew 1961). Consequently, they are thought to be facultative hyperparasitoids (Shibata and Ito 2005), but more details are not known. The Cecidomyiidae sp. is thought to be an inquiline (K. Yamagishi, personal communication). The *Leptacis* sp. parasitizes larvae of Cecidomyiidae sp. (K. Yamagishi, personal communication). Fig. 2 summarizes the possible food web of insects inhabiting bamboo galls based on Takahashi and Mizuta (1971) and Shibata and Ito (2005).

The percentage of galls attacked by parasitoids in July was low and then increased (Shibata and Ito 2005), probably because the parasitoids emerge from overwintering galls later than *A. rhopaloides* and *D. aiolomorphi* (Shibata and Ito 2005). Moreover, they also emerge from June to November (Shibata and Ito 2005) and can attack larvae of *A. rhopaloides* and *D. aiolomorphi* during the emergence periods. The percentage of the inquiline *D. aiolomorphi* emerging from the overwintering galls was relatively high (1999: 45.2%, 2000: 39.7%, 2001: 30.8%) in Nagoya City (E. Shibata, unpublished data), suggesting that parasitism of inquilines might be a key mortality factor of bamboo galls.

17.3 Life-history Traits

Bamboo galls were distinguished visually in early June and had been completed in early July (Takahashi and Mizuta 1971). There was no difference in the diameter of galls with *A. rhopaloides* or *D. aiolomorphi* in July (Shibata and Ito 2005). Galls are made when insects interrupt the natural development of growing tissue as a result of feeding by nymphs or larvae, forming a chamber where they live (Speight et al. 1999). Thus, the larval feeding of the gall maker *A. rhopaloides* and the inquiline *D. aiolomorphi* might stimulate the tissue of the young buds of the bamboo branchlets, despite the fact that larvae of *D. aiolomorphi* kill larvae of *A. rhopaloides* until July (Takahashi and Mizuta 1971). The diameter of empty galls and galls with parasitoids was smaller than that of galls containing *A. rhopaloides* and *D. aiolomorphi*, suggesting that parasitism of larvae of *A. rhopaloides* and *D. aiolormorphi* arrests stimulation of tissue growth and results in gall abortion (Shibata and Ito 2005).

Life-history traits of insects in bamboo galls were detected using soft X-ray photography (Shibata and Ito 2005; Shibata et al. 2004). Larvae of *A. rhopaloides* were observed in 18.7% of galls in July and then percentages

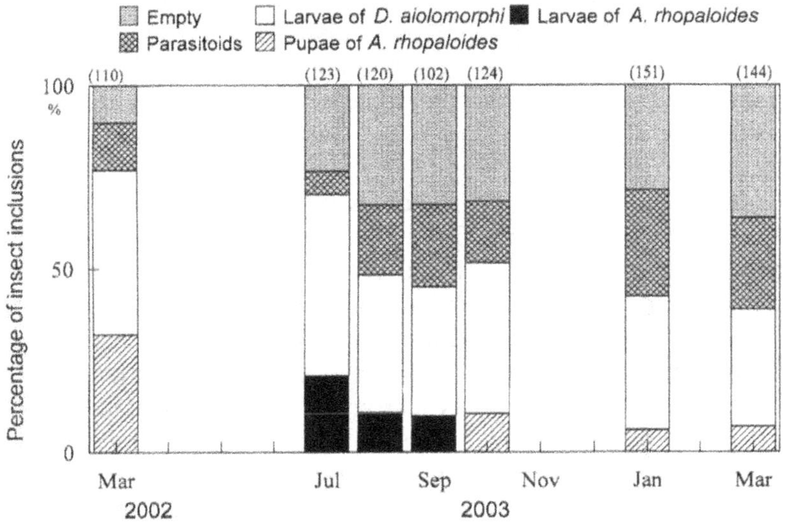

Fig. 3. Seasonal changes in percentage of insect inclusions in bamboo galls in Nagoya City, central Japan (modified from Shibata and Ito [2005]). *Numbers of galls examined* were shown in parenthesis.

decreased from 10.8% in August to 9.8% in September (Fig. 3). Pupae of *A. rhopaloides* appeared in October (10.5%) and percentages of pupae in January and March were 6.5% and 6.9%, respectively (Shibata and Ito 2005). *A. rhopaloides* overwinters in galls as pupae and starts emerging from mid-April to early May. These observations confirm that *A. rhopaloides* is univoltine.

17.4 Shoot Elongation and Emergence

As explained above, *A. rhopaloides* emerges from mid-April to early May. Bud burst of *P. heterocycla*, where females lay eggs, starts in late March and buds elongate until early June (Shibata 2001). Thus, *A. rhopaloides* starts emerging from galls just after bud burst and continues during bud elongation, showing concurrence of emergence and bamboo phenology. This synchronization of adult emergence with host plant phenology is known between the gall midge *Tokiwadiplosis matecola* (Diptera: Cecidomyiidae) and *Lithocarpus edulis* (Fagaceae) (Okuda and Yukawa 2000), and between the gall aphid *Dinipponaphis autumna* (Homoptera: Aphididae) and *Distylium racemosum* (Hamamelidacecae) (Ngakan and Yukawa 1997). Asynchronism of adult emergence of *A. rhopaloides* and bud burst of branches must affect the spatial distribution of galls within bamboo (Shibata 2005) and between bamboos (Shibata 2003). Phenological asynchrony between adult emergence and bud burst may determine the population dynamics of *A. rhopaloides*; a slight advance or delay in emergence may reduce suitable oviposition sites, causing population fluctuations (Shibata 2001).

For gall makers, reproductive success and larval survival depend on successful oviposition in plant tissue suitable for production of nutritious and protective galls (Abrahamson and Weis 1987). *A. rhopaloides* females have mature eggs at emergence and can take advantage of newly elongated buds of bamboo, which have a higher nutritional content than previous-year foliage (Shibata 2001). Galls induced during this period provide a richer resource for larvae. Thus, synchrony between gall-maker emergence and host plant phenology might be advantageous for larval development. In addition, galls on growing buds may grow earlier than on other parts because differentiation is earlier and larvae are enclosed earlier. Hence, oviposition of gall makers on growing parts may enhance protection from abiotic factors (Wool 1977) and natural enemies (Weis and Abrahamson 1985).

17.5 Fecundity

Male and female longevities under rearing conditions were ca. 10 days and ca. 12 days, respectively. There is no significant difference in adult longevity between sexes (Shibata 2002). Counting fully developed eggs in ovaries gives an estimate of the potential maximum fitness of individual females (Sitch et al. 1988). Emerging females of *A. rhopaloides* had ca. 80 mature eggs in their ovaries, which were similar after rearing and death (Shibata 2002). Thus, *A. rhopaloides* is a pro-ovigenic species (Shibata 2002) that emerges with a full complement of mature or nearly mature eggs (Flanders 1950).

It is known that fecundity in most insects varies with female body size (Honek 1993). The body size of *A. rhopaloides* is positively correlated with fecundity (Shibata 2002) as has been previously reported for the gall midge *Dastineura marginemtorquens* on willow (Glynn and Larsson 1994). This suggests that production of larger females might improve fitness. The size of *A. rhopaloides* females is positively correlated with gall diameter. Thus, gall size might be a good predictor of female fecundity, as recently shown for several adelgid gallers (Sopow and Quiring 2001).

17.6 Oviposition Site Preference

Gall size varies with gall position on the host plant (Ngakan and Yukawa 1997; Sitch et al. 1988). Therefore, oviposition site might affect gall size and fitness of gall insects. More bamboo galls induced by *A. rhopaloides* are found on longer branches, similar to galls formed by sawflies (Price 1991; Price and Ohgushi 1995) and by cynipid wasps (Caouette and Price 1989; Ito and Hijii 2001; Pires and Price 2000), suggesting the possibility that module size plays some part in oviposition site preference by female *A. rhopaloides* that utilize larger branches more frequently. However, there is no significant difference in survival rates of *A. rhopaloides* between branch lengths (Shibata 2005). Thus, selection of longer branches for oviposition does not result in better survival of offspring of *A. rhopaloides.*

17.7 Conclusion

The number of galls in stands of *P. heterocycla* bamboo in Nagoya City fluctuates yearly (E. Shibata, unpublished data). As mentioned above, mortality attributable to inquilines as a top-down force and phenological asynchrony between adult emergence and bud burst as a bottom-up effect may have large effects on the population fluctuation of *A. rhopaloides*. However, the relative importance of the two factors in the population dynamics of *A. rhopaloides* remains unclear. Further investigations of mortality factors are needed.

17.8 Acknowledgements

I thank K. Kamijo and K. Yamagishi for insect identification, and M. Ito and N. Ikai for invaluable suggestions. Thanks are also due to the members of the Laboratory of Forest Protection, Nagoya University for their help with the fieldwork. This study was supported in part a Grand-in-Aid for Scientific Research (No. 11460068) from the Ministry of Education, Science and Culture, Japan.

17.9 References

Abrahamson WG, Weis AE (1987) Nutritional ecology of arthropod gall maker In: Slansky Jr F, Rodriguez JG (eds) Nutrition ecology of insects, mites, spiders, and related invertebrates, Wiley, New York, pp 235-258

Askew RR (1961) On the biology of the inhabitants of oak galls of Cynipidae (Hymenoptera) in Britain. Transactions of the Society for British Entomology 14:237-268

Askew RR (1980) The diversity of insect communities in leaf-mines and plant galls. Journal of Animal Ecology 49:817-829

Caouette MR, Price PW (1989) Growth of Arizona rose and attack and establishment of gall wasps *Diplolepis fusiformans* and *D. spinosa* (Hymenoptera: Cynipidae). Environmental Entomology 18:822-828

Honek A (1993) Intraspecific variation in body size and fecundity in insects: A general relationship. Oikos 66:483-492

Ito M, Hijii N (2001) Effect of shoot size and phenological variation of host plants on the spatial patterns of cynipid galls. Journal of Forest Research 6:147-151

Kobayashi F (1994) Bamboo gall chalcid (in Japanese). In: Kobayashi F, Taketani A (eds) Forest insects, Yokendo, Tokyo, pp 523-524

Flanders SE (1950) Regulation of ovulation and egg dispersal in the parasitic Hymenoptera. Canadian Entomologist 82:134-140

Fox CW, Waddell KJ, Groeters FR, Mousseau TA (1997) Variation in budbreak phenology affects the distribution of a leaf-mining beetle (*Brachys tessellates*) on turkey oak (*Quercus laevis*). Écoscience 4:480-489

Glynn S, Larsson S (1994) Gall initiation success and fecundity of *Dasineura marginemtorquens* on variable *Salix viminalis* host plants. Entomologia Experimentalis et Applicata 73:11-17

Ngakan PO, Yukawa J (1997) Synchronization with host plant phenology and gall site preference of *Dinipponaphis autumna* (Homoptera: Aphididae). Applied Entomology and Zoology 32:81-90

Okuda S, Yukawa J (2000) Life history strategy of *Tokiwadiplosis matecola* (Diptera: Cecidomyiidae) relying upon the lammas shoots of *Lithocarpus edulis* (Fagaceae). Entomological Science 3:47-56

Pires CSS, Price PW (2000) Patterns of host plant growth and attack and establishment of gall-inducing wasp (Hymenoptera: Cynipidae). Environmental Entomology 29:49-54

Price PW (1991) The plant vigor hypothesis and herbivore attack. Oikos 62:244-251

Price PW, Martinsen GD (1994) Biological pest control. Biomass Bioenergy 6:93-101

Price PW, Ohgushi T (1995) Preference and performance linkage in a *Phyllocolpa* sawfly on the willow, *Salix miyabeana*, on Hokkaido. Researches on Population Ecology 37:23-28

Shibata E (2001) Synchronization of shoot elongation in the bamboo *Phyllostachys heterocycla* (Monocotyledoneae: Gramineae) and emergence of the gall maker *Aiolomorphus rhopaloides* (Hymenoptera: Eurytomidae) and its inquiline *Diomorus aiolomorphi* (Hymenoptera: Torymidae). Environmental Entomology 30:1098-1102

Shibata E (2002) Potential fecundity of the bamboo gall maker, *Aiolomorphus rhopaloides* (Hymenoptera: Eurytomidae), and its inquiline, *Diomorus aiolomorphi* (hymenoptera: Torymidae), in relation to gall size and body size. Journal of Forest Research 7:117-120

Shibata E (2003) Sampling procedure for density estimation of bamboo galls induced by *Aiolomorphus rhopaloides* (Hymenoptera: Eurytomidae) in a bamboo stand. Journal of Forest Research 8:123-126

Shibata E (2005) Oviposition site preference of bamboo gall maker, *Aiolomorphus rhopaloides* (Hymenoptera: Eurytomidae), on bamboo in terms of plant-vigor hypothesis. Applied Entomology and Zoology 40:631-636

Shibata E, Ito M (2005) Life-history traits in insect inclusions associated with bamboo galls. Insect Science 12:143-150

Shibata E, Ito M, Yoshida M (2004) Detection of insect inclusions and size estimation of bamboo galls using soft X-rays. Nagoya University Forest Science 23:15-17

Sitch TA, Grewcock DA, Gilbert FS (1988) Factors affecting components of fitness in a gall-making wasp (*Cynips divisa* Hartig.). Oecologia 76:371-375

Sopow SL, Quiring DT (2001) Is gall size a good indicator of adelgid fitness? Entomologia Experimentalis et Applicata 99:267-271

Speight MR, Hunter MD, Watt AD (1999) Ecology of insects, concept and applications. Blackwell Science, London

Takahashi F, Mizuta K (1971) Life cycles of a *Eurytomid* wasp, *Aiolomorphous rhopaloides*, and three species of wasps parasitic on it (in Japanese with English summary). Japanese Journal of Applied Entomology and Zoology 15:36-43

Yukawa J, Masuda H (1996) Insect and mite galls of Japan in colors (in Japanese with English explanations for color plates). Zenkoku Nôson Kyôiku Kyôkai, Tokyo

Washburn JO, Cornell HV (1981) Parasitoids, patches, and phenology: their possible role in the local extinction of a cynipid gall wasp population. Ecology 63:1602-1605

Weis AE, Abrahamson WG (1985) Potential selective pressures by parasitoids on a plant–herbivore interaction. Ecology 66:1261-1269

Wool D (1977) Genetic and environmental components of morphological variation in gall-forming aphids (Homoptera, Aphididae, Fordinae) in relation to climate. Journal of Animal Ecology 46:875-889

18 Effects of Host-tree Traits on the Species Composition and Density of Galling Insects on two Oak Species, *Quercus crispula* and *Quercus serrata* (Fagaceae)

Noriyuki Ikai and Naoki Hijii

Laboratory of Forest Protection, Graduate School of Bioagricultural Sciences, Nagoya University, Nagoya 464-8601, Japan

Summary. To clarify the factors that determine the community structure of galling insects, we examined the effects of host-tree traits such as phenology and leaf quality on the species composition and densities of galling insects on sympatric oak species, *Quercus crispula* and *Quercus serrata*. Host-tree traits differed between species. In total, we recorded 23 sorts of gall: 18 from *Q. crispula*, 15 from *Q. serrata*, and 10 from both. Species composition and gall densities varied with tree species. There were no significant correlations between bud-burst phenology and the gall density on individual trees in either species for cynipids, but there were significant correlations for cecidomyiids. Synchronization of insect life history with bud-burst phenology is likely to be more critical for cecidomyiids than for cynipids, probably due to the shorter life span of cecidomyiid adults. There were no significant correlations between gall density and water content, total nitrogen concentration, or tannin astringency for any insect. These results suggest that the difference in bud-burst phenology between tree species and the difference in the response to bud-burst phenology among galling insects can cause the observed difference between tree species in the community structure of galling insects.

Key words. Bud-burst phenology, Cynipid, Cecidomyiid, Gall density, Leaf chemistry

18.1 Introduction

Oviposition and survivorship of insect herbivores are mainly governed by host-plant traits such as phenology and foliar quality. Synchronization with host-plant phenology is a critical event for insect herbivores, because a time lag in the synchronization can adversely affect the quality and quantity of available food resources (Yukawa 2000). Plant water, nitrogen, and secondary compounds such as tannins can have positive or negative effects on the oviposition and survivorship of insect herbivores (Slansky and Rodriguez 1987). Because these host-tree traits may differ among tree species, the density of herbivorous insect species may differ among host trees.

The response to these host-plant traits may differ among insect species. Ayres et al. (1997) showed that the effects of the same condensed tannin on the growth and survivorship of larvae of four leaf beetles and two swallowtail butterflies differed among the insect species. In oak trees, on which large numbers of galling insect species have been recorded, a difference in host-tree traits among tree species and a different response to the host-tree traits among galling insects can cause a difference between tree species in the species composition (i.e., one aspect of community structure) of these insects.

The aim of the present study is to clarify the effects of some host-tree traits, such as bud-burst phenology, leaf water content, total nitrogen concentration, and tannin astringency, on species composition and the densities of the galling insects on *Quercus crispula* Blume and *Quercus serrata* Thunb. (Fagaceae).

18.2 Materials and Methods

The study was carried out in the Nagoya University Experimental Forest at Inabu, central Japan (about 1000 m a.s.l.; 35°11'N, 137°33'E). We selected six *Q. crispula* trees and seven *Q. serrata* trees (each 3 to 5 m tall) in the forest for periodic surveys.

We categorized the bud-burst process into 10 stages ranging from "before flushing" (stage 0) to "leaf-opening finished" (stage 9) (Ito and Hijii 2001). We randomly selected four branches in each tree and the shoots from the previous year at the tip of each branch. On 2 May 2002 and 7 May 2003, we determined the leaf stage for current-year shoots on each shoot from the previous year.

To measure the leaf water content, total nitrogen concentration, and tannin astringency, we collected five shoots from the upper branches of each tree on 28 May, 28 July, and 2 October in 2003, weighed and lyophilized (14 hours) one leaf on the tip of each shoot that had little herbivore damage to determine water content, and then measured the total nitrogen concentration for these lyophilized samples with a CN corder (Macro Corder

JM1000CN, J-Science Lab., Kyoto, Japan). We extracted tannins from these lyophilized samples with 70% aqueous acetone, then employed the radial diffusion method to obtain an index of the tannin astringency (Hagerman 1987). We prepared the standard curve using tannic acid (Wako Pure Chemical Industries, Osaka, Japan) and expressed the value of the tannin astringency as the tannic acid equivalent (TAE, mg) per 100 mg of leaf dry mass.

To clarify the insect species composition and gall density, we estimated the numbers of bud galls per shoot and the number of leaf galls per leaf for the four branches on which we measured the bud-burst phenology every week from May to October in 2002 and 2003. We analyzed differences in gall density (i.e., the number of bud galls per shoot or leaf galls per leaf) between tree species by means of nested ANOVA. We treated tree species as a fixed effect, and the individual tree as a random effect.

To detect the effects of host-tree traits on gall density, we tested correlations between gall density and bud-burst stage on each individual tree using Spearman's rank correlation coefficient, and tested correlations between gall density and foliar quality of each individual tree during the periods of gall emergence using Pearson's correlation coefficient. Significance was set at $P < 0.05$. These statistical analyses were performed using the SPSS v. 11.5.1J (SPSS, Chicago, USA).

18.3 Results

Bud burst occurred earlier in *Q. crispula* than in *Q. serrata* for most trees surveyed (data not shown) and bud-burst stages differed most greatly between tree species on 2 May 2002 and 7 May 2003 (Fig. 1). The quality of the leaves tended to differ between the tree species (Fig. 2).

We recorded 23 sorts of gall: 18 from *Q. crispula*, 15 from *Q. serrata*, and 10 from both tree species (Table 1). Cynipids were responsible for 19 sorts of gall versus only 4 for cecidomyiids. In *Q. crispula*, the galls of Cecidomyiidae sp. 1 and sp. 3 were dominant, whereas the cynipid galls formed by sexual and agamic generations of *Aphelonyx glanduliferae* Mukaigawa and the agamic generation of *Trigonaspis* sp. were dominant in *Q. serrata*. There were many galls of *Andricus symbioticus* Kovalev, *Andricus* sp., and Cynipidae sp. 2 in *Q. crispula*, but these galls were excluded from further analysis because they were only observed on one or two trees.

The densities of these dominant cynipid galls did not differ significantly between tree species (Fig. 1, nested ANOVA, $P > 0.05$), whereas the densities of both cecidomyiid galls were much higher on *Q. crispula* than on

Q. serrata in 2002 (nested ANOVA; Cecidomyiidae sp. 1: $F = 5.95$, d.f. = 1, $P < 0.05$; Cecidomyiidae sp. 3: $F = 9.57$, d.f. = 1, $P < 0.01$) and 2003 (Cecidomyiidae sp. 1: $F = 6.91$, d.f. = 1, $P < 0.05$; Cecidomyiidae sp. 3: $F = 3.00$, d.f. = 1, $P = 0.08$).

Table 1. Sorts and numbers of galls on *Q. crispula* (Qc) and *Q. serrata* (Qs)

Species[a]	Generation[b]	Number of galls			
		2002		2003	
		Qc	Qs	Qc	Qs
Cynipidae					
Andricus moriokae	Sexual	20	-	33	-
Andricus mukaigawae	Sexual	6	-	10	-
Andricus symbioticus	Agamic	1	6	-	34
	Sexual	-	9	-	65
Andricus sp. (C-190)	Sexual	163	-	440	-
Aphelonyx glanduliferae	Sexual	30	31	10	3
	Agamic	21	98	115	21
Trigonaspis sp. (C-191?)	Agamic	48	329	120	273
Neuroterus moriokensis	Agamic	-	11	-	-
Cynipidae sp. 1 (C-130)	Agamic	-	-	-	4
Cynipidae sp. 2 (C-136)	Agamic	22	6	368	-
Cynipidae sp. 3 (C-140)	Agamic	-	-	3	-
Cynipidae sp. 4 (C-141)	Thelytoky	14	7	4	-
Cynipidae sp. 5 (C-147)	Thelytoky?	-	4	-	5
Cynipidae sp. 6 (C-186)	Agamic	3	10	-	3
Cynipidae sp. 7*	Agamic	-	1	-	-
Cynipidae sp. 8*	Sexual	-	-	5	-
Cynipidae sp. 9*	Sexual	20	-	-	-
Cynipidae sp. 10*	Agamic	5	-	10	1
Cecidomyiidae					
Cecidomyiidae sp. 1 (C-067)		10971	8	12372	53
Cecidomyiidae sp. 2 (C-189)		7	-	44	-
Cecidomyiidae sp. 3*		532	17	1034	5
Cecidomyiidae sp. 4**		-	-	11	-

[a]Numbers in parentheses are those in Yukawa and Masuda (1996), and only provided for undescribed species.
[b]Provided only for cynipids.
*Undescribed in Yukawa and Masuda (1996).
**Previously known as *Silvestrina quercifoliae* but its generic position is doubtful (Yukawa and Masuda 1996).
- Not observed.

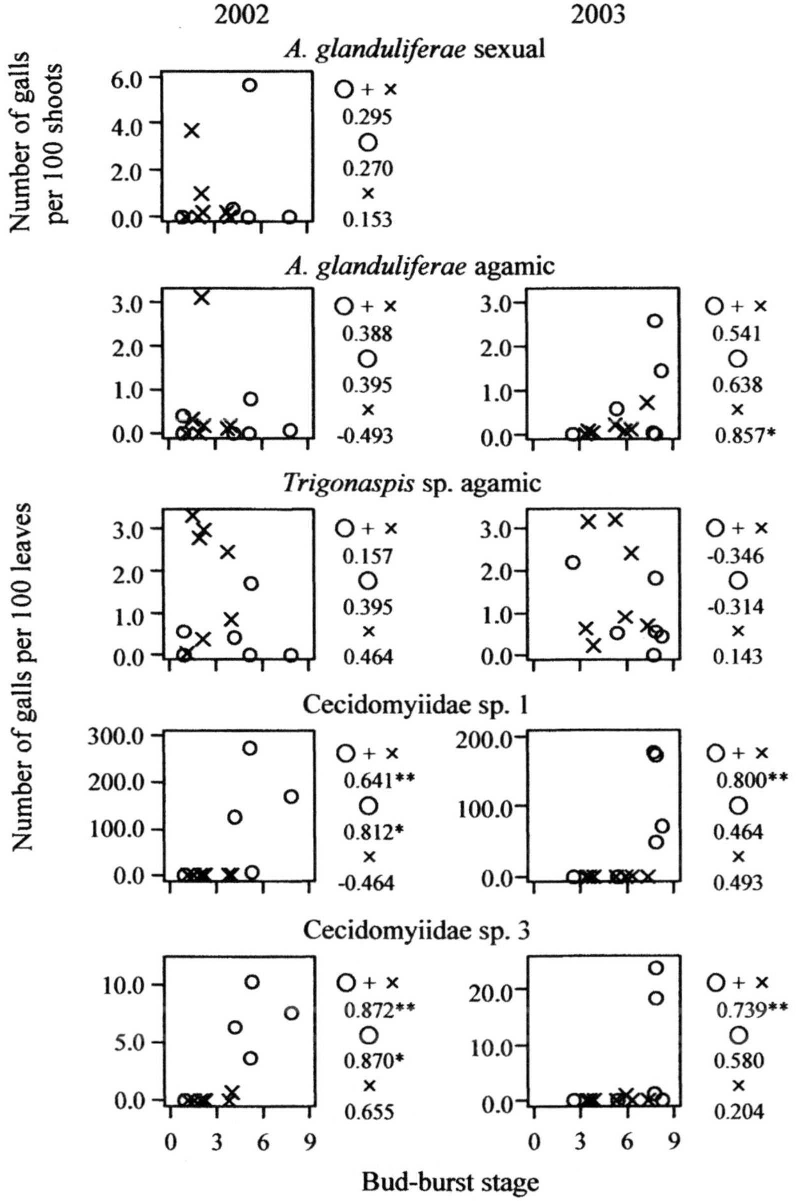

Fig. 1. Gall densities as a function of bud-burst stage in individual trees of *Q. crispula* (○) and *Q. serrata* (×). At the right side of each graph, we have presented the correlations between the gall densities and bud-burst stage of *Q. crispula* and *Q. serrata* combined (○ + ×), *Q. crispula* alone (○), and *Q. serrata* alone (×), with significance tested using Spearman's correlation coefficient (*$P < 0.05$; **$P < 0.01$). In 2003, the sexual generation of *A. glanduliferae* was not analyzed because these galls were rare.

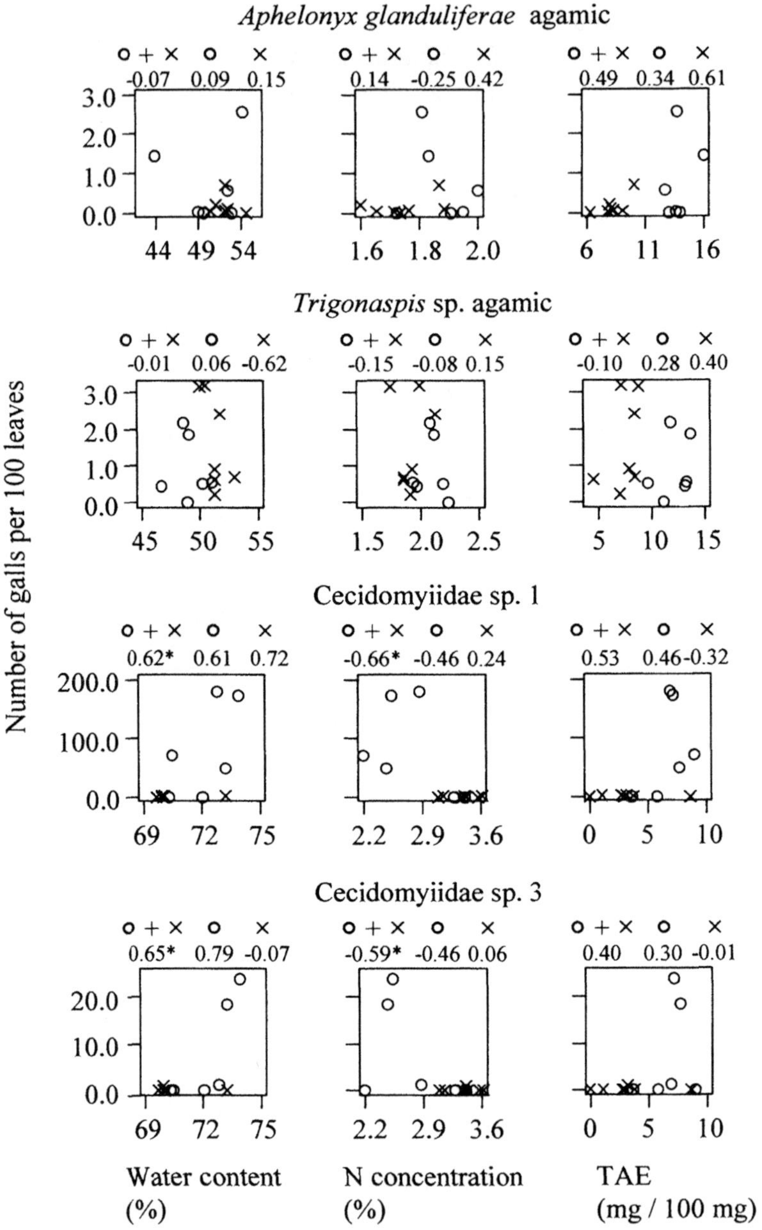

Fig. 2. Gall densities as a function of leaf quality in individual trees of *Q. crispula* (○) and *Q. serrata* (×) during the periods of gall emergence in 2003. Above each graph, we have presented the correlations between the gall densities and each leaf quality parameter for *Q. crispula* and *Q. serrata* combined (○ + ×), *Q. crispula* alone (○), and *Q. serrata* alone (×), with significance tested using Pearson's correlation coefficient (**P* < 0.05).

There were no significant correlations between bud-burst stage and the gall densities of the cynipids (Fig. 1), whereas there were positive correlations between bud-burst stage and the gall densities of cecidomyiids for *Q. crispula* and *Q. serrata* combined and for *Q. crispula* alone in 2002. However, there were no significant correlations observed for either *Q. crispula* alone or *Q. serrata* alone in 2003. This may be because bud-burst phenology did not differ greatly among the individual trees in either tree species. For all the types of galls, there were few significant correlations between gall densities and the quality of the leaves (Fig. 2).

18.4 Discussion

The gall densities of the cynipids were not affected significantly by any of the host-tree traits, thus the gall density may not differ greatly between tree species (Figs. 1, 2). In the cecidomyiids, the gall densities were not affected significantly by leaf quality (Fig. 2), although the gall densities were higher on the trees with an earlier bud-burst phenology in 2002 (Fig. 1). *Q. crispula* had an earlier bud-burst than *Q. serrata*, and the densities of cecidomyiid galls were higher on *Q. crispula* than on *Q. serrata*. These results suggest that differences in the bud-burst phenology between the two tree species and in the response of the insect species to bud-burst phenology are responsible for the difference in the species composition between the two trees (Table 1).

Synchronization of the life history of galling insects with bud-burst phenology is likely to be more critical for cecidomyiids than for cynipids because of the shorter life span of adult cecidomyiids. Adult cynipids live for from 1 week to 1 month (Yukawa and Masuda 1996), and even if they are unable to synchronize their life cycles with host-tree phenology, they are able to survive until resource availability for oviposition increases. In contrast, adults of many cecidomyiids live for 1 or 2 days (Yukawa and Masuda 1996) and can oviposit on leaves only during leaf extension (N. Ikai, unpublished data).

There were few significant relationships between any of the parameters of leaf quality and gall density (Fig. 2). Thus, these parameters are also unlikely to have affected either the oviposition preference of adults or the survivorship of the galling insects before gall induction. Nyman and Julkunen-Tiitto (2000) showed that the composition of phenolic compounds in the leaves of a willow, *Salix reticulata*, differed among the individual trees, whereas the composition in the gall tissue produced by a sawfly, *Pontania reticulatae*, on the willow did not differ significantly among trees. Therefore, the performance of galling insects may not be affected by variations in the quality of the host leaves, and thus the oviposition preference of adult galling insects may also be unaffected by variations in leaf quality, because in many insects there is a linkage between oviposition preference and the performance of their offspring (Price 1997).

18.5 Acknowledgments

We thank Dr. Y. Abe, Kyoto Prefectural University, for identification of the cynipid species. Thanks are also extended to T. Shimada, Forestry and Forest Products Research Institute, for his help in the tannin analysis and to all members of the Laboratory of Forest Protection, Nagoya University, for helpful suggestions.

18.6 References

Ayres MP, Clausen TP, MacLean SF, Redman AM, Reichardt PB (1997) Diversity of structure and antiherbivore activity in condensed tannin. Ecology 78:1696-1712

Hagerman AE (1987) Radial diffusion method for determining tannin in plant extracts. Journal of Chemical Ecology 13:437-449

Ito M, Hijii N (2001) Effect of shoot size and phenological variation of host plants on the spatial patterns of cynipid galls. Journal of Forest Research 6:147-151

Nyman T, Julkunen-Tiitto R (2000) Manipulation of the phenolic chemistry of willows by gall-inducing sawflies. Proceedings of the National Academy of Science, USA 97:13184-13187

Price PW (1997) Insect ecology, 3rd ed. Wiley, New York

Slansky F, Rodriguez JG (1987) Nutritional ecology of insects, mites, spiders, and related invertebrates. Wiley, New York

Yukawa J (2000) Synchronization of gallers with host plant phenology. Population Ecology 42:105-113

Yukawa J, Masuda H (1996) Insect and mite galls of Japan in colors (in Japanese with English explanations for color plates). Zenkoku Nôson Kyôiku Kyôkai, Tokyo

4. Indirect Effects of Galling Arthropods

19 Positive Indirect Effects of Biotic- and Abiotic-mediated Changes in Plant Traits on Herbivory

Masahiro Nakamura

Tomakomai Research Station, Field Science Center for Northern Biosphere, Hokkaido University, Takaoka, Tomakomai 053-0035, Japan

Summary. Plants are subject to damage by a wide variety of threats and to varying degrees of destruction. Sustained damage induces changes in plant traits, and the effects of such damage-induced responses on the abundance and impact of insect herbivores is well documented. Although such plant responses may indirectly influence other insect herbivores either positively or negatively, little is known about positive indirect effects. This chapter focuses on the indirect effects of damage by three different threats: natural disturbances, mammals, and insects. In some cases, damage may have a positive indirect effect on insect herbivores by producing a new source of food, a new microhabitat, and/or changing plant defense chemicals. Therefore, positive indirect effects may potentially enhance the biodiversity of insect communities on terrestrial plants.

Key words. Insects, Mammals, Natural disturbances, Regrowth-mediated, Shelter-mediated

19.1 Introduction

Damage and stress often induce changes in plant traits. The importance of these damage-induced changes on the abundance and impact of insect herbivores is well documented (Faeth 1988, 1991; Ohgushi 2005). For example, insect herbivory increases secondary defense chemicals and/or decreases nutritional quality of host plants (Karban and Myers 1989; Schultz and Baldwin 1982); these changes may prevent further herbivory, resulting in a negative indirect impact of plant trait changes on future insect herbivores (Faeth 1988, 1991).

Plants are subject to damage by a wide variety of threats and to varying degrees of destruction. Insects typically damage plants at the leaf or shoot level, mammals at the branch level (Bryant et al. 1991; Roininen et al. 1994), and natural disturbances (e.g., fire, hurricane, flood) at the tree level (Del Tredici 2001). Such large-scale destruction also induces changes in plant traits (Danell and Huss-Danell 1985; Nakamura et al. 2006; Spiller and Agrawal 2003), which may in turn have indirect effects on insect herbivores. Therefore, both biotic and abiotic damage should be considered potential initiators of indirect effects.

Damage may induce both nutritional and structural changes in plants. In general, the destruction of terminal shoots stimulates growth in other areas of the plant by terminating apical dominance (Mopper et al. 1991); this process, called compensatory regrowth, tends to increase both the structural complexity (Mopper et al. 1991) and nutritional quality (Nakamura et al. 2003; Pilson 1992) of host plants. These changes, in turn, may provide new food and habitat for other herbivorous insects. Although insect herbivory may have either positive or negative indirect impacts on other insect herbivores (reviews in Denno et al. 1995; Faeth 1991; Ohgushi 2005), most studies have focused on the negative effects; thus, little is known about the potential positive indirect effects (Faeth 1991).

This chapter focuses on the indirect effects of damage by three different sources: natural disturbances, mammals, and insects. The effects of the gall midge *Rabdophaga rigidae* on the willow *Salix eriocarpa* is used to illustrate a positive indirect effect of herbivory on other insect herbivores.

19.2 Natural Disturbances

Large-scale destruction by a natural disturbance may stimulate compensatory regrowth in trees and shrubs (Bond and Midgley 2001; Del Tredici 2001) by removing aboveground biomass and a large proportion of apical buds. This trimming or pruning effect terminates apical dominance, which physiologically suppresses shoot growth (Cline et al. 1997), thereby leading to the rapid sprouting of shoots from dormant buds (Bond and Midgley 2001; Del Tredici 2001).

Price (1991) suggested that many herbivorous insects, particularly endophytic species (e.g., gall-producing, mining, and boring insects), prefer fast-growing plants. Many arthropod populations grow quickly due to high reproductive rates and colonization abilities (Spiller et al. 1998). Thus, disturbances that lead to rapid plant growth may have a positive indirect effect on the abundance and impact of herbivorous arthropods. Vieira et al. (1996) demonstrated that many young leaves of *Palicourea rigida* appeared following fire in the Cerrado savanna of Brazil. The regrowth resulted in a larger population of a leaf-galling midge in the genus *Contarinia* (Diptera Cecidomyiidae). Like endophytic insects, free-living insects (e.g., lepidopteran larvae and leaf beetles) respond quickly to plant regrowth following disturbance (Bailey and Whitham 2002; Nakamura et al. 2005; Radho-Toly et al. 2001; Spiller and Agrawal 2003; Stein et al. 1992). For example, Spiller and Agrawal (2003) reported that buttonwood mangrove trees (*Conocarpus erectus*) on an island hit by a hurricane suffered major leaf damage by moth larvae. They also experimentally demonstrated that the leaves of pruned trees on the island had higher nitrogen content and lower toughness and trichome density than the leaves of control trees.

Pioneer woody plants (e.g., eucalypts and willows) typically compensate for damage with vigorous growth (Del Tredici 2001; Price 1991). Therefore, bottom-up forces, such as sprouting following a natural disturbance, on insect communities are more likely to occur in pioneer woody plants.

19.3 Mammals

Any type of physical or biological damage that removes a large proportion of apical buds can stimulate sprouting (Del Tredici 2001). In the boreal forest, winter browsing by mammals stimulates sprouting of woody plants and positively affects insect herbivores (Danell and Huss-Danell 1985; Olofsson and Strengbom 2000; Roininen et al. 1994). For example, Danell and Huss-Danell (1985) reported that sucking, chewing, mining, and galling insects preferred the birch trees *Betula pendula* and *B. pubescens* following browsing by moose (*Alces alces*). The newly sprouted, young plant tissues had relatively low concentrations of defense chemicals and high nutritional quality.

However, mammal browsing does not always lead to high-quality resources. In some cases, a plant may respond to browsing by increasing defense chemicals to repel generalist herbivores (Bryant et al. 1991). At the same time, however, an increase in defense chemicals may benefit some specialist insect herbivores. For example, Martinsen et al. (1998) reported that sprouts from the stumps and roots of beaver-cut cottonwoods (*Populus* sp.) contained higher levels of secondary chemicals (e.g., phenolic glycosides) that repel mammal herbivores. These same chemicals benefited leaf beetles (*Chrysomela confluens*), which sequester the substances for use in their own defense; beetles fed cottonwood sprouts were better defended against predators (e.g., ants) than those fed non-sprout growth. This finding implies that mammal-induced changes in plant traits can have positive indirect effects on specialist insect herbivores by not only increasing the nutritional quality of the host plant but also increasing defense chemicals important to the insect.

19.4 Insects

In response to insect herbivory, plants may increase the secondary chemical content and/or decrease the nutritional quality of new foliage (Karban and Myers 1989; Schultz and Baldwin 1982). Both responses have detrimental effects on the survival and growth of some insect herbivores that appear later in the season (Karban and Myers 1989). However, such insect-induced plant responses also may have positive indirect effects on other insects (reviews in Denno et al. 1995; Faeth 1991; Ohgushi 2005). Here, I introduce two types of such positive effects: shelter-mediated and regrowth-mediated indirect effects.

In addition to food, plants provide insects with habitats that offer protection against natural enemies and/or environmental stress (Hunter and West 1990; Strong et al. 1984). Herbivorous lepidopterans, weevils, sawflies, and aphids construct leaf shelters, such as leaf rolls and galls, on a wide variety of plants, from trees and shrubs to herbs and even ferns (Martinsen et al. 2000). Such leaf shelters may later be reused by other arthropods (review in Fukui 2001). For example, Cappuccino (1993) showed that the leaf shelters constructed by various species of birch-feeding lepidopteran larvae are later colonized by other lepidopterans. Jones et al. (1994) defined "ecosystem engineers" as organisms that directly and indirectly modulate the availability of resources to other species by modifying, maintaining, and creating habitats. Thus, shelter-making insects can be considered ecosystem engineers because they create microhabitats for other insect species (Fukui 2001).

Furthermore, insect herbivory that destroys the apical meristems of shoots may induce regrowth responses in host plants (Mopper et al. 1991; Whitham et al. 1991). Whitham and Mopper (1985) demonstrated that the destruction of a plant's terminal shoots stimulates dormant lateral bud development adjacent to the site of attack in a process called "lateral branching." The regrowth is dependent on the plant species, as well as the timing and type of insect herbivory (Mopper et al. 1991), and may enhance the availability of food resources to other insect herbivores (Craig et al. 1986; Damman 1989; Pilson 1992). Indeed, Pilson (1992) showed that herbivory by stem and rosette gallers, moth caterpillars, and beetles stimulated lateral branching in the goldenrod *Solidago altissima* and that aphids and spittlebugs were more abundant on these plants than control plants.

19.5 Case Study: Galls Indirectly Benefit Other Insects

Nakamura et al. (2003) presented an example of regrowth-mediated interactions between a gall midge *R. rigidae* (Diptera: Cecidomyiidae) and other insect herbivores on a willow (*S. eriocarpa*) located on the Yasu River (35° N, 136° E) in Shiga Prefecture, central Japan. *Salix eriocarpa* is commonly attacked by the gall midge, which induces stem galls on the apical regions of current-year shoots. After gall initiation in mid-May, the plant vigorously develops lateral shoots from the leaf axils below the galls (Fig. 1). The regrowth produces new plant tissues that some insect herbivores depend on for successful development (Feeny 1970; Rausher 1981). Thus, Nakamura et al. (2003) predicted that gall initiation would have a positive indirect effect on the feeding preferences of other insect herbivores via plant regrowth. The study investigated whether gall initiation would induce lateral branching and subsequent leaf flush, how these regrowth responses would affect the nutritional and physical properties of the willow, and how the responses would influence the feeding preferences of the aphid *Aphis farinosa* (Hemiptera: Aphididae) and the leaf beetles *Plagiodera versicolora* (Coleoptera: Chrysomelidae) and *Smaragdina semiaurantiaca* (Coleoptera: Chrysomelidae). These three insect species are frequently observed on *S. eriocarpa* during July.

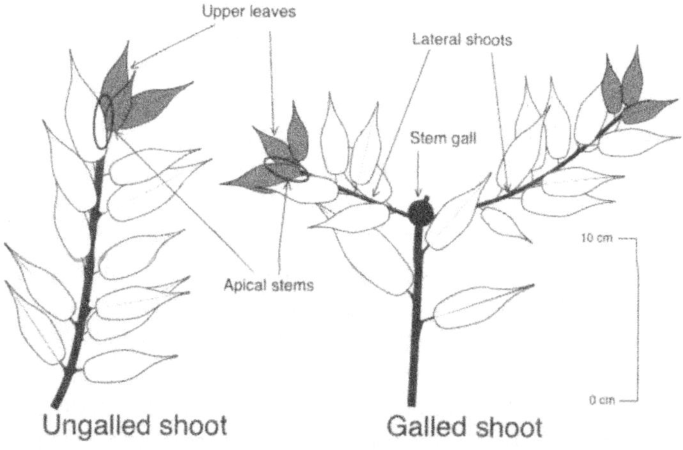

Fig. 1. Illustration showing ungalled and galled shoots of *Salix eriocarpa*. Modified from Nakamura et al. (2003); reprinted by permission of Blackwell Publishing.

Gall midges stimulated the development of lateral shoots in *S. eriocarpa*, and galled shoots produced five and six times more lateral shoots than ungalled shoots in 2000 and 2001, respectively ($P = 0.03$ for 2000; $P < 0.01$ for 2001). Moreover, gall midges affected temporal patterns of leaf flush in *S. eriocarpa*. In mid-June, a secondary leaf flush occurred rapidly on galled shoots and continued until late July. In contrast, the cumulative number of leaves produced on ungalled shoots increased gradually until late July. Consequently, galled shoots produced seven times more newly expanded leaves per week in mid-July than ungalled shoots ($P = 0.04$).

Because aphids colonize apical shoot stems and the two leaf beetles feed on leaves on the upper parts of shoots, nitrogen and water content and tissue toughness of apical stems and upper leaves (Fig. 1) were measured. Apical stems of lateral shoots were more tender and had higher nitrogen and water content than those of ungalled shoots (toughness, $P < 0.01$; nitrogen content, $P < 0.01$; water content, $P < 0.01$). Also, the quality of upper leaves differed significantly. Those on lateral shoots were more tender and had higher nitrogen and water content than those on ungalled shoots (toughness, $P < 0.01$; nitrogen content, $P < 0.01$; water content, $P < 0.01$). The results indicated that gall midges increased the nutritional and physical properties of their host plant by inducing a regrowth response; this finding contradicts most previous studies, which report that insect herbivory decreases the nutritional status of damaged parts in host plants (Karban and Myers 1989; Schultz and Baldwin 1982).

New plant tissues on galled shoots were highly attractive to other insect herbivores. The aphid colonization rate was significantly higher on galled shoots than on ungalled shoots in both 2000 and 2001 ($P < 0.01$ for 2000; $P = 0.03$ for 2001), and aphids preferentially colonized the apical stems of lateral shoots. The difference in the number of adult *P. versicolora* on ungalled and galled shoots was marginally significant in 2000 ($P = 0.08$; Fig. 2a), although no difference was detected in 2001 ($P = 0.20$). In contrast, the number of *S. semiaurantiaca* adults was ten times greater on galled shoots than on ungalled shoots in 2001 ($P < 0.01$; Fig. 2b). Adults of both beetle species aggregated and fed on young leaves produced during the secondary leaf flush of galled shoots. This suggests that the increased water and nitrogen content and tenderness of plant tissues following gall initiation resulted in increased numbers of *A. farinosa*, *P. versicolora*, and *S. semiaurantiaca* on galled shoots.

The regrowth responses of host plants to insect herbivory have long been discounted in studies on indirect effects, probably because of the assumption that host plants respond in a way that limits further herbivory. However, many studies of insect–plant interactions have reported that insect herbivory often stimulates plant regrowth, depending on the plant spe-

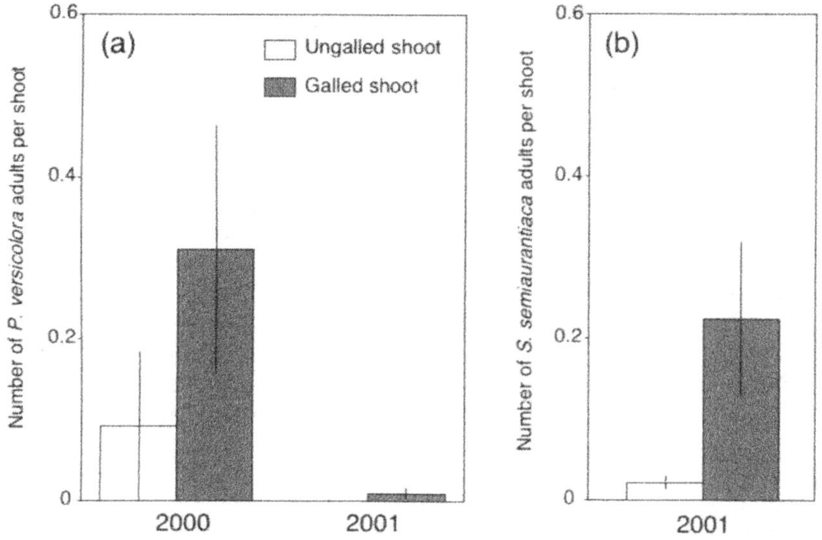

Fig. 2. Number of adult **(a)** *P. versicolora* and **(b)** *S. semiaurantiaca* on ungalled and galled shoots in mid-July (mean ± SE). Modified from Nakamura et al. (2003); reprinted by permission of Blackwell Publishing.

cies as well as the timing and type of herbivory (Mopper et al. 1991; Whitham et al. 1991). This implies that indirect effects of insect herbivory can be either negative or positive (Denno et al. 1995; Faeth 1991; Ohgushi 2005), and that damage on different scales can lead to regrowth-mediated indirect effects.

In conclusion, both biotic and abiotic damage of varying degrees can have positive indirect effects on the abundance and impact of insect herbivores by inducing changes in host plants that provide new food resources, microhabitats, and defense chemicals important to those insect herbivores. In this way, positive indirect effects may potentially enhance the biodiversity of insect communities (e.g., Bailey and Whitham 2002; Martinsen et al. 2000; Nakamura et al. 2006). Therefore, to fully understand the mechanisms that lead to diverse ecological communities, further studies on interactions with positive indirect effects are needed.

19.6 Acknowledgements

I would like to thank T. Hirao for his valuable comments on earlier drafts of this manuscript. This study was partly supported by the Ministry of

Education, Culture, Sports, Science and Technology Grant-in-Aid for Scientific Research (A-15207003) to T. Ohgushi, and the 21st Century COE Program (A14).

19.7 References

Bailey JK, Whitham TG (2002) Interactions among fire, aspen, and elk affect insect diversity: reversal of a community response. Ecology 83:1701-1712

Bond WJ, Midgley JJ (2001) Ecology of sprouting in woody plants: the persistence niche. Trends in Ecology & Evolution 16:45-51.

Bryant JP, Danell K, Provenza F, Reichardt PB, Clausen TA, Werner RA (1991) Effects of mammal browsing on the chemistry of deciduous woody plants. In: Tallamy DW, Raupp MJ (eds) Phytochemical induction by herbivores. John Wiley, New York, pp 135-154

Cappuccino N (1993) Mutual use of leaf-shelters by lepidopteran larvae on paper birch. Ecological Entomology 18:287-292

Cline M, Wessel T, Iwamura H (1997) Cytokinin/auxin control of apical dominance in *Ipomoea nil.* Plant Cell Physiology 38:659-667

Craig TP, Price PW, Itami JK (1986) Resource regulation by a stem-galling sawfly on the arroyo willow. Ecology 67:419-425

Damman H (1989) Facilitative interactions between two lepidopteran herbivores of *Asimina.* Oecologia 78:214-219

Danell K, Huss-Danell K (1985) Feeding by insects and hares on birches earlier affected by moose browsing. Oikos 44:75-81

Del Tredici P (2001) Sprouting in temperate trees: a morphological and ecological review. Botanical Review 67:121-140

Denno RF, McClure MS, Ott JR (1995) Interspecific interactions in phytophagous insects: competition reexamined and resurrected. Annual Review of Entomology 40:297-331

Faeth SH (1988) Plant-mediated interactions between seasonal herbivores: enough for evolution or coevolution? In: Spencer KC (ed) Chemical mediation of coevolution. Academic Press, New York, pp 391-414

Faeth SH (1991) Variable induced responses: direct and indirect effects on oak folivores. In: Tallamy DW, Raupp MJ (eds) Phytochemical induction by herbivores. John Wiley, New York, pp 293-323

Feeny P (1970) Seasonal changes in oak leaf tannins and nutrients as a cause of spring feeding by winter moth caterpillars. Ecology 51:565-581

Fukui A (2001) Indirect interactions mediated by leaf shelters in animal–plant communities. Population Ecology 43:31-40

Hunter MD, West C (1990) Variation in the effects of spring defoliation on the late season phytophagous insects of *Quercus robur*. In: Watt AD, Leather SR, Hunter MD, Kidd NAC (eds) Population dynamics of forest insects. Intercept, Edinburgh, pp 123-135

Jones CG, Lawton JH, Shachak M (1994) Organisms as ecosystem engineers. Oikos 69:373-386

Karban R, Myers JH (1989) Induced plant responses to herbivory. Annual Review of Ecology and Systematics 20:331-348

Martinsen GD, Driebe EM, Whitham TG (1998) Indirect interactions mediated by changing plant chemistry: beaver browsing benefits beetles. Ecology 79:192-200

Martinsen GD, Floate KD, Waltz AM, Wimp GM, Whitham TG (2000) Positive interactions between leafrollers and other arthropods enhance biodiversity on hybrid cottonwoods. Oecologia 123:82-89

Mopper S, Maschinski J, Cobb N, Whitham TG (1991) A new look at habitat structure: consequences of herbivore-modified plant architecture. In: Bell SS, McCoy ED, Mushinsky HR (eds) Habitat structure. Chapman and Hall, London, pp 260-280

Nakamura M, Miyamoto Y, Ohgushi T (2003) Gall initiation enhances the availability of food resources for herbivorous insects. Functional Ecology 17:851-857

Nakamura M, Utsumi S, Miki T, Ohgushi T (2005) Flood initiates bottom-up cascades in a tri-trophic system: host plant regrowth increases densities of a leaf beetle and its predators. Journal of Animal Ecology 74:683-691

Nakamura M, Kagata H, Ohgushi T (2006) Trunk cutting initiates bottom-up cascades in a tri-trophic system: sprouting increases biodiversity of herbivorous and predaceous arthropods on willows. Oikos (in press)

Ohgushi T (2005) Indirect interaction webs: herbivore-induced effects through trait change in plants. Annual Review of Ecology, Evolution and Systematics 36:81-105

Olofsson J, Strengbom J (2000) Response of galling invertebrates on *Salix lanata* to reindeer herbivory. Oikos 91:493-498

Pilson D (1992) Aphid distribution and the evolution of goldenrod resistance. Evolution 46:1358-1372

Price PW (1991) The plant vigor hypothesis and herbivore attack. Oikos 62:244-251

Radho-Toly S, Majer JD, Yates C (2001) Impact of fire on leaf nutrients, arthropod fauna and herbivory of native and exotic eucalypts in Kings Park, Perth, Western Australia. Austral Ecology 26:500-506

Rausher MD (1981) Host plant selection by *Battus philenor* butterflies: the roles of predation, nutrition, and plant chemistry. Ecological Monographs 51:1-20

Roininen H, Price PW, Tahvanainen J (1994) Does the willow bud galler, *Euura mucronata*, benefit from hare browsing on its host plant? In: Price PW, Baranchikov Y, Mattson WJ (eds) The ecology, physiology, and evolution of gall forming insects. General Technical Report NC-174. USDA Forest Service, St Paul, pp 12-26

Schultz JC, Baldwin IT (1982) Oak leaf quality declines in response to defoliation by gypsy moth larvae. Science 217:149-151

Spiller DA, Agrawal AA (2003) Intense disturbance enhances plant susceptibility to herbivory: natural and experimental evidence. Ecology 84:890-897

Spiller DA, Losos JB, Schoener TW (1998) Impact of a catastrophic hurricane on island populations. Science 281:695-697

Stein SJ, Price PW, Abrahamson WG, Sacchi CF (1992) The effect of fire on stimulating willow regrowth and subsequent attack by grasshoppers and elk. Oikos 65:190-196

Strong DR, Lawton JH, Southwood TRE (1984) Insects on plant: community patterns and mechanisms. Blackwell Scientific Publications, Oxford

Vieira EM, Andrade I, Price PW (1996) Fire effects on a *Palicourea rigida* (Rubiaceae) gall midge: a test of the plant vigor hypothesis. Biotropica 28:210-217

Whitham TG, Mopper S (1985) Chronic herbivory: impacts on architecture and sex expression of pinyon pine. Science 228:1089-1091

Whitham TG, Maschinski J, Larson KC, Paige KN (1991) Plant responses to herbivory: the continuum from negative to positive and underlying physiological mechanisms. In: Lewinsohn TM, Fernandes GW, Benson WW, Price PW (eds) Plant-animal interactions: evolutionary ecology in tropical and temperate regions. John Wiley, New York, pp 227-256

20 Deer Browsing on Dwarf Bamboo Affects the Interspecies Relationships among the Parasitoids Associated with a Gall Midge

Akira Ueda[1], Teruaki Hino[2], and Ken Tabuchi[1]

[1]Hokkaido Research Center, Forestry and Forest Products Research Institute, 7 Hitsujigaoka, Toyohira, Sapporo 062-8516, Japan
[2]Kansai Research Center, Forestry and Forest Products Research Institute, 68 Nagaikyutaro, Fushimi, Kyoto 612-0855, Japan

Summary. We found that deer browsing alters the species composition of parasitoids of gall-forming insects via its effects on the host plants. At Mt. Ôdaigahara, in west-central Japan, we compared the species composition of two parasitoid wasps, *Pediobius* sp. (Eulophidae) and *Torymus* sp. (Torymidae), on an unidentified gall midge (tribe Oligotrophini) that forms galls on dwarf bamboo (*Sasa nipponica* Makino et Shibata), the major forage for Sika deer (*Cervus nippon* Temminck). Gall width was larger inside deer exclosures, where the bamboo culms were longer and thicker due to their escape from browsing. The parasitism rate by *Pediobius* sp. was lower inside the exclosures, where parasitism concentrated on the smaller galls. In contrast, the parasitism rate by *Torymus* sp. was higher inside the exclosures and concentrated on the larger galls. *Torymus* sp. emerge earlier and have a longer ovipositor than *Pediobius* sp., thus it should be able to oviposit throughout all gall developmental stages. Because *Torymus* sp. may be hyperparasitized by *Pediobius* sp., more *Torymus* sp. larvae survived in larger galls that *Pediobius* sp. could not penetrate with its shorter ovipositor. Thus, deer browsing indirectly favors *Pediobius* sp. by reducing gall size and thereby improving access to host.

Key words. Deer browsing, Gall midge, Indirect effect, Parasitoid, *Sasa nipponica*

20.1 Introduction

Browsing by mammals can have indirect positive or negative effects on insects (Bailey and Whitham 2002; Baines et al. 1994; Danell and Huss-

Danell 1985; Suominen et al. 1999a, b, 2003). With gall-forming insects, several studies have reported that browsing can increase their number (Danell and Huss-Danell 1985; Olofsson and Strengbom 2000; Roininen et al. 1997). However, no studies have reported the effects of browsing on parasitoids of herbivorous insects, including gall-forming insects.

Forest declines resulting from an overabundance of deer have occurred worldwide in recent decades (Côté et al. 2004). The direct effects of browsing by deer on vegetation and on the performance and nutrient content of plants, as well as the indirect effects on other organisms or soil physical chemistry, have been studied to clarify the influence of deer overabundance (Côté et al. 2004). The browsing pressure sometimes drives small animals to large decrease indirectly (Baines et al. 1994). Sudden forest decline caused by an overabundance of deer in Japan has required managers to construct deer exclosures to protect severely damaged forests without any consideration of the influences of these exclosures on other organisms and on soils. To develop effective methods for facilitating forest regeneration without adversely affecting the diversity of other organisms, we established deer exclosures at Mt. Ôdaigahara, an area in Japan where forest decline has been caused by an overabundance of Sika deer (*Cervus nippon* Temminck). At this site, we have studied the influence of deer browsing on the dynamics of the understory vegetation, invertebrates, abundance of mycorrhizae, and soil physical chemistry (Furusawa et al. 2001, 2003, 2005; Hino et al. 2003; Itô and Hino 2004).

In the present study, we observed the influence of deer browsing on the performance of a gall-forming insect and its parasitoids on *Sasa nipponica* Makino et Shibata, a kind of dwarf bamboo that is the major forage for deer at Mt. Ôdaigahara (Yokoyama et al. 1996). Herein, we describe the indirect effects of deer browsing on the gall-forming insect and the species composition of the parasitoids in the galls that form on the dwarf bamboo. We discuss how the present species composition arises and consider the possibility of large population decreases of small animals as a result of the overabundance of deer.

20.2 Materials and Methods

20.2.1 Study Site

The present study was carried out in a forest at Mt. Ôdaigahara, on the Kii Peninsula of mid-western Japan (34°11'N, 136°06'E, 1540 m a.s.l.). The dominant vegetation is a temperate mixed forest of evergreen coniferous and broad-leaved deciduous trees. *Fagus crenata* Blume was the most dominant in the basal area of live trees (48.2%), followed by *Abies homolepis* Sieb. et Zucc. (23.9%), *Quercus crispula* Blume (12.3%), and *Acer shirasawanum* Koidz. (8.4%) (Itô and Hino 2004). The forest floor was dominated by dwarf bamboo (*Sasa nipponica*), although its biomass was kept at low levels by deer browsing, as this species is a major forage for the deer (Yokoyama and Shibata 1998). Maeji et al. (1999) estimated the population density of deer at 22.3 to 34.1 per km^2 from 1996 to 1997 near our study site.

We established five deer exclosures in November 1996 in a 90 × 150 m study site. Each exclosure size was 10 × 20 m, and the deer fences (2 m tall) were constructed from steel pipe (5 cm in diameter), wire, and polyethylene netting. Then we treated inside the exclosures as the deer exclusion areas and within 5 m outside the exclosures as deer browsing areas.

20.2.2 Biology of the Gall Midge and Its Parasitoids

The unidentified gall midge (tribe Oligotrophini) forms bean-shaped, multilocular galls parallel to the length of the culm by transforming the tissues of the culm at or above the second joint from the top of the bamboo shoot. As many as eight oval larval chambers form in a line, with their long axis parallel to the longitudinal center line of the gall. The gall midges overwinter in the gall as mature larvae or pupae, and adults emerge from mid-May to early June. A few mature larvae observed in August do not emerge in that year as a result of prolonged diapause and may overwinter again. This species of gall midge has only been reported at Mt. Ôdaigahara. Two parasitoid wasps, *Pediobius* sp. (Eulophidae) and *Torymus* sp. (Torymidae), attack the larvae of the gall midge and overwinter in the galls as larvae. *Pediobius* sp. pupates from April to July of the following year and may emerge from mid-June to mid-August. *Torymus* sp. pupates in early June of the following year and may emerge from mid-June to early July.

20.2.3 Data Collection

We collected data on dwarf bamboo size, gall size, and occupancy of the larval chambers so as to clarify the direct effects of deer browsing on the dwarf bamboo, and the indirect effects of browsing on the gall midge and its parasitoids. In late September or early October from 2001 to 2003, we collected about 20 culms per year of dwarf bamboo on which the gall midge had formed galls inside the five exclosures (deer exclusion areas). We also simultaneously collected about 20 culms per year that contained the gall and that were found within 5 m of the exclosures (deer browsing areas).

In 2004, we collected about 10 culms with galls inside and outside each of the five exclosures in early October, because the density of galls was too low for us to find the same number of galls as in previous years. The mean density of the galls outside the exclosures ranged from 13.8 to 20.1 galls/m2 from 2001 to 2003 but averaged only 1.1 galls/m2 in 2004 (A. Ueda, unpublished data). We used the measured culm length and thickness as indicators of the size of the bamboo. We measured the culm thickness at the point of minimum thickness immediately below the second joint from the tip (i.e., immediately below the gall). We also measured the maximum width of the gall and defined this as the gall size.

To determine occupancy of the larval chambers, we cut the galls open to observe their chambers and classified the chambers into four categories: chambers occupied by the gall midge (either a mature larva or pupa), by *Pediobius* sp. (larva), or by *Torymus* sp. (larva), and empty chambers. We could easily distinguish between the larvae of the parasitoids and the mature larvae and pupae of the gall midge because the body color of the gall midges is yellow and that of the parasitoids is white. We could also easily distinguish between the larvae of *Pediobius* sp. and *Torymus* sp. because the former were hairless and glossy, whereas the latter were densely hirsute and less glossy. When we measured the galls in 2004, we also recorded the insect species in each chamber so we could compare the chamber size among the species.

Gall size and wall thickness affect the species composition of the parasitoids, as these factors determine which parasitoids will be capable of successfully attacking the gall-forming insects (Ito and Hijii 2002, 2004; Plantard and Hochberg 1998; Price and Clancy 1986; Schönrogge et al. 1996; Washburn and Cornell 1981). To determine whether this was the case with our study species, we measured the diameter of the larval chambers in the galls collected in 2002 and 2004 and subtracted half of that value from half of the gall width to provide an gall wall thickness (Ito and Hijii 2004).

The period of adult emergence also affects the species composition of the parasitoids, because adults that emerge early can oviposit in smaller, more immature galls whose walls are thinner (Plantard and Hochberg 1998). To determine whether this was the case in our study area, we collected galls on 24 April and 4 June 2003 and stored them in the dark at 23°C. We counted the number of adult gall midges and parasitoids that emerged at intervals of 1 to 5 days. Ovipositor length also affects the species composition of the parasitoids, since a long ovipositor lets the parasitoid oviposit even on hosts protected by thicker gall walls (Ito and Hijii 2004; Plantard and Hochberg 1998). We measured the ovipositor length of emerged adult female parasitoids. We also measured their body length, excluding the ovipositor, to calculate an ovipositor length index (the ratio of ovipositor length to body length).

20.2.4 Data Analysis

We performed our statistical analyses with the JMP (ver. 5.1.1) software (SAS Institute 2004). To examine the effects of deer browsing on the bamboo size (culm length and thickness), the gall width, and the gall wall thickness, we calculated the averages for these measurements respectively on the outside and inside of a exclosure at each plot in a year. Then, we compared the averages between outside and inside of the exclosures in respective years using Wilcoxon signed-rank test. We tested the relationships between bamboo size (culm length and thickness) and gall wall thickness using Pearson's correlation coefficient (r). To examine the effects of deer browsing on occupancy of the larval chambers, we used the Wald test in logistic regression analysis for the number of chambers with each type of insect and for empty chambers between years and treatments (with and without deer browsing). To clarify the relationships between gall size and whether the galls harbored each type of insect or empty chambers, we compared the mean gall widths using the Tukey–Kramer HSD test with the combined data of four years. We used the Kruskal–Wallis test to compare chamber diameter and gall wall thickness for chambers occupied by each type of insect and for empty chambers with the data in 2004 (We could not analyze the data in 2002, because we measured the chamber diameter without considering the tenant of each chamber). If a significant difference was found, we compared the data using the Tukey–Kramer HSD test. We also used the Kruskal–Wallis test to compare both ovipositor length and ovipositor length index (the ratio of ovipositor length to body length) between *Pediobius* sp. and *Torymus* sp.

20.3 Results

20.3.1 Effect of Deer Browsing on Dwarf Bamboo, Gall Midges, and Parasitoids

Deer browsing significantly reduced all measured values for culm length, culm thickness, gall width, chamber diameter, and the index of gall wall thickness (Fig. 1). There were significant relationships between bamboo size and gall wall thickness (culm length vs. index of gall wall thickness: r

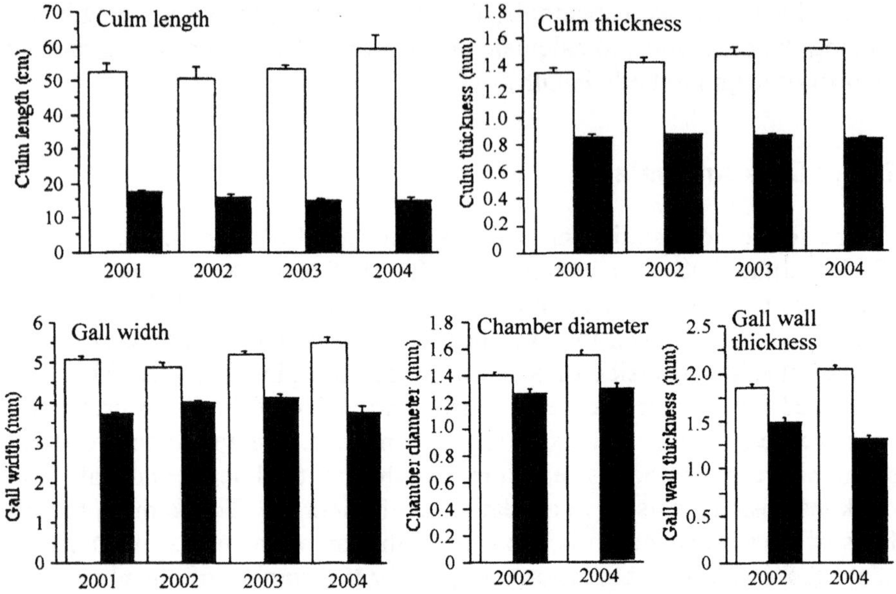

Fig. 1. Means of culm length and thickness of dwarf bamboo, gall width, chamber diameter, and gall wall thickness inside the deer exclosures (white bars) and outside the exclosures (black bars) each year. Vertical range lines represent standard errors. All measurements differed significantly between inside and outside of the exclosures in respective years with the same value of Wilcoxon signed-ranks test (P = 0.031, n = 5). The culm thickness was measured immediately below the bamboo joint where the gall formed. The gall wall thickness is the value of subtracted half of chamber diameter from half of the gall width

Table 1. Significance level (P value) in the Wald test in logistic regression analysis for chambers occupied by each type of insect and for empty chambers

	Gall midge	*Pediobius* sp.	*Torymus* sp.	Empty chamber
Year	<0.0001	0.0001	0.052	0.40
Deer	<0.0001	<0.0001	0.0002	0.36
Year × Deer	0.0048	0.0019	0.21	0.085

= 0.597, n = 428, P < 0.0001; culm thickness *vs* index of gall wall thickness: r = 0.662, n = 609, P < 0.0001).

Deer browsing significantly reduced the proportions of the chambers occupied by the gall midge and by *Torymus* sp. in every year (Fig. 2, Table 1). In contrast, deer browsing significantly increased the proportion of the chambers occupied by *Pediobius* sp. in every year (Fig. 2, Table 1). No

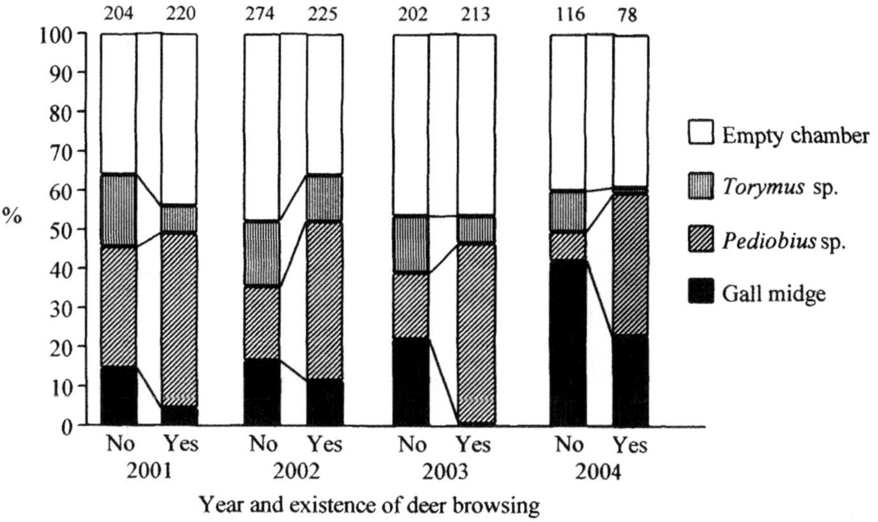

Fig. 2. Percentages of chambers occupied by each type of insect and of empty chambers under deer browsing *(yes)* and deer exclusion *(no)* from 2001 to 2004. *Numbers above the bars* represent the number of chambers observed.

significant or consistent pattern was seen in the proportion of empty chambers (Fig. 2, Table 1).

Inside the exclosures, galls that harbored gall midges or *Torymus* sp. were significantly larger than those that harbored *Pediobius* sp. and those with empty chambers (Fig. 3). The width of galls that harbored *Pediobius* sp. was also significantly smaller then that of galls with empty chambers (Fig. 3). The chamber diameter did not differ significantly among the chamber categories ($P = 0.63$). Gall wall thickness differed significantly among these categories ($P = 0.0005$), with the thickness for *Torymus* sp. larger than those for *Pediobius* sp. and for empty chambers (Fig. 3). Thus, the thickness for chambers occupied by the gall midge and for empty chambers did not differ significantly, but both were significantly larger than that for chambers occupied by *Pediobius* sp. (Fig. 3).

Outside the exclosures, galls that harbored gall midges were significantly larger than galls that harbored *Pediobius* sp. and galls with empty chambers, but galls that harbored *Pediobius* sp. did not differ significantly in width from those harboring *Torymus* sp. and those with empty chambers (Fig. 3). Neither the chamber diameter ($P = 0.90$) nor gall wall thickness ($P = 0.20$) differed significantly among the chamber categories (Fig. 3).

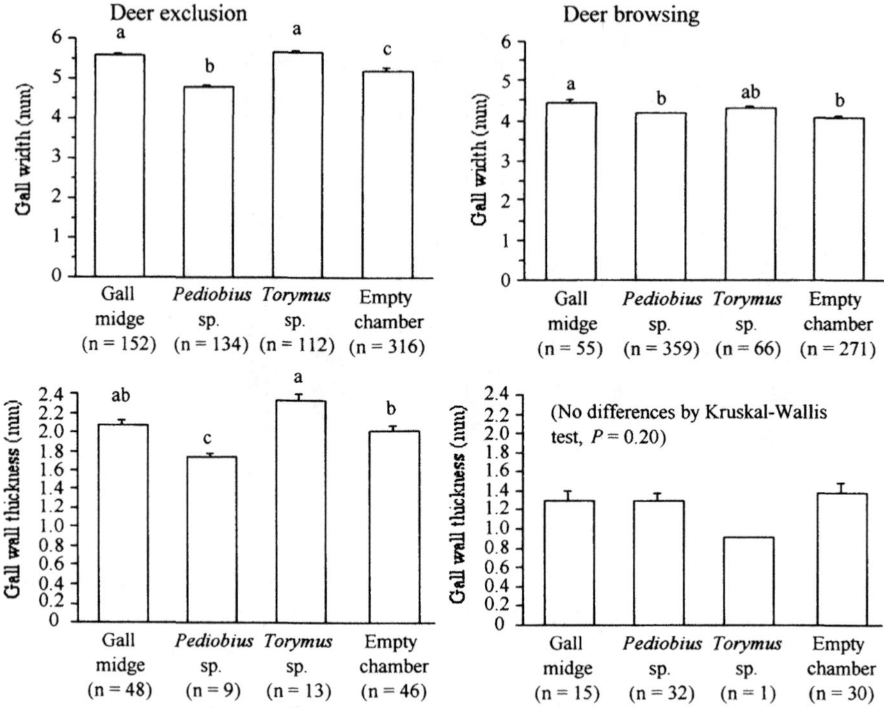

Fig. 3. Mean gall width (upper panels) and mean gall wall thickness (lower panels) inside the deer exclosures (left panels) and outside the exclosures (right panels). Vertical range lines represent standard errors. Bars within a graph labeled with the same letter do not differ significantly (Tukey–Kramer HSD test, $P > 0.05$).

20.3.2 Period of Adult Emergence and Ovipositor Length of Parasitoids

Adult gall midges emerged from the galls earlier than adult parasitoids (Fig. 4). *Pediobius* sp. and *Torymus* sp. began to emerge simultaneously about 10 days after the gall midges began to emerge, but *Torymus* sp. completed its emergence within 20 days while *Pediobius* sp. prolonged its emergence for 60 days (Fig. 4). There were no differences between male and female parasitoids in terms of emergence timing, so the data in Fig. 4 represent the results for both sexes combined. The ovipositor was significantly longer on *Torymus* sp. than on *Pediobius* sp. (Fig. 5), and the ovipositor length index (the ratio of ovipositor length to body length) was also significantly larger for *Torymus* sp. than for *Pediobius* sp. (Fig. 5).

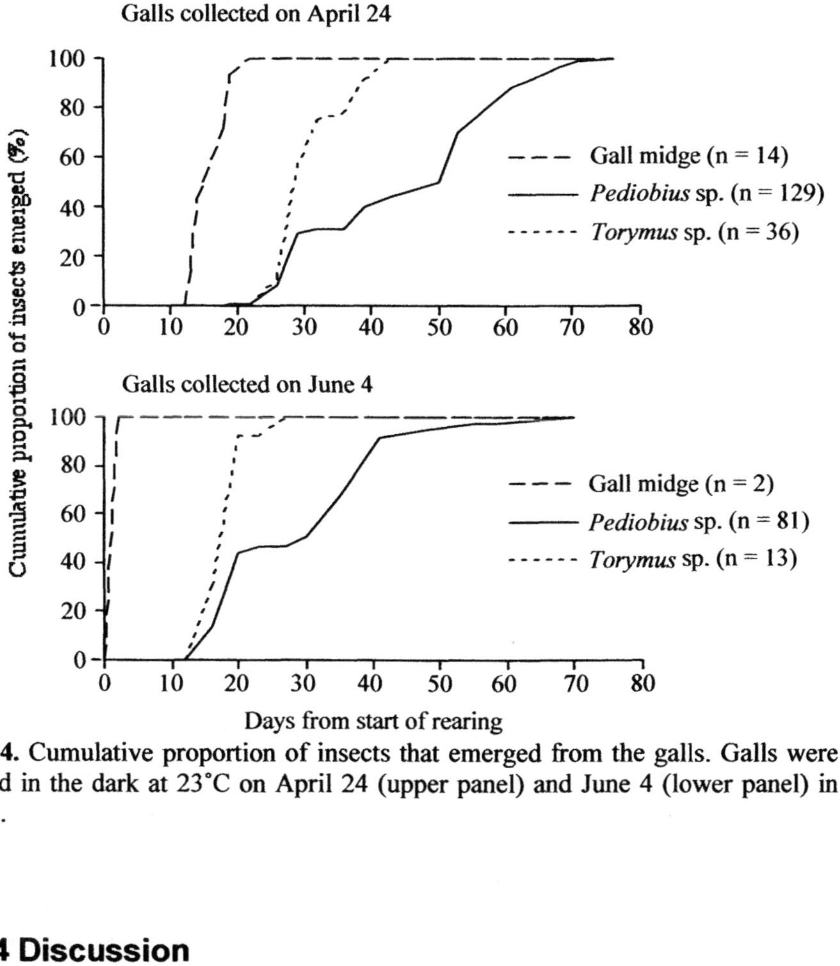

Fig. 4. Cumulative proportion of insects that emerged from the galls. Galls were stored in the dark at 23°C on April 24 (upper panel) and June 4 (lower panel) in 2003.

20.4 Discussion

Deer browsing directly reduced the size of the dwarf bam boo and indirectly reduced gall width and gall wall thickness. Moreover, deer browsing increased the rate of attack by *Pediobius* sp., and reduced both the survival rate of the gall midge and the rate of attack by *Torymus* sp. Inside the exclosures, where no deer browsing occurred, *Pediobius* sp. was concentrated in smaller galls. This can be explained by the interaction between the parasitoid's shorter ovipositor (compared with that of *Torymus*) and the thickness of the gall wall (Ito and Hijii 2004; Plantard and Hochberg 1998).

The more concentrated attack by *Pediobius* sp. in the smaller galls may result from both the prolonged emergence of the adult parasitoids and their short ovipositors. *Pediobius* sp. adults that emerge late may be unable to

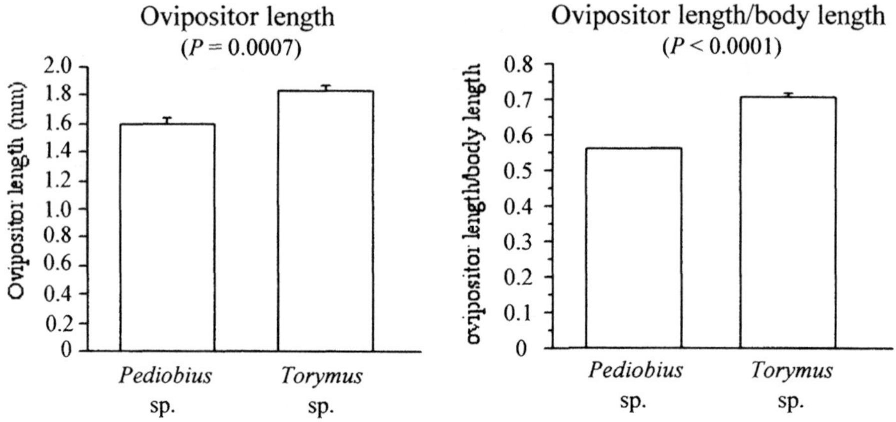

Fig. 5. Mean ovipositor length (left) and mean value of the ovipositor length index (the ratio of ovipositor length to body length) for *Pediobius* sp. (n = 20) and *Torymus* sp. (n = 16). Vertical range lines represent standard errors. The significance levels in the Kruskal–Wallis test are presented above the bars.

successfully oviposit in galls that have already developed thick walls. As a result, successful oviposition by *Pediobius* sp. may be limited to small galls with thinner walls. During the period of emergence of *Torymus* sp., the galls are small and the species that has a relatively long ovipositor may be able to oviposit in gall midges in galls of all sizes, but in small galls they may be hyperparasitized by *Pediobius* sp. This conclusion is supported by the lack of a significant difference in gall width or gall wall thickness between chambers occupied by gall midges and those occupied by *Torymus* sp.

Outside the exclosures, where deer browsing occurred, *Pediobius* sp. was able to oviposit in galls of almost all sizes because deer browsing reduced gall size. As a result, the rate of attack by *Pediobius* sp. increased and the survival rate of the gall midge decreased. Because the rate of attack by *Torymus* sp. decreased outside the exclosures, this species may have been hyperparasitized by *Pediobius* sp. even if *Torymus* sp. was able to oviposit in galls of all sizes, at rates similar to those inside the exclosures. As a result, the net rate of attack by *Torymus* sp. decreased.

Plants and animals at Mt. Ôdaigahara are currently exposed to overpopulation of deer (Maeji et al. 1999; Yokoyama and Shibata 1998). The present study showed that this overpopulation could greatly decrease *Torymus* sp. Currently, many areas of Japan have been exposed to overpopulations of deer, and the resultant browsing pressure may drive small animals other than the parasitoids in the present study to suffer large population decreases. To identify and mitigate population decreases, we must carefully monitor the population dynamics of small animals in areas with high deer populations.

20.5 Acknowledgments

We thank K. Kamijo (Bibai City) and S. Sato (Kyushu University) for their identification of the parasitoid and gall midge species. We are grateful to H. Furusawa, Y. Takahata, T. Shimada, S. Chikaguchi, and S. Narayama of the Kansai Research Center, Forestry and Forest Products Research Institute, for their support; to the staff of the Mt. Ôdaigahara Visitors' Center and to S. and M. Tagaito from the Ôdai Shrine for their kind cooperation. Thanks are also due to E. Shibata (Nagoya University), T. Nakashizuka (Research Institute for Human and Nature), and M. Ito (Hokkaido Research Center, Forestry and Forest Products Research Institute) for their useful advices. This study was supported by a grant from the Ministry of the Environment of Japan (Environmental Research by National Research Institutes of Government Ministries and Agencies) and by a Grant-in-Aid for Scientific Research from the Ministry of Education, Culture, Sports, Science and Technology, Japan (No. 14206019).

20.6 References

Bailey JK, Whitham TG (2002) Interactions among fire, aspen, and elk affect insect diversity: reversal of a community response. Ecology:1701-1712

Baines D, Sage RB, Baines MM (1994) The implications of red deer grazing to ground vegetation and invertebrate communities of Scottish native pinewoods. Journal of Applied Ecology 31:776-783

Côte SD, Rooney TP, Tremblay JP, Dussault C, Waller DM (2004) Ecological impacts of deer overabundance. Annual Review of Ecology, Evolution, and Systematics 35:113-147

Danell K, Huss-Danell K (1985) Feeding by insects and hares on birches earlier affected by moose browsing. Oikos 44:75-81

Furusawa H, Araki M, Hino T (2001) Effects of sika deer and sasa on water potential in surface soil – A case study at Ohdaigahara. Applied Forest Science 10(1):31-36 (in Japanese with English abstract)

Furusawa H, Miyanishi H, Kaneko S, Hino T (2003) Movement of soil and litter on the floor of a temperate mixed forest with an impoverished understory grazed by deer (*Cervus nippon centralis* Temminck). Journal of the Japanese Forest Society 85:318-325 (in Japanese with English abstract)

Furusawa H, Hino T, Kaneko S, Araki M (2005) Effects of dwarf bamboo (*Sasa nipponica*) and deer (*Cervus nippon centralis*) on the chemical properties of soil and microbial biomass in a forest at Ohdaigahara, central Japan. Bulletin of the Forestry and Forest Products Research Institute 4:157-165

Hino T, Furusawa H, Ito H, Ueda A, Takahata Y, Ito M (2003) Forest ecosystem management based on an interaction network in Ohdaigahara. Japanese Journal of Conservation Ecology 8: 145-158 (in Japanese with English abstract)

Itô H, Hino T (2004) Effects of deer, mice and dwarf bamboo on the emergence, survival and growth of *Abies homolepis* (Piceaceae) seedlings. Ecological Research 19:217-223

Ito M, Hijii N (2002) Factors affecting refuge from parasitoid attack in a cynipid wasp, *Aphelonyx glanduliferae*. Population Ecology 44:23-32

Ito M, Hijii N (2004) Relationships among abundance of galls, survivorship, and mortality factors in a cynipid wasp, *Andricus moriokae* (Hymenoptera: Cynipidae). Journal of Forest Research 9:355-359

Maeji I, Yokoyama S, Shibata E (1999) Population density and range use of sika deer, *Cervus nippon*, on Mt. Ohdaigahara, central Japan. Journal of Forest Research 4:235-239

Olofsson J, Strengbom J (2000) Response of galling invertebrates on *Salix lanata* to reindeer herbivory. OIKOS 91:493-498

Plantard O, Hochberg ME (1998) Factors affecting parasitism in the oak-galler *Neuroterus quercusbaccarum* (Hymenoptera: Cynipidae). Oikos 81:289-298

Price PW, Clancy KM (1986) Interactions among three trophic levels: gall size and parasitoid attack. Ecology 67:1593-1600

Roininen H, Price PW, Bryant JP (1997) Response of galling insects to natural browsing by mammals in Alaska. Oikos 80:481-486

SAS Institute (2004) JMP software ver. 5.1.1. SAS Institute, Cary

Schönrogge K, Stone GN, Crawley MJ (1996) Abundance patterns and species richness of the parasitoids and inquilines of the alien gall-former *Andricus quercuscalicis* (Hymenoptera: Cynipidae). Oikos 77:507-518

Suominen O, Danell K, Bergström R (1999a) Moose, trees and ground-living invertebrates: indirect interactions in Swedish pine forests. Oikos 84:215-226

Suominen O, Danell K, Bryant JP (1999b) Indirect effects of mammalian browsers on vegetation and ground-dwelling insects in an Alaskan floodplain. Ecoscience 6:505-510

Suominen O, Niemelä J, Martikainen P, Niemelä P, Kojola I (2003) Impact of reindeer grazing on ground-dwelling Carabidae and Curculionidae assemblages in Lapland. Ecography 26:503-513

Yokoyama S, Shibata E (1998) The effects of sika-deer browsing on the biomass and morphology of a dwarf bamboo, *Sasa nipponica*, in Mt. Ohdaigahara, central Japan. Forest Ecology and Management 103:49-56

Yokoyama S, Koizumi T, Shibata E (1996) Food habits of sika deer as assessed by fecal analysis in Mt. Ohdaigahara, central Japan. Journal of Forest Research 1:161-164

Washburn JO, Cornell HV (1981) Parasitoids, patches, and phenology: their possible role in the local extinction of a cynipid gall wasp population. Ecology 62:1597-1607

21 Influence of the Population Dynamics of a Gall-inducing Cecidomyiid and Its Parasitoids on the Abundance of a Successor, *Lasioptera yadokariae* (Diptera: Cecidomyiidae)

Junichi Yukawa[1], Shigekazu Haitsuka[2], Katsuhiko Miyaji[3], and Takahiro Kamikado[4]

[1]Kyushu University, Fukuoka 812-8581, Japan
[2]Saga Prefectural Agriculture Research Center, Saga 840-2205, Japan
[3]Agricultural Management Division, Kagoshima Prefectural Agricultural Experiment Station, Kagoshima 891-0116, Japan
[4]Kagoshima Prefectural Plant Protection Office, Kagoshima 891-0116, Japan

Summary. *Lasioptera yadokariae* (Diptera: Cecidomyiidae) is a successor in galls induced by three gall midge species. Its larvae inhabit vacated leaf galls after the primary gall inducers and their parasitoids have departed. The successor is fundamentally univoltine and its larvae feed on fungal mycelium within the galls. From 1973 to 1991, we surveyed the population dynamics and the emergence season of *Pseudasphondylia neolitseae*, which is one of the three gall midge species, and its parasitoids, *Bracon tamabae* (Braconidae) and *Gastrancistrus* sp. (Pteromalidae) in southern Kyushu, Japan, in order to assess the number of vacated galls that are available for the successor. Then, the proportion of vacated galls utilized by the successor was evaluated. Our data suggest that when the density of *P. neolitseae* or *B. tamabae* is high, *L. yadokariae* increases its population number by utilizing the abundant vacated galls after their emergence, whereas the high density of *Gastrancistrus* sp. does not contribute to an increase of the successor because of its delayed emergence. The high density of *B. tamabae*, however, would decrease the number of *L. yadokariae* in the following generation, because the high percentage parasitism will reduce the density of *P. neolitseae*, resulting in a shortage of vacated galls.

Key words. *Lasioptera yadokariae*, Population dynamics, Vacated gall, Successor, Cecidomyiidae

21.1 Introduction

Successors inhabit galls after gall-inducing organisms and their associates, including parasitoids, predators, and inquilines, have departed (Mani 1964, as 'successori'). As the galls usually do not decay for some time after the departure of these organisms, the vacant galls provide shelter to a variety of other arthropods such as ants, aphids, thrips, psocids, and mites, and become a suitable substrate for fungi to grow (Mani 1964; Yukawa and Rohfritsch 2005).

Lasioptera yadokariae Yukawa and Haitsuka, 1994 (Diptera: Cecidomyiidae) is the only known Japanese gall midge that is specialized to live in leaf galls after the gall-inducing cecidomyiids have departed. It is known to follow univoltine cecidomyiids such as *Pseudasphondylia neolitseae* Yukawa on *Neolitsea sericea* (Blume) Koidz. (Lauraceae) (Fig. 1), *Daphnephila* sp. on *Machilus japonica* Sieb. and Zucc. (Lauraceae), and *Masakimyia pustulae* Yukawa and Sunose on *Euonymus japonicus* Thunb. (Celastraceae) (Yukawa and Haitsuka 1994). *L. yadokariae* is fundamentally univoltine and in April and early May the females lay their eggs in vacated galls from which the gall-inducing cecidomyiids and their parasitoids have departed. Larvae of *L. yadokariae* live in the galls, feeding on the fungal mycelium of the genus *Pestalotia* (Fungi Imperfecti: Melanoconiales: Melanoconiaceae) that grows in the vacated galls (Yukawa and Haitsuka 1994).

The density of gall-inducing species and their parasitoids, which fluctuates independently of the density of the successor, is considered to influence the abundance and survival of *L. yadokariae*. In addition, synchronization of the emergence season of *L. yadokariae* adults with that of gall-inducing cecidomyiids and their parasitoids is an important factor in determining the density of the successor because its adult life span is very short and the females choose fresh vacant galls for oviposition (Yukawa and Haitsuka 1994). In this study, we surveyed the population dynamics and the emergence season of *P. neolitseae* and its parasitoids, *Bracon tamabae* Maetô (Braconidae) and *Gastrancistrus* sp. (Pteromalidae), to assess the number of vacated galls from which they had emerged. Then, the proportion of vacated galls utilized by *L. yadokariae* was evaluated. Based on these data, we discuss the influence of the gall midge and parasitoids on the abundance of the successor.

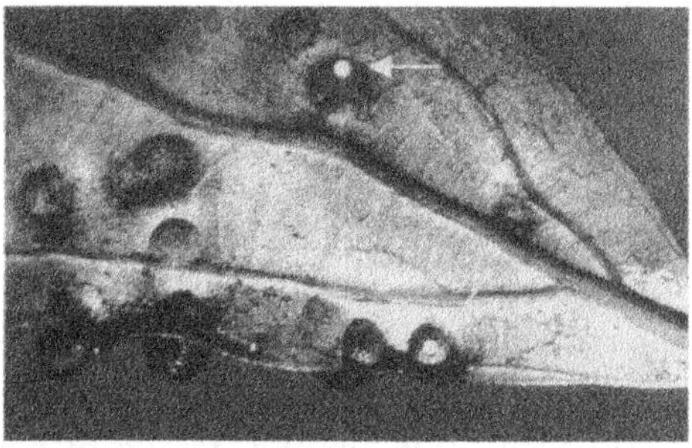

Fig. 1. Leaf galls induced by *P. neolitseae* on *N. sericea.* Galls inhabited by larvae of a successor, *L. yadokariae*, are recognized by the presence of white mycelia plugging exit holes of the gall midge. Arrows point toward the galls inhabited by *L. yadokariae* larvae.

21.2 Materials and Methods

21.2.1 Insects Studied

Successor: *L. yadokariae* utilizes leaf galls of the aforementioned three species of gall midges (Yukawa and Haitsuka 1994), but we investigated galls of only *P. neolitseae*, because the two others did not coexist in the census field. In southern Kyushu, Japan, the larvae mature and quit the galls to drop to the ground by mid May, pass through the summer, autumn, and winter in the soil, and pupate in the following spring. In addition to most univoltine individuals, some individuals seem to require two years to complete one generation (Yukawa and Haitsuka 1994).

Gall-inducing cecidomyiid: *P. neolitseae* is monophagous and fundamentally univoltine (Yukawa 1974; Yukawa et al. 1976). In southern Kyushu, Japan, larvae pass through summer as first instars, molt into second instars in October, and overwinter as full-grown larvae in the leaf galls on the host plant. Pupation takes place in February or March. Adults emerge in late March or April. Some individuals are known to require two years to complete one generation (Takasu and Yukawa 1984).

Parasitoids: *B. tamabae* (previously identified as *Ipobracon scurra*) and *Gastrancistrus* sp. attack *P. neolitseae* larvae (Yukawa 1983). *B. tamabae* is a multivoltine and polyphagous ectoparasitoid, attacking full-grown larvae or pupae of host gall midges. From *P. neolitseae* galls, adults usually emerge during the period from March to April, occasionally to May. *Gastrancistrus* sp. is a univoltine and possibly monophagous endoparasitoid of

P. neolitseae. In May, the females lay their eggs into the first instars and their larvae do not develop until the host larvae become full-grown in the following spring. Adults emerge from the midge galls in April and May.

21.2.2 Fieldwork

We conducted fieldwork from 1973 to 1991 at Mt. Shiroyama, Kagoshima City, Kyushu, Japan (31°35′47″N, 130°30′00″E, about 100 m asl). In 1970, we selected 100 saplings (1–3 m in height) of *N. sericea* for field census (see Yukawa and Akimoto 2006 for the details of census saplings). The daily emergence of adult gall midges and parasitoids was surveyed every day from early March to late May by recording the number of exit holes produced by them on the under surface of galls. The exit hole of *P. neolitseae* was easily distinguished from those of the parasitoids as it retains a pupal case of the gall midge for a few days (Fig. 2a). *B. tamabae* produced smaller exit holes than *Gastrancistrus* sp. (Fig. 2b, c). After recording the number, emergence holes were marked with a felt pen to avoid double counting.

We could not record the daily number of *L. yadokariae* adults that emerged from the ground. Instead, we counted galls inhabited by the successor at intervals of three to four days by examining the presence of white fungal mycelia plugging the exit holes of vacated galls (Fig. 1). Because the mycelia appeared several days after oviposition by the females, we could roughly determine the time of *L. yadokariae* emergence.

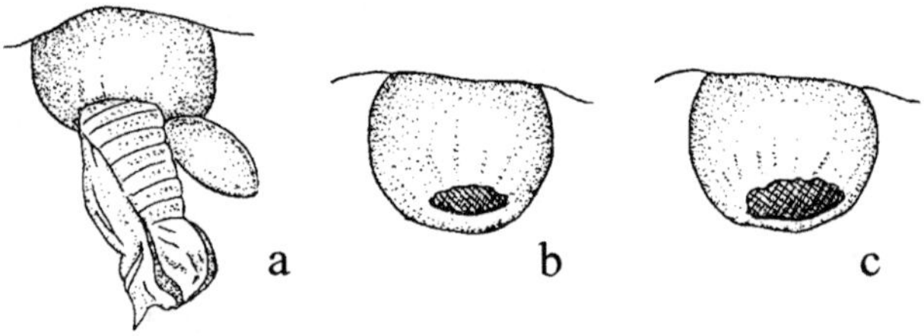

Fig. 2. Three different kinds of vacated *P. neolitseae* galls. **a** Vacated gall with pupal exuviae of *P. neolitseae* and trapdoor, which dehisce within several days. **b** Vacated gall from which *B. tamabae* emerged. **c** Vacated gall from which *Gastrancistrus* sp. emerged. The last two vacated galls are distinguishable by the shape and diameter of exit hole.

21.3 Results

We gathered 16-year data for the population fluctuation of the gall midge and its parasitoids at the time of emergence (Fig. 3). The data indicate that the gall midge has been parasitized more severely by *B. tamabae* than *Gastrancistrus* sp. The figure also shows that the numbers of vacated galls after the emergence of *P. neolitseae*, *B. tamabae*, and *Gastrancistrus* sp. fluctuated largely from year to year and vacated galls from which *B. tamabae* had emerged were most abundant throughout the survey period except in 1983 and 1985. The accumulated number of vacated galls from which *B. tamabae* had emerged for 16 years was 3022, and followed by 887 of the vacated galls after the emergence of *P. neolitseae*. The accumulated number of vacated galls from which *Gastrancistrus* sp. had emerged was 495, fewest among the three kinds.

Fig. 4 shows the annual changes in the total number of three kinds of vacated galls and the number of vacated galls utilized by *L. yadokariae*, of which the latter represents the population dynamics of the successor. It fluctuated from 9 in 1979 up to 108 in 1989. The proportion of vacated galls utilized by *L. yadokariae* fluctuated from year to year between 3.1% in 1985 and 36.0% in 1987. Because the proportion varied yearly among the three kinds of vacated galls, we compared the proportions based on the mean percentage of 16-year data (Table 1). The vacated galls from which *B. tamabae* had emerged were more frequently utilized by *L. yadokariae* than vacated galls from which *P. neolitseae* or *Gastrancistrus* sp. had

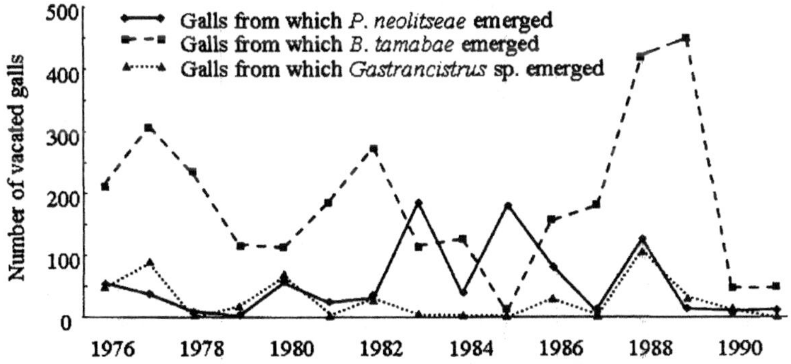

Fig. 3. Population dynamics of *P. neolitseae* and its two parasitoid species, *B. tamabae* and *Gastrancistrus* sp. at the time of emergence, indicating also annual changes in the number of vacated galls available for *L. yadokariae* larvae to inhabit.

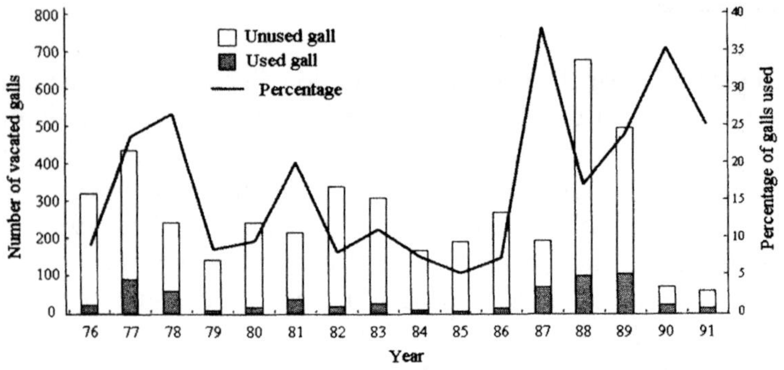

Fig. 4. Annual changes in the numbers of unused and used vacated galls and the proportion of vacated galls used by *L. yadokariae.*

Table 1. Comparison between vacated galls from which *P. neolitseae*, *B. tamabae*, and *Gastrancistrus* sp. emerged in the proportion of galls used by *L. yadokariae*

Gall midge and parasitoids departed from the galls	Vacated galls available	Vacated galls used	%*
Pseudasphondylia neolitseae	887	67	7.6 [a]
Bracon tamabae	3022	532	17.6 [b]
Gastrancistrus sp.	495	40	8.1 [a]
Total	4404	639	14.5

* The mean percentage of 16-year data combined together. Different letters in this column indicate a significant difference ($\chi^2 > \chi^2_{0.001}$).

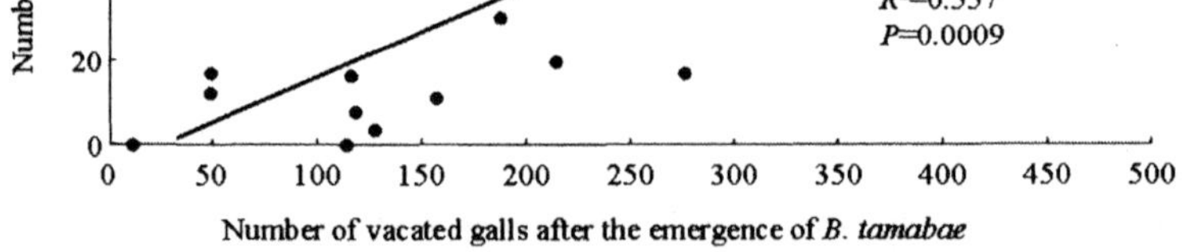

Fig. 5. Relationship between the number of vacated galls after the emergence of *B. tamabae* and the number of galls utilized by *L. yadokariae.*

emerged. The number of *L. yadokariae* larvae increased when the number of vacated galls increased. The relationship was most clearly demonstrated for vacated galls from which *B. tamabae* had emerged (Fig. 5), followed by those from which *P. neolitseae* had emerged ($R^2 = 0.410$), but it was less clear for vacated galls from which *Gastrancistrus* sp. had emerged ($R^2 = 0.364$).

The time and length of emergence period for the gall midge and parasitoids was expressed as the duration in days from the first to the last emerg-

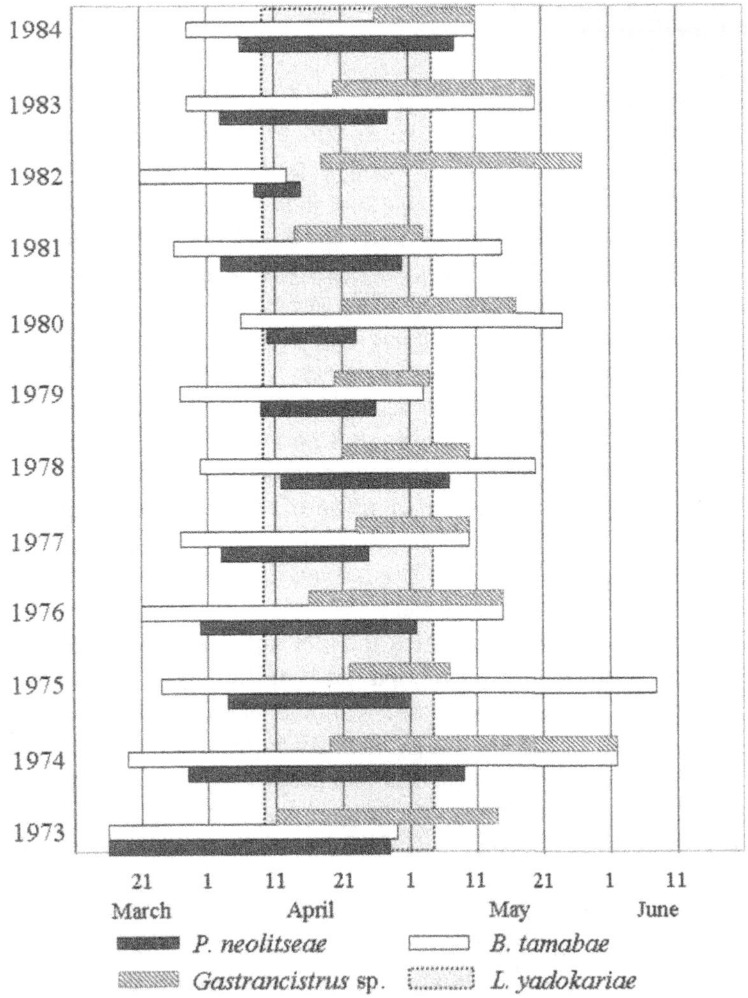

Fig. 6. Emergence season of *P. neolitseae, B. tamabae, Gastrancistrus* sp., and *L. yadokariae.*

ence date (Fig. 6). The time of emergence season for *L. yadokariae* was roughly indicated based on the data in Yukawa and Haitsuka (1994). *P. neolitseae* started to emerge from March or April and the emergence lasted 7 to 42 days until late April or early May. The emergence period of *B. tamabae* was longest among them throughout the surveys except in 1982, and lasted 22 to 74 days. The emergence of *Gastrancistrus* sp. started from mid or late April and lasted 15 to 43 days until early to mid May.

21.4 Discussion

Although many vacated galls were available in the field, at most 36% of them were utilized by *L. yadokariae* (Fig. 4). Sometimes two or more *L. yadokariae* eggs are found in a vacated gall, although only one mature larva can develop in a gall (Yukawa and Haitsuka 1994). These data show that the number of freshly vacated galls may be insufficient for *L. yadokariae* during the time of their oviposition season. In fact, the number of *L. yadokariae* larvae increased when the number of vacated galls from which *B. tamabae* and *P. neolitseae* had emerged increased (Fig. 5). As pointed out for gall-inducing cecidomyiids (Yukawa 2000; Yukawa and Akimoto 2006), synchronization of the *L. yadokariae* oviposition season with the emergence of gall inducer and parasitoids would also affect the quality and quantity of available vacated galls and determines the density of the short-lived successor. The duration of the *L. yadokariae* oviposition season is shorter than the total length of the emergence season of the gall midge and parasitoids (Fig. 6). This means that *L. yadokariae* adults can utilize only part of all vacated galls.

Although *L. yadokariae* females did not show any preference between the exit holes with or without a pupal case of the gall midge (Yukawa and Haitsuka 1994), vacated galls from which *B. tamabae* had emerged were utilized by the successor at a significantly higher rate than vacated galls from which *P. neolitseae* or *Gastrancistrus* sp. had emerged (Table 1). This was possibly caused by the higher population density and longer emergence season of *B. tamabae* than *P. neolitseae* and *Gastrancistrus* sp. (Figs. 3, 6).

These data suggest that when the density of *P. neolitseae* or *B. tamabae* is high, *L. yadokariae* numbers increase in the current generation using abundant vacated galls after their emergence, whereas the high density of *Gastrancistrus* sp. does not contribute to an increase of the successor because of its delayed emergence. The high density of *B. tamabae*, however, would decrease the number of *L. yadokariae* in the following generation, because high percentage parasitism by *T. tamabae* will reduce the density of *P. neolitseae*, resulting in the shortage of vacated galls. A possible existence of some *L. yadokariae* individuals that require two years to complete one generation (Yukawa and Haitsuka 1994) may play an important role in diversifying the risk of low vacated gall density and asynchrony with gall midge and parasitoid emergence.

21.5 Acknowledgements

We thank past students of the Entomological Laboratory, Faculty of Agriculture, Kagoshima University for their help in the field survey. Our thanks are also due to Dr. K. M. Harris (former Director of IIE, UK) for his suggestions and comments on an early draft.

21.6 References

Mani MS (1964) Ecology of plant galls. Dr. W. Junk, The Hague

Takasu K, Yukawa J (1984) Two-year life history of the neolitsea leaf gall midge, *Pseudasphondylia neolitseae* Yukawa (Diptera, Cecidomyiidae). Kontyû 52:596-604

Yukawa J (1974) Descriptions of new Japanese gall midges (Diptera, Cecidomyiidae, Asphondyliidi) causing leaf galls on Lauraceae. Kontyû 42:293-304

Yukawa J (1983) Arthropod community centred upon the neolitsea leaf gall midge, *Pseudasphondylia neolitseae* Yukawa (Diptera, Cecidomyiidae) and its host plant, *Neolitsea sericea* (Blume) Koidz. (Lauraceae). Memoirs of the Faculty of Agriculture, Kagoshima University 19:89-96

Yukawa J (2000) Synchronization of gallers with host plant phenology. Population Ecology 42:105-113

Yukawa J, Akimoto K (2006) Influence of synchronization between adult emergence and host plant phenology on the population density of *Pseudasphondylia neolitseae* (Diptera: Cecidomyiidae) inducing leaf galls on *Neolitseae sericea* (Lauraceae). Population Ecology 48:13-21

Yukawa J, Haitsuka S (1994) A new cecidomyiid successor (Diptera) inhabiting empty midge galls. Japanese Journal of Entomology 62:709-718

Yukawa J, Rohfritsch O (2005) Biology and ecology of gall-inducing Cecidomyiidae (Diptera). In: Raman A, Schaefer CW, Withers TM (eds) Biology, ecology, and evolution of gall-inducing arthropods. Science Publishers, Enfield, pp 273-304

Yukawa J, Takahashi K, Ohsaki N (1976) Population behaviour of the neolitsea leaf gall midge, *Pseudasphondylia neolitseae* Yukawa (Diptera, Cecidomyiidae). Kontyû 44:358-365

5. Evolution and Taxonomy

22 Evolution of Wing Pigmentation Patterns in a Tephritid Gallmaker: Divergence and Hybridization

Jonathan M. Brown and Idelle Cooper

Department of Biology, Grinnell College, Grinnell, IA 50112, USA

Summary. Wing pigmentation patterns are commonly used for taxonomic identification in many groups of gall-making flies in the family Tephritidae; however, neither the functional significance nor the mechanisms of evolutionary change in these characters have been well studied. We applied a quantitative image analysis approach to measure wing-pattern variation and differentiation in populations of the goldenrod gall fly *Eurosta solidaginis.* A large body of work by Abrahamson and coworkers has established that "host races" have emerged from within the eastern U.S. subspecies, *E. solidaginis solidaginis,* via a shift from an ancestral host-plant species *(Solidago altissima* [Asteraceae]) to a derived host species (*S. gigantea*). The image analysis demonstrated that host races are significantly differentiated in wing patterns at a sympatric site. We also quantified wing-pattern variation across a hybrid zone in Iowa between the named subspecies of *E. solidaginis,* which are distinguished taxonomically solely based on wing-pattern differences. The presence of intermediate wing patterns suggests hybridization between wing forms. These results suggest that host shifts provide barriers strong enough to allow populations to diverge in wing patterns, but that neither sympatric host shifts nor wing pattern differences evolved during periods of geographic isolation provide complete barriers to gene flow.

Key words. Speciation, Tephritidae, *Eurosta solidaginis*, Wing patterns, Image analysis

22.1 Introduction

Wing pigmentation patterns are used as a primary character in species identification in many groups of flies, including many gall-making taxa in the family Tephritidae (e.g., see Foote et al. 1993), but their function is poorly studied. This is surprising given the taxonomic importance of these characters, as well as the common suggestions in the tephritid literature that wing and body color patterns may play a role in signaling during courtship. Experimental evidence for the significance of a particular wing pattern has only been demonstrated for tephritids that use them to mimic jumping spider threat displays (Greene et al. 1987; Mather and Roitberg 1987); indirect evidence for their role in visual signaling during mating comes from (1) the association of patterned wings with stylized wing movements during courtship (Headrick and Goeden 1999), and (2) the *lack* of patterned wings in the tephritoid family Lonchaeidae, in which mating commonly occurs in aerial swarms, where visual cues should be ineffective (Sivinski et al. 1999).

If wing patterns are involved in signaling during courtship, understanding their evolution may be important for elucidating mechanisms of diversification. For example, in his paper developing the argument for sympatric speciation via host shifts in the tephritid *Rhagoletis pomonella* group, Bush (1969) contrasted this group of host-associated lineages with another *Rhagoletis* species group, the *suavis* group, all of whose lineages share the same host (*Juglans*), but vary in mating behavior and wing and body melanization patterns. He suggested a model of allopatric speciation for the *suavis* group, in which diversification of wing patterns displayed during courtship have developed via reinforcement (i.e., selection against hybrids between locally-adapted forms that diverged during previous periods of allopatry). As with other mating characters, diversification of wing patterns does not require reinforcement; if genetic drift and/or sexual selection causes patterns to diverge whenever barriers to gene flow arise, divergence can occur either in sympatry (when there are host shifts) or allopatry. The study of wing pattern variation, function and evolution thus can contribute to our understanding of how barriers to gene flow arise and are maintained during the complex process of speciation.

In order to address such questions, wing patterns need to be treated as a quantitative character that varies continuously within and between populations, rather than a discrete character that characterizes different named taxa. Although there are no studies in tephritids, developmental genetic studies of *Drosophila* species (True et al. 1999) suggest that genetic and environmental changes in the timing of gene regulatory events could explain variability in their wing patterns. Condon and Norboom (1994) describe such variation in a radiation of tropical tephritids (*Blephaneura*), but Faust and Brown (1998) and Brown et al. (in press) are the only published studies to quantify such variation.

Here we apply the image analysis techniques described in Brown et al. (in press) to populations and subspecies of the tephritid fly *Eurosta soli-*

daginis, which induce galls on multiple species of goldenrods (Asteraceae: *Solidago*). This has become a model system for the investigation of plant-herbivore interactions (see Abrahamson and Weis 1997) and, in the last decade, herbivore speciation via host shifts (see review in Abrahamson et al. 2003). Populations of flies infesting two host-plant species in eastern North America, *S. altissima* and *S. gigantea,* have been shown to be reproductively isolated due to (1) allochronic emergence and strict host preference, both of which lead to assortative mating (Craig et al. 1994; How et al. 1993), and (2) divergent adaptation to their host plants, as demonstrated by lower fitness of hybrids on both host plants (Craig et al. 1997, 2001). Allozyme (Itami et al. 1998; Waring et al. 1990) and mtDNA (Brown et al. 1996) studies have established that these populations are genetically differentiated; furthermore, higher genetic diversity and paraphyly of mtDNA haplotypes suggest that *S. altissima* is the ancestral and *S. gigantea* the derived (or novel) host. New England (northeastern N. American) was identified as the putative location of the host shift by phylogeographic patterns of mtDNA variation (Brown et al. 1996).

Here, we report on wing pattern differentiation from sympatric host race populations from New England. While no morphological distinctions between host races have been previously identified, interruption of gene flow between host races predicts that traits involved in mate recognition should begin to diverge by drift or sexual selection. In addition, we analyzed wing pattern variation across a series of populations in the Midwestern United States where intermediates have been reported (Faust and Brown 1998; Foote et al. 1993) between two geographic subspecies of *E. solidaginis. E. s. solidaginis* and *E .s. fascipennis* are morphologically differentiated solely by wing pattern differences (see Fig. 1) and are distributed east and west (respectively) of the tall-grass prairie region of central North America (Foote et al. 1993). Both host races are found in our study area. If the gigantea host race had a single origin in New England populations before spreading westwards, it should exhibit the wing pattern of *E. s. solidaginis*, unless there is gene flow between host races. The results of our analyses suggest that wing pattern does evolve as reproductive barriers form between populations and host races, but that neither host association nor the different wing patterns of the geographic subspecies constitute an absolute barrier to gene flow.

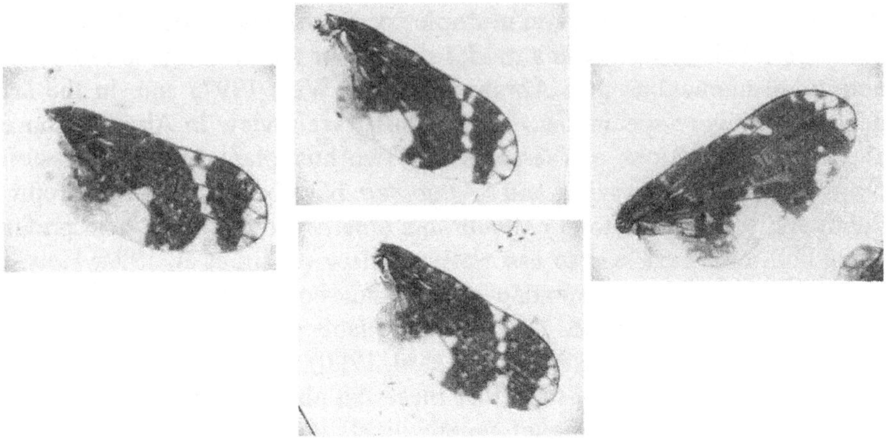

Fig. 1. Wings of *E. s. solidaginis (left)*, *E. s. fascipennis (right)* and intermediates from Iowa, USA.

22.2 Methods

Galls formed by larval *E. solidaginis* were collected in fall 1999, 2000 and 2001 after they entered winter diapause (between October and March), stored at –20°C, and then reared to adults by placing them at room temperature, which causes larvae to break diapause. Following 2–3 days, emerged adults were frozen at –80°C until wings were removed and, slide-mounted in Permount™ medium. Population locations and sample sizes are shown in Table 1.

We captured digital images of mounted wings through a Leica MZ8 microscope under identical light conditions using a video camera and digital framegrabber on a Powermac 8500. Using Wingmeasure (O'Fallon 2001), a plug-in applet of ImageJ 1.16f (Rasband 2001), we then defined for each wing a set of 13 homologous landmarks, i.e., apices or intersections of the veins defining wing cells. The program defined seven polygons formed by lines connecting these landmarks (see Fig. 2). In the Iowa subspecies transect analysis, each polygon was further subdivided by lines connecting the midpoints of opposite sides, forming 28 homologous areas. The plug-in calculated the average of each area's grayscale values, which define the brightness of each pixel in the image and thus measure the relative amount of melanin deposition in that area.

We compared host races and populations after reducing the number of variables using two multivariate techniques: (1) Principal components analysis (PCA) identified orthogonal axes of greatest variation in multivar-

Table 1. Location of sampled populations

Region	Location	Latitude	Longitude	altissima flies (M:F)	gigantea flies (M:F)
New England	Huntington, VT	44:22.2	73:03.2	31(20:11)	41 (26:15)
Iowa	Mineral Point, WI	42:51.7	90:10.2	11	
	Dubuque, IA	42:25.2	90:41.9	8	
	Hale, IA	42:00.8	91:04.7	6	12
	E of Central City, IA	42:12.5	91:25.4	8	
	Sugar Hollow Road, IA	41:45.0	91:26.0	10	
	Central City, IA	42:11.8	91:35.8	8	
	Casey's Paha SP, IA	42:16	92:14		11
	Krumm Preserve, Grinnell, IA	41:43	92:47	16	10
	Conard Env. Res. Area, IA	41:41	92:52	9	
	Kellogg, IA	41:43.7	92:57.6	6	
	Kellogg, IA	41:44.9	92:57.6		12
	Colo Bogs, IA	42:01.1	93:15.2	11	14
	Colfax, IA	41:40.3	93:17.1		5
	Midvale, IA	41:46	93:37	5	
	Albion, IA	42:08.2	93:33.9		24
	Doolittle Prairie, IA	42:09	93:36	5	
	West Des Moines, IA	42:01	93:47	8	
	Ames, IA	41:32.5	93:46.7	15	
	Van Meter, IA	41:31.1	93:56.4		10
	DeSoto, IA	41:31.1	93:58.6	14	
	Pilot Mound, IA	42:09	93:58		15
	Don Williams County Park, IA	42:08	94:01		11
	Casey, IA	41:34.5	94.29.4	16	
	Prairie Rose SP, IA	41:36	95:13	13	
	Denison, IA	42:02	95:19	9	
	Dunlap, IA	41:51	95:34	8	
	Soldier, IA	42:00	95:51	18	
	Sylvan Runkle SP, IA	42:07	95:59	18	

iate space. For the Vermont sympatric host races, we detected significant pattern variation for *HOST* and *SEX* main effects and their interaction using ANOVA on the first three PC scores. For the Iowa subspecies transect, we plotted the mean values for PC1 vs. longitude, to see if a transition zone between the subspecies could be located. (2) For the Vermont sympatric host races, we also used Discriminant Function Analysis (DFA) to test the ability to correctly place each individual into each host race or into each sex. All statistical analyses were performed using Minitab v14.

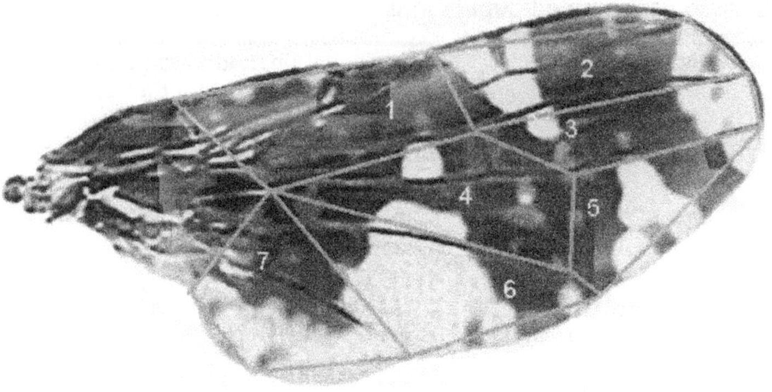

Fig. 2. Wing of *E. s. solidaginis,* with seven wing regions defined by landmarks.

Table 2. Percent variation explained and loadings of first three principal components for the Vermont host race populations

	PC1	PC2	PC3
% variation (cum.)	56.6 (56.6)	17.8 (74.5)	9.3 (83.8)
Region 1	–0.37	–0.35	0.61
Region 2	–0.40	–0.38	0.18
Region 3	–0.41	–0.24	–0.42
Region 4	–0.40	–0.19	–0.03
Region 5	–0.40	0.09	–0.56
Region 6	–0.37	0.49	0.07
Region 7	–0.27	0.65	0.32

22.3 Results

22.3.1 Vermont Sympatric Host Races

Multivariate analysis discovered highly significant differentiation between the sympatric host races and significant differentiation between sexes. Table 2 shows the loadings of each PC score on each variable (the mean grayscale values of the seven wing regions). PC1 measures variation in the overall brightness of the wing, but ANOVA showed no significant *HOST*, *SEX* or interaction effects on this measure. PC2 reflects most strongly the variation in pigmentation in regions 6 and 7 (cell cual and anal lobe) and

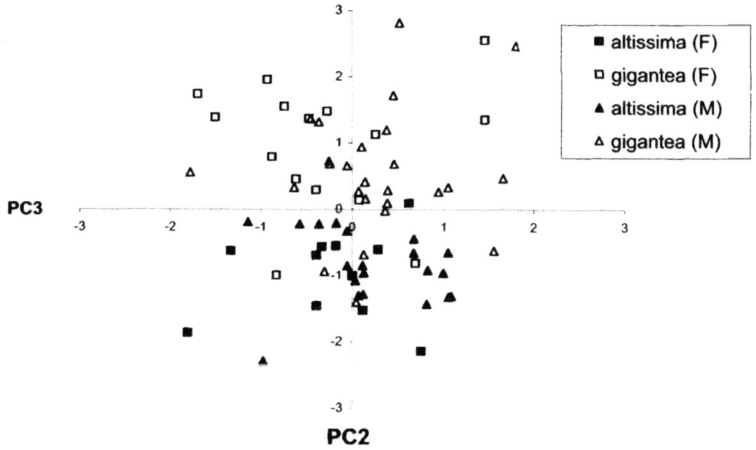

Fig. 3. Individual variation in wing pattern in the Vermont host race populations.

showed a highly significant *HOST* effect ($F = 67.4$, $P < 0.001$). PC3 reflects most strongly the variation in regions 1 and 5 (the pterostigma and cell m) and showed a significant *SEX* effect ($F = 7.1$, $P = 0.01$; Fig. 3).

Discriminant Function Analysis correctly classified 30/31 (97%) altissima flies and 40/45 (89%) gigantea flies when *HOST* was used as a grouping variable (test for difference: $\chi^2 = 1.57$, $P = NS$). 32/46 (70%) males and 17/26 (65%) females when *SEX* was used as the grouping variable.

22.3.2 Iowa Subspecies Transect

PC1 explained 57% of total variation and loaded positively on all 28 area means with coefficients from 0.12 to 0.21. Fig. 4 shows the transition in population means ($n_{population} > 5$) between those typical of the eastern subspecies in eastern Iowa (positive PC1 score) to those typical of the western subspecies (negative PC1 score) in north Central Iowa at longitudes higher than 92.8. Note that some populations in SW Iowa (between longitude 95 and 96) have eastern wings, so any potential hybrid zone is mosaic. Intermediacy in values of means reflects both mixtures of wings typical of the two subspecies and intermediates. Note that gigantea host race populations appear to acquire more western subspecies-like wing forms (lower means for PC1) at longitudes with western altissima populations.

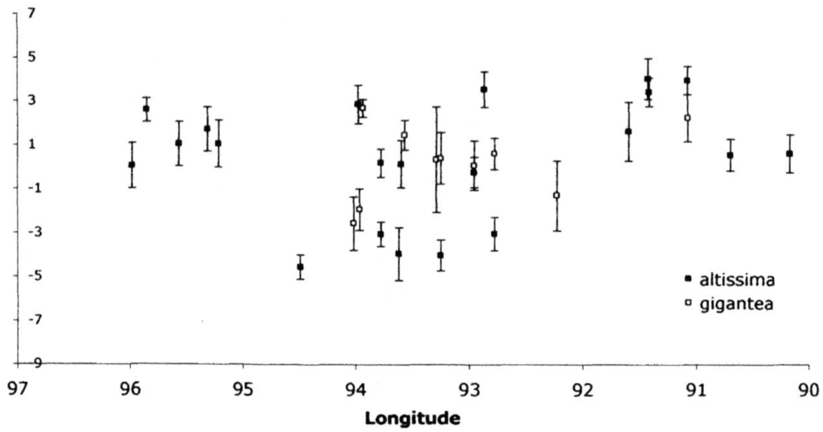

Fig. 4. Mean scores (± SE) for host race populations in Iowa, plotted vs. longitude.

22.4 Discussion

Quantitative analysis of wing-pattern variation illustrated that host race formation in these gall-making flies provides enough of a barrier to gene flow to allow for morphological differentiation. This is not surprising, since these host races are known to differ in other heritable traits such as behavior and adaptation to host plants (Craig et al. 1994, 1997, 1999). This is the first indication, however, that morphological features can be used to distinguish them. A larger study of multiple populations of sympatric host races (J. Brown, unpublished) illustrates that differentiation is consistent and often larger than reported here. While it is not known how much the variation seen in this population is heritable, the presence of individuals that are intermediate in wing pattern between the eastern and western subspecies of *E. solidaginis* suggests that it is possible. In addition, True et al. (1999) and J. Brown and D. Price (unpublished) have illustrated that relatively minor variations in degree of wing melanization in *Drosophila* species are heritable. The results are consistent with a model of wing pattern evolution by which patterns used primarily as mating signals diverge by genetic drift and/or sexual selection when populations are subdivided.

Despite the presence of significant variation in wing patterns between host races in the Vermont population, wing patterns were not so differentiated that host race could be assigned without error by discriminant function analysis. This is consistent with evidence that host races are connected by low levels of gene flow: Itami et al. (1998) present evidence from allozyme studies that suggest geographic variation in neutral differentiation (F_{ST}) between host races and discuss the ecological conditions, e.g., suitability of host plants for hybrids, degree of sympatry, and degree of allochronic isolation, that could lead to local and temporal fluctuations in gene flow. Gigantea flies had a non-significantly higher rate of error in placement by DFA, which could reflect a bias in the rate of gene flow from altissima to gigantea host races; studies of the inheritance of host fidelity and adaptation in these host races (Craig et al. 2001) argue for such a bias based on (a) the relatively lower level of host fidelity of gigantea males, which leads to (b) hybrid females who have high host fidelity to *S. gigantea* and thus (c) backcross with gigantea males, all of which would introgress altissima alleles into the gigantea host race.

The Iowa subspecies transect also illustrates the potential for gene flow between host races. The gigantea host race is primarily distributed in areas east of the zone of subspecies overlap, and an analysis of mtDNA haplotype diversity (Brown et al. 1996) suggested a far eastern (New England) origin for the host shift to gigantea. Thus, the gigantea host race should retain an eastern subspecies wing form, if host races are genetically isolated. In contrast, gigantea host race populations in Iowa acquire more western subspecies wing phenotypes when they reach the zone of overlap between the two (Fig. 4). Since eastern and western altissima flies use the same host plants, it would be informative to compare the levels of differentiation between host races and subspecies in the same locations. Better estimates of gene flow are needed for this region where the two subspecies come into contact. Mating experiments using both host races and subspecies may shed light on the relative strengths of host-related vs. morphological/behavioral barriers, and thus the relative importance of ecological vs. morphological barriers in processes of speciation.

22.5 Acknowledgments

Amy Whipple and Warren Abrahamson kindly provided wings from the Vermont sympatric host race populations. Research support was provided by Grinnell College. This is publication #10 of the Conard Environmental Research Area of Grinnell College.

22.6 References

Abrahamson WG, Weis AE (1997) Evolutionary ecology across three trophic levels: goldenrods, gallmakers, and natural enemies. Monographs in Population Biology 29, Princeton University Press, Princeton

Abrahamson WG, Blair CP, Eubanks MD, Morehead SA (2003) Sequential radiation of unrelated organisms: the gall fly *Eurosta solidaginis* and the tumbling flower beetle *Mordellistena convicta.* Journal of Evolutionary Biology 16:781-789

Brown JM, Abrahamson WG, Way PA. (1996) mtDNA phylogeography of host races of the goldenrod ball gallmaker (Diptera: Tephritidae: *Eurosta solidaginis*). Evolution 50:777-786

Brown JM, Todd-Thompson M, McCord A, O'Brien A, O'Fallon B (in press) Phylogeny, host association, and wing pattern variation in the endemic Hawaiian tephritids. (Tephritidae: Tephritini). Instrumentas Biodiversitatis

Bush GL (1969) Sympatric host race formation and speciation in frugivorous flies of the genus *Rhagoletis* (Diptera: Tephritidae). Evolution 23:237-251.

Condon ML, Norboom AL (1994) Three sympatric species of *Blephaneura* (Diptera: Tephritidae) on a single species of host (*Guarania spinulosa*, Cucurbitaceae): new species and new taxonomic methods. Systematic Entomology 19:279-304

Craig TP, Itami JK, Abrahamson WG, Horner JD (1994) Behavioral evidence for host-race formation in *Eurosta solidaginis*. Evolution 47:1696-1710

Craig TP, Horner JD, Itami JK (1997) Hybridization studies on the host races of *Eurosta solidaginis*: implications for sympatric speciation. Evolution 51:1552-1560

Craig TP, Abrahamson WG, Itami JK, Horner JD (1999) Oviposition preference and offspring performance of *Eurosta solidaginis* on genotypes of *Solidago altissima.* Oikos 86:119-128

Craig TP, Itami JK, Horner JD (2001) Genetics, experience, and host-plant preference in *Eurosta solidaginis*: implications for host shifts and speciation. Evolution 55:773-782

Faust L, Brown JM (1998) Sexual selection via female choice in the gall-making fly *Eurosta solidaginis* Fitch (Diptera: Tephritidae). In: Csoka G, Mattson WJ, Stone GN, Price PW (eds) The biology of gall-inducing arthropods. General Technical Report NC-199. USDA Forest Service, North Central Research Station, St. Paul, pp 82-89

Foote RH, Blanc FL, Norrbom AL (1993) Handbook of the fruit flies (Diptera: Tephritidae) of America north of Mexico. Cornell University Press, Ithaca

Greene E, Orsak LJ, Whitman DW (1987) A tephritid fly mimics the territorial displays of it jumping spider predators. Science 236:310-312

Headrick DH, Goeden RD (1999) Behavior of flies in the subfamily Tephritinae. In: Aluja M, Norboom AL (eds) Fruit flies (Tephritidae): phylogeny and evolution of behavior. CRC Press, Boca Raton, pp 671-710

How ST, Abrahamson WG, Craig TP (1993) Role of host plant phenology in host use by *Eurosta solidaginis* (Diptera: Tephritidae) on *Solidago* (Compositae). Environmental Entomology 22:388-396

Itami JK, Craig TP, Horner JD (1998) Factors affecting gene flow between the host races of *Eurosta solidaginis*. In: Mopper S, Strauss SY (eds) Genetic structure and local adaptation in natural insect populations. Chapman and Hall, New York, pp 375-407

Mather MH, Roitberg, BD (1987) A sheep in wolf's clothing: tephritid flies mimic spider predators. Science 236:308-310

O'Fallon, B (2001) Wingmeasure: a plug-in for ImageJ. Available at http://web.grinnell.edu/individuals/brownj/wingmeasure/index.html

Rasband, W (2001) ImageJ: Image processing and analysis in Java. Available at http://rsb.info.nih.gov/ij/

Sivinski J, Aluja M, Dodson GN, Friedberg A, Headrick DH, Kaneshiro KY, Landholt PJ (1999) Topics in the evolution of sexual behavior in the Tephritidae. In: Aluja M, Norboom AL (eds) Fruit flies (Tephritidae): phylogeny and evolution of behavior. CRC Press, Boca Raton, pp 751-792

True JR, Edwards KA, Yamamoto D, Carroll SB (1999) *Drosophila* wing melanin patterns form by vein-dependent elaboration of enzymatic prepatterns. Current Biology 9: 1382-1391

Waring GL, Abrahamson WG, Howard DJ (1990) Genetic differentiation among host-associated populations of the gallmaker *Eurosta solidaginis* (Diptera: Tephritidae). Evolution 44:1648-1655

23 The Evolution of Gall Traits in the Fordinae (Homoptera)

Moshe Inbar[1]

Department of Evolutionary & Environmental Biology, University of Haifa, Mount Carmel, Haifa 31905, Israel

Summary. The evolutionary divergence of the galling habit of aphids (Homoptera: Fordinae) that induce different gall types on *Pistacia* spp. (Anacardiaceae) trees in the Mediterranean region was examined. The phylogenetic cladogram of the aphids that was based on sequences of mitochondrial genes (COI and COII) was constructed. Placing gall traits on the single parsimony cladogram suggests that gall types evolved gradually from simple to complex structure and higher reproductive success. The importance of improved nutrition (sink strength) and defense in the evolution and maintenance of gall divergence in the Fordinae is discussed.

Key words. Aphids, Molecular phylogenetics, *Pistacia*

23.1 Introduction

One of the most striking characteristics in many groups of gall-forming insects is the variability in gall position, morphology, and structural complexity. Although the mechanism of gall formation remains unknown, it seems that the insects somehow control the process (Abrahamson and Weis 1997). Several studies suggested a close association between the insect and the morphology of the galls (Crespi and Worobey 1998; Nyman et al. 2000; Stern 1995).

The galling habit probably evolved from related free-feeding insects. Usually, galling is preceded by leaf folding or simple pseudogalls (Crespi and Worobey 1998; Price and Roininen 1993). Within a group, ancestral galls are usually simple with a single chamber (Fukatsu et al. 1994), although complex ancestral galls have also been reported (e.g., Dorchin et al. 2004).

---

[1] Former surname, Burstein

Natural selection may create and maintain gall divergence. Gall traits may be related to pressure imposed by natural enemies (Abrahamson and Weis 1997; Cornell 1983; Price et al. 1987) and competition among gall formers for galling sites (often time limited) and nutrients (Akimoto 1988; Inbar et al. 1995; Yukawa 2000). Furthermore, it has been suggested that gall structure in some social lineages of aphids and thrips is affected by the evolution of sociality (Crespi and Worobey 1998; Stern 1995).

In the Mediterranean region, a group of aphids (Homoptera: Fordinae) induces remarkably variable galls on wild pistachio, *Pistacia* (Anacardiaceae). The galls are formed on various host organs and differ in size, shape, and phenology (Koach and Wool 1997). They therefore provide an important opportunity to trace and understand the evolution of gall traits (see also Inbar et al. 2004). Molecular tools were used to establish phylogenetic relationships among the species. Then, possible evolutionary pathways and driving forces were suggested based on biological ecological information.

23.2 Materials and Methods

Approximately sixteen gall-forming aphids (Fordinae) are found in Israel, widely distributed in the Mediterranean and Irano-Turanien type habitats. Each species induces a characteristic gall on a specific *Pistacia* host (Koach and Wool 1977; Table 1). The Fordinae have been divided into two tribes, Fordini and Baizongiini. Their complex life cycle includes sexual and parthenogenetic reproduction and alternation between *Pistacia* and roots (without gall induction) of non-specific secondary hosts (Wool 1984). In Israel, the galls are formed on *P. palaestina, P. atlantica* (deciduous trees), and *P. lentiscus* (evergreen shrub). Galls are induced in the spring by the fundatrix and the migrating aphids leave the galls in the fall (details in Wool 1984). The Fordinae, as mentioned above, induce remarkably different gall types defined as (Table 1): pea, margin, bag, spherical, and bud (Inbar et al. 2004). Furthermore, four species induce two different galls; the fundatrix induces pea ('temporary') galls on the leaflet midvein whereas the offspring induce ('final') galls on the leaflet margin (Wool and Burstein 1991b).

The phylogeny of the aphids was based on DNA that was extracted from species collected in Israel. Sequences of COI and COII (1952 nt) were analyzed with PAUP* 4.0.10 (Inbar et al. 2004). The aphids are phloem feeders that divert assimilates from the host. The ability of the aphids (gall) to create a physiological sink was measured with a ^{14}C labeling technique;

Table 1. Life history traits of the Fordinae. Pea galls (induced by the fundatrix) are only found in species that have two galls

Gall type	Description	Source of assimilates	Aphids/gall	Genera
Pea	Unsealed, lentil-shaped (~ 5 mm long), located on the leaflet midvein.	Galled leaflet	< 20	*Forda*, *Smynthurodes*
Margin	Unsealed, elongate galls (~ 20 mm long) located on the leaflet margin (adaxial). The gall openings are near the leaflet midvein.	Galled leaflet	20 - 100	*Fordini* spp., *Forda*, *Paracletus*, *Smynthurodes*
Bag	Unsealed galls located on the upper (adaxial) leaflet surface, occupying most of the leaflet. The gall openings are near the leaflet midvein.	?	Several hundreds	*Aploneura*, *Asiphonella*
Spherical	Completely sealed globular galls (volume ~ 4 cm^3) located on the lower (abaxial) side of the leaflet or leaf midveins. Recently, a new *Geoica* species was found on the midvein of the leaf (Remaudière et al. 2004).	Galled leaflets & leaf. Leaves on the same shoot.	~ 1000	*Geoica*, *Rectinasus*
Bud	Large and completely sealed galls located on the buds initially induced on the main vein of young leaflet, and eventually take over the entire bud. May have a size and shape of a banana or cauliflower.	Galled shoot and leaves on neighboring shoots	Several thousand	*Baizongia*, *Slavum*

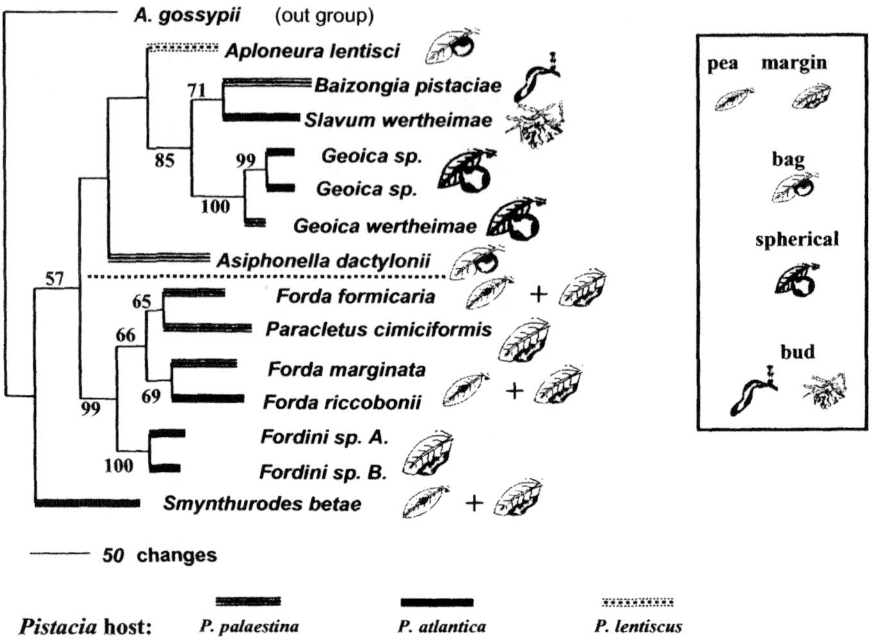

Fig. 1. Single most parsimonious tree of the Fordinae with *Aphis gossypii* (Homoptera: Aphididae) as an outgroup. Bootstrap values (>55) are shown next to the branches. Both *Geoica* spp. were collected, probably from different galls on *P. atlantica. Dashed horizontal line* divides the cladogram between the Baizongiini (up) and Fordini. Gall types are indicated with charts *(box)*.

the data was adopted from Burstein et al. (1994) and Inbar et al. (1995).

23.3 Results

The parsimony analysis yielded a single cladogram (Fig. 1) that divided (as the morphological systematics) the Fordinae to Fordini and Baizongiini tribes, but placed *Smynthurodes betae* in a new ancestral group. The Fordini (99% bootstrap support), includes the *Forda*, *Paracletus,* and the undescribed species (Fordini sp. A & sp. B). Because of the shared pea and margin gall types, *S. betae* is probably more closely related to the Fordini. The Baizongiini (85% bootstrap support) is composed of the *Geoica* (spherical galls) and the *Slavum–Baizongia* clades (bud galls). The position of *Aploneura* and *Asiphonella* (bag galls) is not clear.

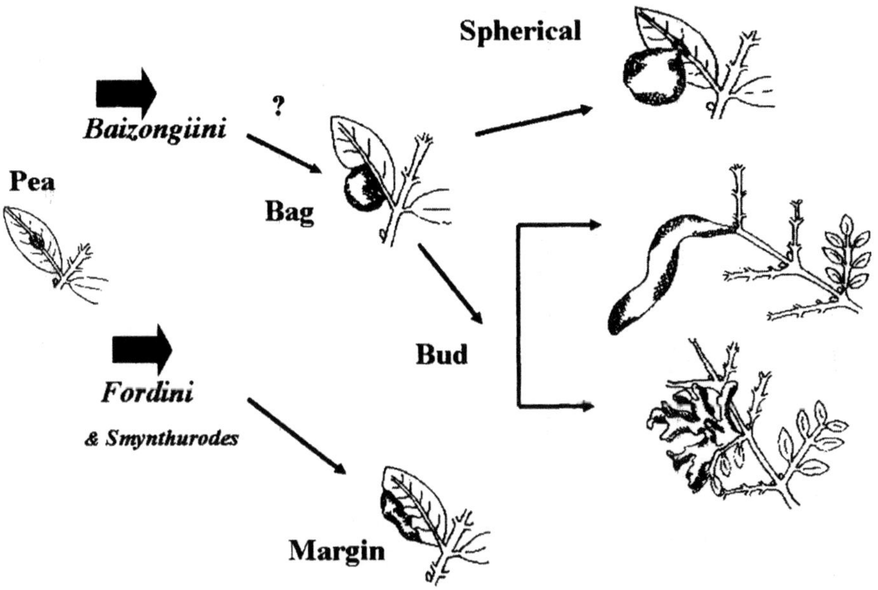

Fig. 2. The evolutionary scenario of the Fordinae gall type. Scaling was not maintained; see actual sizes in Table 1 and Fig 3. The position of the bag galls is questionable. Drawing by Adi Ne'eman.

The cladogram does not correspond with host plant specialization, types on different hosts maintained their shape and galling sites. The two-galltrait evolved early in the evolution of the Fordinae with the formation of margin galls (*Smynthurodes* and *Forda*). If this scenario is true, then the two-gall trait was lost in *Paracletus* and the Fordini spp. The Baizongiini induce large galls associated with the midrib and the bud. Sealed galls also developed once in the Baizongiini (bud and spherical galls). There is a strong association between gall type, sink strength, and the reproductive success of the aphid (Fig. 1, Table 1). Gall sink strength and aphid reproductive success increase in the following order: pea < margin < bag ≤ spherical < bud.

23.4 Discussion

The data clearly indicate that gall structure is controlled by the aphids. Repeated shifts between the two hosts, at least once in each aphid clade, are the most likely explanation for the similarity in gall types on *P. atlantica* and *P. palaestina* that was reported by Inbar and Wool (1995). It is likely that the aphids first shifted host (to sympatric *Pistacia* host), while

Fig. 3. Galls on *P. palaestina* demonstrating the differences between the Fordini and Baizongiini. *Left*, Final galls of *Forda formicaria* (20mm long). *Right*, The bud galls of *Baizongia pistaciae* (note *the scaling hand*).

maintaining gall characteristics, and then speciated. Similar pattern of initial shifting (timing) followed by later speciation (allochronic speciation) has been recently detected in the related gall-forming aphids (*Pemphigus*) in North America (Abott and Withgott 2004). Gall types were retained during host shifts, indicating that the aphids control gall trait.

In *Smynthurodes* (the ancestral group) and *Forda*, the fundatrix's offspring can also induce "final" margin galls (Table 1). Because, in most aphids, only the fundatrix induces galls and because the entire life cycle can be completed in the 'temporary' pea galls (as in *S. betae*; Wool and Burstein 1991b), it is likely that the primitive Fordinae had a single pea type gall on the midvein. From this point, the evolution of gall type in the Fordinae may have developed in two parallel lines as follows (Fig. 2; see also Inbar et al. 2004): First, in the *Smynthurodes* and Fordini lines, higher reproductive success was achieved by the ability of the fundatrix's offspring to induce their own margin galls with a slightly stronger sink. Consequently, a single fundatrix line can potentially continue in several galls that also spread the risk of destruction (Wool and Burstein 1991b). This trait was lost in *Paracelsus* and Fordini spp. In the second route, the Baizongiini found the way to induce larger galls on the midvein, possibly an open bag type. Next the sealed spherical and bud galls evolved. This route was correlated with increasing ability to manipulate the midvein, increase sink strength and produce thousands of aphids (Fig. 3).

Increasing sink strength seems to be an important factor in the evolution of gall types in the Fordinae, but what about other selective factors such as natural enemies, sociality, and competition? Natural enemies are thought to affect gall traits (Stone et al. 2002). This seems unlikely in the Fordinae. Although many hymenopteran parasitoids are known from free-living aphids, only one, *Monoctonia pistaciaecola* Stary, attacks the Fordinae before the galls are completely formed (Wool and Burstein 1991a). In addition, several predators attack the aphids in all gall types (Wool and Steinitz, unpublished). The production of soldiers in social aphids is associated with a small entrance and low surface area/gall volume (Stern 1995; see also Crespi and Worobey 1998). Nevertheless, eusociality was not discovered among the Fordinae (Inbar 1998). Potentially, competition may also cause shifts in galling site and shape. Although competition is common among gall-formers, its role in shaping community structure is questionable. In the Fordinae competition is weak due to clear niche separation (Inbar and Wool 1995; Inbar et al. 1995).

In conclusion, the molecular phylogenetics and the biological data available suggest that the remarkable divergence in gall size is most probably a result of selection for stronger host plant manipulation that results in superior nutrient supply and higher reproductive success.

23.5 Acknowledgments

I thank D. Wool and M. Wink for their fruitful collaboration. The comments of D. Graur, S. Lev Yadun, and P.W. Price are greatly appreciated.

23.6 References

Abott P, Withgott JH (2004) Phylogenetic and molecular evidence for allochronic speciation in gall-forming aphids (*Pemphigus*). Evolution 58:539-553

Abrahamson WG, Weis AE (1997) Evolutionary ecology across three trophic levels: goldenrods, gallmakers, and natural enemies. Princeton Univ. Press, Cambridge

Akimoto S (1988) Competition and niche relationships among *Eriosoma* aphids occurring on the Japanese elm. Oecologia 75:44-53

Burstein M, Wool D, Eshel A (1994) Sink strength and clone size of sympatric, gall forming aphids. European Journal of Entomology 91:57-61

Crespi BJ, Worobey M (1998) Comparative analysis of gall morphology in Australian gall thrips: the evolution of extended phenotypes. Evolution 52:1686-1696

Cornell HV (1983) The secondary chemistry and complex morphology of galls formed by the Cynipinae (Hymenoptera): why and how? American Midland Naturalist 110:225-234

Dorchin N, Freidberg A, Mokady O (2004) Phylogeny of the Baldratiina (Diptera : Cecidomyiidae) inferred from morphological, ecological and molecular data sources, and evolutionary patterns in plant-galler relationships. Molecular Phylogenetics and Evolution 30:503-515

Fukatsu T, Aoki S, Kurosu U, Ishikawa H (1994) Phylogeny of Cerataphidini aphids revealed by their symbiotic microorganisms and basic structure of their galls: implications for host-symbiont coevolution and evolution of sterile soldier castes. Zoological Science 11:613-623

Inbar M (1998) Competition, territoriality and maternal defense in a gall-forming aphid. Ethology, Ecology and Evolution 10:159-170

Inbar M, Wool D (1995) Phloem-feeding specialists sharing a host tree: resource partitioning minimizes interference competition among galling aphid species. Oikos 73:109-119

Inbar M, Eshel A, Wool D (1995) Interspecific competition among phloem-feeding insects mediated by induced host-plant sinks. Ecology 76:1506-1515

Inbar M, Wink M, Wool D (2004) The evolution of host plant manipulation by insects: molecular and ecological evidence from gall-forming aphids on *Pistacia*. Molecular Phylogenetics and Evolution 32:504-511

Koach J, Wool D (1977) Geographic distribution and host specificity of gall-forming aphids (Homoptera, Fordinae) on *Pistacia* trees in Israel. Marcellia 40:207-216

Nyman T, Widmer A, Roininen H (2000) Evolution of gall morphology and host-plant relationships in willow-feeding sawflies (Hymenoptera: Tenthredinidae). Evolution 54:526-533

Price PW, Roininen H (1993) Adaptive radiation in gall induction. In: Wagner MR, Raffa KF (eds) Sawfly life history adaptations to woody plants. Academic Press, San Diego, pp 229-257

Price, W, Fernandes GW, Waring GL (1987) Adaptive nature of insect galls. Environmental Entomology 16:15-24

Remaudière G, Inbar M, Menier JJ, Shmida A (2004) Un nouveau geoica gallicole sur *Pistacia atlantica* en jordanie (Hemiptera, Aphididae, Eriosomatinae, Fordini). Revue Francaise d'Entomologie 26:37-42

Stern DL (1995) Phylogenetic evidence that aphids, rather than plants, determine gall morphology. Proceedings of the Royal Society of London. B Biological Sciences. 260:85-89

Stone GN, Schönrogge K, Atkinson RJ, Bellido D, Pujade-Villar J (2002) The population biology of oak gall wasps (Hymenoptera: Cynipidae). Annual Review of Entomology 47:633-688

Wool, D (1984) Gall forming aphids. In: Ananthakrishnan TN (ed) Biology of gall insects. Oxford & IBH, New Delhi, pp 11-58

Wool D, Burstein M (1991a) Parasitoids of the gall-forming aphid *Smynthurodes betae* Westw. (Aphidoidea, Fordinae) in Israel. Entomophaga 36:531-538

Wool D, Burstein M (1991b) A galling aphid with extra life cycle complexity: population ecology and evolutionary considerations. Researches on Population Ecology 33:307-322

Yukawa J (2000) Synchronization of gallers with host plant phenology. Population Ecology 42:105-113

24 Life History Patterns and Host Ranges of the Genus *Asphondylia* (Diptera: Cecidomyiidae)

Nami Uechi[1] and Junichi Yukawa[2]

[1]Okinawa Prefectural Agricultural Experiment Station, 4-222 Sakiyama-cho, Naha, Okinawa 903-0814, Japan
[2]Kyushu University, Fukuoka 812-8581, Japan

Summary. Based on ecological, morphological, distributional, and molecular data, the Japanese *Asphondylia* gall midges (Diptera: Cecidomyiidae) were classified into the following five groups in terms of life history patterns and host ranges: (I) univoltine and monophagous or oligophagous species; (II) bivoltine species, which can complete their annual life cycle on a single host plant species by alternating between different host organs; (III) multivoltine species on one organ of a single host plant; (IV) multivoltine species, which alternate between different host plants seasonally to complete their annual life cycle; (V) univoltine and oligophagous species but partly bi- or multivoltine by utilizing occasional alternate hosts. *Asphondylia* species seem to be less constrained by the phenology of their host plants than other leaf gallers and to have flexible potential for adaptation of their life history strategy to the habitat and life style of their host plants. The potential may be derived from strong flight ability in search for suitable plants and the existence of a fungal symbiont on whose hyphae the larvae feed within galls. The categorization of life history patterns could contribute to future evolutionary studies of *Asphondylia* life histories when phylogenetic relationships among the species are analyzed.

Key words. *Asphondylia*, Cecidomyiidae, Life history, Host range, Gall

24.1 Introduction

The genus *Asphondylia* (Diptera: Cecidomyiidae) contains about 270 nominal species in the world that mostly gall flowers and prevent fruiting in various plant species and families (Gagné 2004). *Asphondylia* species are usually monophagous or oligophagous within a single plant genus as are most other gall-inducing cecidomyiids (Gagné 2004). In *Asphondylia*, polyphagy across different plant families has been known in only a few species, including *Asphondylia websteri* Felt (Gagné and Wuensche 1986), *Asphondylia gennadii* (Marchal) (Orphanides 1975), *Asphondylia yushimai* Yukawa and Uechi (Yukawa et al. 2003), and *Asphondylia baca* Monzen (Uechi et al. 2004), of which the latter three exhibit host alternation (Uechi et al. 2004, 2005; Yukawa et al. 2003).

In Japan, six species and 13 unidentified segregates of *Asphondylia* were known to occur on 25 plant genera of 17 families (Yukawa and Masuda 1996). The segregates were left unnamed due to their morphological similarity and the lack of information on host ranges and annual life cycles. Since 1996 we identified and described some of these segregates and clarified their host ranges and annual life cycles based on DNA sequence data, pupal morphological features, distributional information, and ecological data on the time of their emergence and host plant phenology (e.g., Uechi et al. 2004, 2005; Yukawa et al. 2003). Recent DNA sequence data revealed that a segregate on *Ampelopsis* on Ishigaki Island is different from *A. baca* (N. Uechi et al., unpublished data). As a result, six species and 11 segregates of *Asphondylia* are now known to exist in Japan. Through the aforementioned study of the Japanese *Asphondylia* gall midges, we found that they exhibit diverse patterns in life history and host plant utilization.

In this paper we review life history strategies of *Asphondylia* gall midges and classify the strategies into several categories in terms of voltinism, host range, host alternation, and organs galled. By superimposing the life history patterns on a phylogenetic tree of the Japanese *Asphondylia* gall midges, we consider the evolution of their life history strategies.

24.2 Materials and Methods

We obtained information on host plants, distribution, and other ecological traits of the Japanese *Asphondylia* gall midges mainly from Yukawa and Masuda (1996) and partly from our recent papers (e.g., Uechi et al. 2002, 2004; Yukawa et al. 2003) and unpublished data. Information for exotic congeners was obtained through literature searches. Unfortunately, the information that they provided was limited because detailed life histories have not been described in most cases, although scattered statements on host plants and organs galled were given for most species. The species and segregates treated in this paper are listed in Table 1, together with their host plants, host organs, and other data sources. For convenience, we give

Table 1. *Asphondylia* species and segregates discussed in this paper, their host plants, host organs, and data sources

Species and segregates	Host plant	Host organ	Voltinism	Isolation name	Data sources	Acc.No
Japanese species						
A. aucubae	*Aucuba japonica* (Cornaceae)	Fruit	Uni	Aspauc93FkIno	This paper	AB238595
A. baca	*Ampelopsis brevipedunculata* (Vitaceae), *Weigela* spp. (Caprifoliaceae)	Fruit	Multi	Asp.baca61(Ampelopsis)	Uechi et al. (2004)	AB115589
A. itoi	*Distylium racemosum* (Hamamelidaceae)	Fruit	Uni	DistyOk70	Uechi and Yukawa (2004)	AB162344
A. morivorella	*Morus* spp. (Moraceae)	Leaf bud	Uni	MorHk148	This paper	AB238596
A. sphaera	*Ligustrum* spp. (Oleaceae), *Rhus* spp. (Anacardiaceae)	Fruit	Uni	LigObFrCh155	Uechi and Yukawa unpublished	AB197941
A. yushimai	*Glycine max* and other wild fabaceous plants (Fabaceae), *Prunus zippeliana* (Rosaceae), *Osmanthus heterophyllus* (Oleaceae)	Pod	Multi	SoyFuk15	Yukawa et al. (2003), Uechi et al. (2005)	AB085787
Ampelopsis fruit gall midge on Ishigaki Island	*Ampelopsis brevipedunculata* (Vitaceae)	Fruit	Multi	AmpOki145	This paper	AB238597
Ardisia flower bud gall midge	*Ardisia japonica* (Myrsinaceae)	Fruit	Bi	ArdFrMyg99	Uechi et al. (2005)	AB194465
Hedera flower bud gall midge	*Hedera rhombea* (Araliaceae)	Flower bud	Multi	HedFl99A	Yukawa et al. (2003)	AB085874
Hedera fruit gall midge	*Hedera rhombea* (Araliaceae)	Fruit	Multi	HedFrFkSk107	Yukawa et al. (2003)	AB085882
Sapium leaf bud gall midge	*Sapium japonicum* (Euphorbiaceae)	Leaf bud	Uni	SapLbGnm335	Uechi et al. (2005)	AB194443
European species						
A. gennadii			Multi	Asp.gennadii262(carob)	Uechi et al. (2004)	AB115569
Asphondylia sarothamni	*Cytisus scoparius* (Fabaceae)	Pod, leaf bud	Bi	Asp.sarothamni128	Uechi et al. (2004)	AB115567
Pseudasphondylia matatabi	*Actinidia polygama* (Actinidiaceae)	Fruit		ActFuk30	Yukawa et al. (2003)	AB085873

each unnamed Japanese segregate an English name derived from its host genus and galled plant part.

DNA sequence data of each *Asphondylia* species and segregate were obtained by the methods described in our previous papers (e.g., Uechi et al. 2004; Yukawa et al. 2003). The sequence data were analyzed by the maximum parsimony method, using PAUP* 4.0b10 (Swofford 2002). A maximum parsimony tree was determined under the heuristic search strategy with 100 random addition sequences and the TBR branch swapping algorithm. The resulting tree was evaluated by the bootstrap test (Efron 1982; Felsenstein 1985) based on 1000 replications.

24.3 Results

24.3.1 Life History Patterns

The Japanese *Asphondylia* species and segregates were classified into the following five groups in terms of life history patterns and host ranges: (I) univoltine and monophagous or oligophagous species, which can complete their annual life cycle on one or a few congeneric host species (Fig. 1); (II) bivoltine species, which can complete their annual life cycle on a single host plant species by alternating between different host organs for galling (Fig. 2); (III) multivoltine species, which can complete their annual life cycle on one organ of a single host plant (Fig. 3); (IV) multivoltine species, which alternate between different host plants seasonally to complete their annual life cycle (Fig. 4); (V) univoltine and oligophagous species, which can complete their annual life cycle on one or more host plant species within a genus. However, some individuals are partly bi- or multivoltine by utilizing occasional alternate host plants (Fig. 5).

24.3.2 The Five Life History Patterns

Pattern I (Fig. 1): This category includes *Asphondylia aucubae* Yukawa and Ohsaki, *Asphondylia morivorella* (Naito), *Asphondylia itoi* Uechi and Yukawa (Uechi and Yukawa 2004), and the *Sapium* leaf bud gall midge. The first instars pass through the summer, autumn and winter in the gall and moult into the second instar in the following year. Then, adults emerge when the appropriate host organ becomes available for oviposition.

Pattern II (Fig. 2): This category includes the *Ardisia* fruit gall midge in Japan and other species in Europe and North America. To complete its annual life cycle, *Asphondylia sarothamni* (Loew) alternates between galls

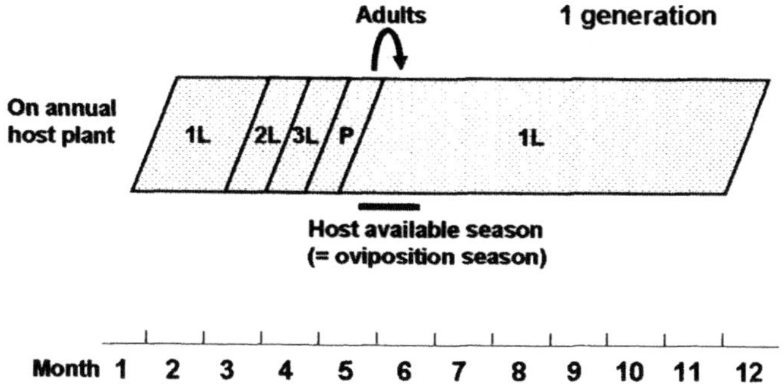

Fig. 1. Schematic representation of life history patterns of the Japanese *Asphondylia* species and segregates: (I) univoltine and monophagous or oligophagous species, which can complete their annual life cycle on a single host plant species.

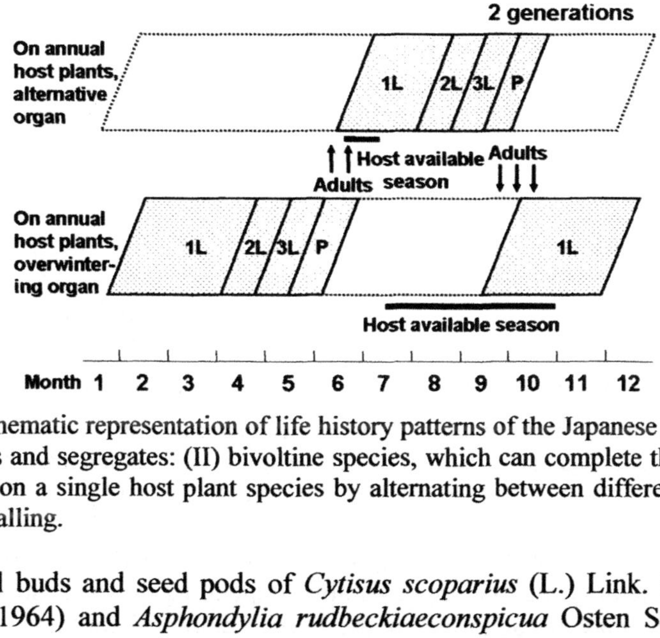

Fig. 2. Schematic representation of life history patterns of the Japanese *Asphondylia* species and segregates: (II) bivoltine species, which can complete their annual life cycle on a single host plant species by alternating between different host organs for galling.

on lateral buds and seed pods of *Cytisus scoparius* (L.) Link. in Europe (Parnell 1964) and *Asphondylia rudbeckiaeconspicua* Osten Sacken between buds at the base of *Rudbeckia laciniata* L. (Asteraceae) in spring and flowers in summer (Plakidas 1988) in North America.

Pattern III (Fig. 3): In Japan, only the *Ampelopsis* fruit gall midge on Ishigaki Island is included in this category. The galled organ is young fruit, which is available throughout the year, hence adult females can lay their eggs whenever they emerge (N. Uechi et al., unpublished data). Adults of *Asphondylia* spp. on creosote bush in North America irregularly appear on many occasions in a year (Gagné and Waring 1990). Their host plant is a perennial, evergreen, and one of the most drought-tolerant plants in the

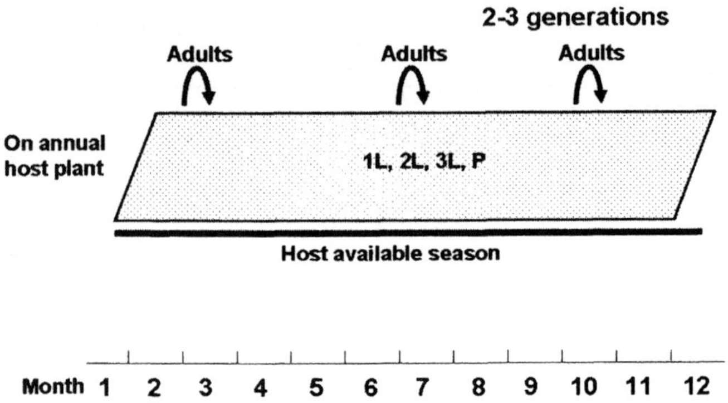

Fig. 3. Schematic representation of life history patterns of the Japanese *Asphondylia* species and segregates: (III) multivoltine species, which can complete their annual life cycle on one organ of a single host plant species.

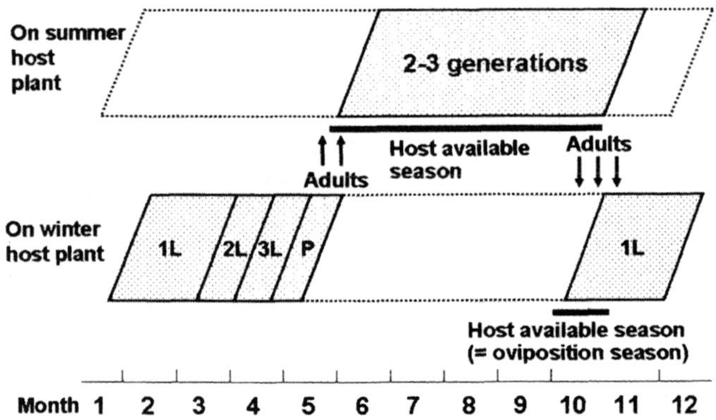

Fig. 4. Schematic representation of life history patterns of the Japanese *Asphondylia* species and segregates: (IV) multivoltine species, which alternate between different host plants seasonally to complete their annual life cycle.

southwestern United States (Gagné and Waring 1990). Because host seasonal availability may be unpredictable, depending on rains, such irregular emergence may be adaptive to the unpredictable phenology of the host plant.

Pattern IV (Fig. 4): This category includes *A. yushimai* (Uechi et al. 2005; Yukawa et al. 2003), *A. baca* (Uechi et al. 2004), the *Hedera* fruit gall midge, and the *Hedera* flower bud gall midge. Usually their spring–summer–autumn hosts die back in late autumn and the emergent summer–autumn generation has nowhere to lay their eggs on those hosts. The emergent winter generation has also nowhere to lay its eggs because its winter hosts do not proffer the proper organ for oviposition at the time of adult

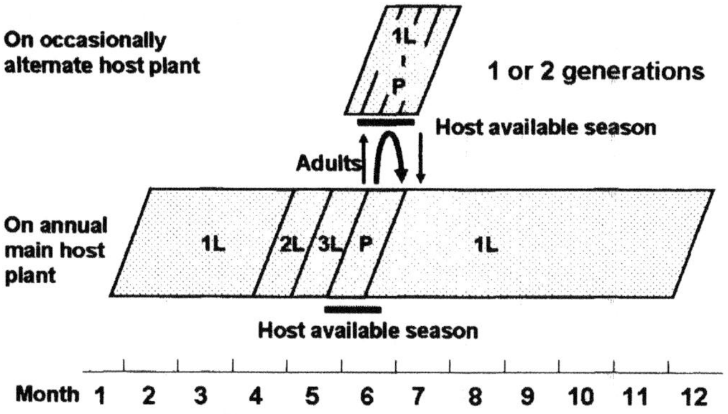

Fig. 5. Schematic representation of life history patterns of the Japanese *Asphondylia* species and segregates: (V) univoltine and oligophagous species, which can complete their annual life cycle on one or more host plant species within a genus. However, some individuals are partly bi- or multivoltine by utilizing occasional alternate host plants.

emergence in spring. Therefore, the multivoltine species needs some alternative hosts on which to lay their eggs. The European species, *A. gennadii*, also alternates between host plants and has a wide host range across four plant families (Orphanides 1975; Uechi et al. 2004).

Pattern V (Fig. 5): At the moment, only *Asphondylia sphaera* Monzen seems to be included in this category (N. Uechi and J. Yukawa, unpublished data). It is univoltine although some individuals utilize flower buds of *Ligustrum japonicum* Thunb. as an occasional alternative host organ (Ohsako et al. 1981). We also suspect that *Rhus* spp. is an occasional alternative host, although it has not yet been confirmed by molecular analysis. However, these individuals may be in a minority in that they correspond to only a small portion of the whole population of spring emergent adults.

24.3.3 Molecular Phylogenetic Inference

The length of the amplified mitochondrial COI gene fragment in all samples analyzed was 439 bp (Uechi et al. 2005; Yukawa et al. 2003). The monophyly of Japanese species, except *A. itoi* and *A. morivorella*, was supported by a 93% bootstrap value in an MP tree (Fig. 6). These phylogenetic relationships did not represent life history pattern, galling part, or host plant differences.

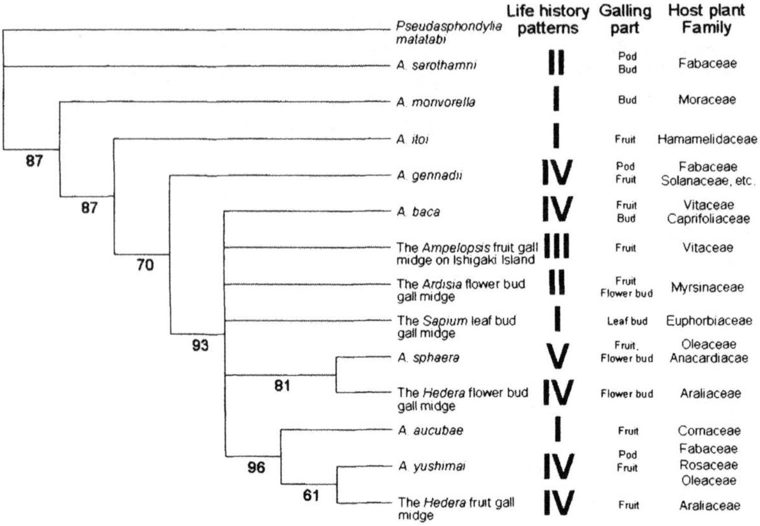

Fig. 6. Strict consensus tree of four equally parsimonious trees (tree length, 350; CI = 0.649, RI = 0.504, RC = 0.327) for *Asphondylia* gall midges, based on 439 bp of the mitochondrial COI gene. *Bootstrap values* are indicated for nodes gaining >60% support (1000 replications). Detailed data of respective species and segregates are shown in Table 1.

24.4 Discussion

Most leaf-gall-inducing cecidomyiids that oviposit in or on rapidly extending shoots have to synchronize strictly with the leaf opening phenology of the host plant (Yukawa 2000; Yukawa and Akimoto 2006). In contrast, synchronization is not always strict for *Asphondylia* gall midges that utilize flower buds, fruit, and overwintering leaf buds because these organs last longer at various stages, hence host seasonal availability for oviposition is longer. The only exception is the case of *A. aucubae*, whose oviposition season is restricted to an extremely short period of time due to physical and physiological conditions of the young fruit of the host plant (Imai and Ohsaki 2005).

The life-history strategy of gall midges has been fundamentally constrained by the phenology, morphology, habitat, and life style of their host plants: annual or perennial, herbs or trees, deciduous or evergreen (Yukawa and Rohfritsch 2005). *Asphondylia* species, however, are less constrained by the phenology of their host plants than other leaf-gall-inducing species. *Asphondylia* species seem to have flexible potential for adaptation of their life history strategy to the habitat and life style of their host plants. The potential may be derived from strong flight ability in search for suitable plants (Yukawa et al. 2003) and the existence of a fungal symbiont on whose hyphae the larvae feed within characteristic 'ambrosia' galls (Yukawa and Rohfritsch 2005).

Host alternation and utilization of occasional alternative hosts have adaptive significance, such as an increase of voltinism that enhances reproductive potential, escape from parasitoid attacks by changing habitats, and seeking fresher and more nutritious host plants. Although host alternation, including utilization of occasional alternative host plants, is quite rare in the life history of Diptera (Uechi et al. 2004), the examples of host alternation in the Japanese and European multivoltine *Asphondylia* species indicate that host alternation has occurred at a number of different locations during the course of speciation in the genus *Asphondylia*. A phylogenetic tree supports this explanation (Fig. 6). Recently possible host alternation by two species of the genus *Pseudasphondylia* (Diptera: Cecidomyiidae) has also been suggested (Tokuda and Yukawa 2005). Morphological similarity among a large majority of Japanese *Asphondylia* species and segregates indicates that they are closely related to each other, but DNA analysis suggests that they are now undergoing host race formation or speciation by expanding host ranges and adapting to new host plants.

Simultaneous polyphagy known in *A. websteri*, the stem-galling habit in *Asphondylia atriplics* Gagné, and the leaf-galling habit in *A. rudbeckiaeconspicua* suggest that there may be other patterns of life history and host utilization in some exotic species in addition to the aforementioned five patterns found in the Japanese *Asphondylia*. Further examinations of life history strategies for exotic species are needed to trace evolutionary processes of life history pattern in *Asphondylia* and related genera.

24.5 Acknowledgements

We are very grateful to Dr. K. M. Harris (former Director of the International Institute of Entomology), for his critical reading of an early draft. This study was partly supported by the Research Fellowship of the Japanese Society for the Promotion of Sciences for Young Scientists to NU.

24.6 References

Efron B (1982) The jackknife, the bootstrap and other resampling plans. Society for Industrial and Applied Mathematics, Philadelphia, PA

Felsenstein J (1985) Confidence-limits on phylogenies - an approach using the bootstrap. Evolution 39:783-791

Gagné RJ (2004) A catalog of the Cecidomyiidae (Diptera) of the world. Memoirs of the Entomological Society of Washington 25:1-408

Gagné RJ, Waring GL (1990) The *Asphondylia* (Cecidomyiidae: Diptera) of creosote bush (*Larrea tridentata*) in North America. Proceeding of the Entomological Society of Washington 92:649-671

Gagné RJ, Wuensche AL (1986) Identity of the *Asphondylia* (Diptera: Cecidomyiidae) on Guar, *Cyamopsis tetragonoloba* (Fabaceae), in the southwestern United States. Annals of the Entomological Society of America 79: 246-250

Imai K, Ohsaki N (2004) Oviposition site of and gall formation by the fruit gall midge *Asphondylia aucubae* (Diptera: Cecidomyiidae) in relation to internal fruit structure. Entomological Science 7:133-137

Ohsako S, Yukawa J, Horikiri M (1981) New data on the life history of the ligustrum fruit midge, *Asphondylia sphaera* Monzen (Diptera, Cecidomyiidae) (in Japanese, with English summery). Proceedings of the Association for Plant Protecton of Kyushu 27:116-118

Orphanides GM (1975) Biology of the carob midge complex, *Asphondylia* spp. (Diptera, Cecidomyiidae), in Cyprus. Bulletin of Entomological Research 65:381-390

Parnell JR (1964) Investigations on the biology and larval morphology of the insects associated with the galls of *Asphondylia sarothamni* H. Loew (Diptera: Cecidomyiidae) on broom (*Sarothamnus scoparius* (L.) Wimmer.). Transactions of the Royal Entomological Society of London 116:255-273

Plakidas JD (1988) The newly discovered spring crown gall of *Asphondylia rudbeckiaeconspicua* (Diptera: Cecidomyiidae) on *Rudbeckia laciniata* (Asteraceae) in Pennsylvania. Proceedings of the Entomological Society of Washington 90:393

Swofford DL (2002) PAUP*. Phylogenetic analysis using parsimony (*and other methods). Version 4. Sinauer, Sunderland, MA

Tokuda M, Yukawa J (2005) Two new and three known Japanese species of genus *Pseudasphondylia* Monzen (Diptera: Cecidomyiidae: Asphondyliini) and their life history strategies. Annals of the Entomological Society of America 98:259-272

Uechi N, Yukawa J (2004) Description of *Asphondylia itoi* sp. n. (Diptera: Cecidomyiidae) inducing fruit galls on *Distylium racemosum* (Hamamelidaceae) in Japan. Esakia 44:27-43

Uechi N, Tokuda M, Yukawa J (2002) Distribution of *Asphondylia* gall midges (Diptera: Cecidomyiidae) in Japan. Esakia 42:1-10

Uechi N, Yukawa J, Yamaguchi D (2004) Host alternation by gall midges of the genus *Asphondylia* (Diptera: Cecidomyiidae). In: Evenhuis NL, Kaneshiro

KY (eds) Contributions to the systematics and evolution of Diptera. D. Elmo Hardy Memorial Volume of Bishop Museum Bulletin in Entomology 12. Bishop Museum Press, Honolulu, pp 53-66.

Uechi N, Yukawa J, Usuba S (2005) Discovery of an additional winter host of the soybean pod gall midge, *Asphondylia yushimai* (Diptera: Cecidomyiidae) in Japan. Applied Entomology and Zoology 40:597-607

Yukawa J (2000) Synchronization of gallers with host plant phenology. Population Ecology 42:105-113

Yukawa J, Akimoto K (2006) Influence of synchronization between adult emergence and host plant phenology on the population density of *Pseudasphondylia neolitseae* (Diptera: Cecidomyiidae) inducing leaf galls on *Neolitseae sericea* (Lauraceae). Population Ecology 48:13-21

Yukawa J, Masuda H (1996) Insect and mite galls of Japan in colors (in Japanese, with English explanation for color plates). Zenkoku Nôson Kyôiku Kyôkai, Tokyo

Yukawa J, Miyamoto K (1979) Redescription of *Asphondylia sphaera* Monzen (Diptera, Cecidomyiidae), with notes on its bionomics. Memoirs of the Faculty of Agriculture, Kagoshima University 15:99-106

Yukawa J, Rohfritsch O (2005) Biology and ecology of gall-inducing Cecidomyiidae (Diptera). In: Raman A, Schaefer CW, Withers TM (eds) Biology, ecology, and evolution of gall-inducing arthropods. Science Publishers, Enfield, pp 273-304

Yukawa J, Uechi N, Horikiri M, Tuda M (2003) Description of the soybean pod gall midge, *Asphondylia yushimai* sp. n. (Diptera: Cecidomyiidae), a major pest of soybean and findings of host alternation. Bulletin of Entomological Research 93:73-86

25 Taxonomic Status of the Genus *Trichagalma* (Hymenoptera: Cynipidae), with Description of the Bisexual Generation

Yoshihisa Abe

Laboratory of Applied Entomology, Graduate School of Agriculture, Kyoto Prefectural University, Kyoto 606-8522, Japan

Summary. Taxonomic considerations of a monotypic genus *Trichagalma* have been based on the unisexual generation only, and its taxonomic status—i.e. whether or not *Trichagalma* is synonymized with *Neuroterus* —has been uncertain. Female and male adults of the bisexual generation of *T. serratae* are described. The adult and gall of the unisexual generation and the gall of the bisexual generation are also redescribed. The present results show that the bisexual generation of *T. serratae* has all the morphological features diagnostic for *Neuroterus*, as does the unisexual generation except for one feature. Chromosomal examination revealed that the haploid chromosome number of this species is 10, as in most oak gall wasps including three *Neuroterus* species. Mitochondrial sequence data obtained in a previous study suggest that *T. serratae* and some species of *Neuroterus* are in the same clade. No possible autapomorphies are found among the morphological, karyological or molecular characteristics of *Trichagalma*. Therefore, *Trichagalma* is synonymous with *Neuroterus*.

Key words. Synonymy, Morphology, Karyotype, Molecular data, *Neuroterus*

25.1 Introduction

Trichagalma serratae was described by Ashmead (1904) under the name *Dryophanta serratae*, and the monotypic genus *Trichagalma* was established by Mayr (1907) for *Trichagalma drouardi* Mayr. Later, Monzen (1929) synonymized *T. drouardi* with *D. serratae*, and treated this species as *T. serratae*. Monzen's taxonomic treatment was subsequently followed by other investigators of gall wasps (e.g., Sakagami 1952; Yukawa and Masuda 1996). Melika and Abrahamson (2002) mentioned that *Tricha-*

galma resembles *Neuroterus* Hartig, 1840, although the former is distinguishable from the latter by its robust size and strongly arched and densely pubescent mesosoma. On the basis of phylogenetic analyses of adult morphology, Lilgeblad (2002) suggested that *Trichagalma* might be synonymous with *Neuroterus*.

Before Masuda (1972) demonstrated that *T. serratae* shows alternation of generations, only the unisexual generation had been known in this species. Masuda (1972) described the bisexual gall of *T. serratae*, and Usuba (1995) confirmed its heterogonic life cycle by a rearing experiment. Nonetheless, the adults of the bisexual generation have not been described. Until now, taxonomic considerations of the genus *Trichagalma* have been based on the unisexual generation alone. Thus, description of the bisexual generation of *T. serratae* is required to determine the taxonomic status of the genus. The present paper describes the bisexual generation of *T. serratae*, and also the karyotype of this species. Based on the adult morphology of both generations, karyotype, and molecular data (Rokas et al. 2003), the taxonomic status of *Trichagalma* is discussed.

25.2 Materials and Methods

25.2.1 External Morphology of Adults

Dry adult specimens of *T. serratae* examined for this paper were reared from galls on *Quercus acutissima* Carruth in Japan. The external structure of the dry-mounted specimens was studied with a Nikon SMZ stereo microscope. Four specimens of both generations were sputter-coated with gold and examined with a JSM-5510LV scanning electron microscope.

The materials examined were as follows. Bisexual generation: 11 females and 21 males, Makioka-cho, Yamanashi Prefecture, IV. 1956 (em. V. 1956) (H. Masuda); 7 females and 1 male, same locality and collector, IV. 1957 (em. V. 1957); 116 females and 103 males, Chojabaru, Oita Prefecture, 9. V. 1984 (em. V. 1984) (Y. Abe).

Unisexual generation: 3 females, Makioka-cho, Yamanashi Prefecture, em. XII. 1956 (H. Masuda); 7 females, same locality and collector, XI. 1988; 5 females, Mt. Oyama, Kanagawa Prefecture, 28. III. 1964 (em. after X. 1964) (Y. Murakami); 14 females, Chojabaru, Oita Prefecture, 29. XI. 1981 (em. XII. 1981) (Y. Abe).

25.2.2 Karyotype

I collected 5 specimens of the unisexual generation of *T. serratae* from *Q. acutissima* on 16 September 1996 in Seika-cho, Kyoto Prefecture, Japan. Mature larvae of these specimens were reared at 25°C under a 15L–9D photoregime until they became pupae. Using an air-drying method (Imai et al. 1988), the ovaries of pupae were prepared for chromosomal examination under a Nikon SMZ-U stereo microscope. The chromosome preparations were stained with Giemsa solution and then rinsed with water. The chromosome number of each specimen was determined by counting at least four good metaphase figures, except for one specimen for which two metaphase figures were counted. The chromosomes observed were classified into the categories defined by Levan et al. (1964).

25.3 Taxonomy

Genus ***Neuroterus*** Hartig, 1840.

Neuroterus Hartig, 1840: 185, 192. Type species by subsequent designation (Ashmead 1903: 151): *Neuroterus politus* Hartig, 1840.

Trichagalma Mayr, 1907: 3. Type species by monotypy: *Trichagalma drouardi* Mayr, 1907. **New Synonymy**.

Diagnosis of *Neuroterus* mentioned by Melika et al. (1999) is cited in **Taxonomic status of *Trichagalma***. Historical review of *Neuroterus* is detailed in Melika and Abrahamson (2002).

Neuroterus serratae (Ashmead, 1904) **n. comb.**
(Figs. 1–4)
Dryophanta serratae Ashmead, 1904: 80.
Trichagalma drouardi Mayr, 1907: 5.
Trichagalma serratae: Monzen, 1929: 347.

25.3.1 Description

Female of bisexual generation: Body almost smooth and bare. Body black; antennae, palpi, tegulae and legs brownish yellow.

Head wider than mesosoma without tegulae in dorsal view. Vertex imbricate. Facial strigae radiating from lateral clypeus weak; the outer adjacent area at ventral margin of head weakly striate. Antenna 14-segmented; relative lengths of flagellar segments 1–12: 10, 8, 7.5, 7, 6.3, 6, 5.7, 5.7, 5.7, 5, 5, 5.7.

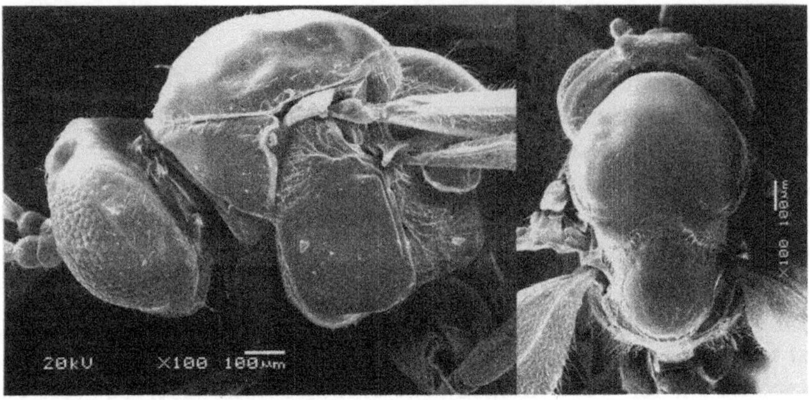

Fig. 1. Head and mesosoma of male adult of bisexual generation of *Neuroterus serratae. Left*, lateral aspect; *right*, dorsal aspect.

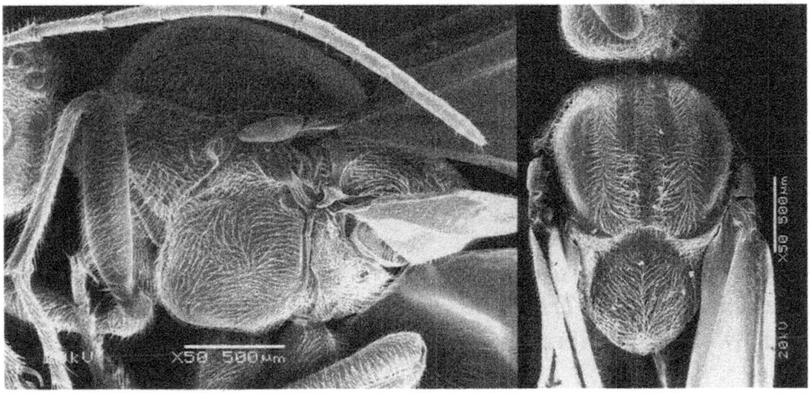

Fig. 2. Head and mesosoma of unisexual female of *Neuroterus serratae. Left*, lateral aspect; *right*, dorsal aspect.

Mesosoma smooth with a few setae; notauli and anteroadmedian and parapsidal signa absent. Transscutal fissure absent medially. Scutellum reticulate rugose, but smooth medially; scutellar foveae absent. Propodeum finely sculptured without median and lateral carinae, sparsely haired laterally.

Wing surface closely ciliated. Smoky spots absent on forewing; marginal cell of forewing elongate, open on anterior margin.

Metasomal tergite 2 with a few setae basolaterally. Hypopygial spine slender in ventral view; length of projecting part approximately two times height of hypopygial spine; apical hairs denser than basal hairs, beyond the apex but not forming hair tuft.

Male of bisexual generation (Fig. 1): Differs from the female as follows. Antenna 15-segmented; relative lengths of flagellar segments 1–13: 10, 8, 7.8, 7.5, 6, 6, 6, 6, 6, 5.6, 5.6, 5.3, 5.3; flagellar segment 1 incised on outer margin. Petiole long.

Unisexual generation (Fig. 2): Head and mesosoma pubescent, metasoma almost bare. Head, mesosoma and metasoma reddish brown; antennae brown. Occiput, surrounding area of ocelli, ventral margin of head, anteroadmedian and parapsidal signa, metapleura, propodeum, outer margins of scutellum and mesopleura black.

Head as wide as mesosoma without tegulae in dorsal view. Facial strigae radiating from lateral clypeus weak; the outer adjacent area at ventral margin of head weakly striate. Antenna 15-segmented; relative lengths of flagellar segments 1–13: 10, 8, 7, 6.8, 5.3, 4.6, 3.8, 3.8, 3.8, 3.8, 3.8, 3.5, 3.5.

Notauli absent. Anteroadmedian and parapsidal signa wide, weakly raised, less pubescent. Transscutal fissure absent medially. Scutellum reticulate rugose, scutellar foveae absent. Propodeum smooth and bare medially, median and lateral propodeal carinae absent.

Wing surface closely ciliated. Smoky spots present on forewing; marginal cell of forewing elongate, open on anterior margin.

Metasomal tergites 2, 3, 4, 5 and 7 pubescent laterally. Hypopygial spine slender in ventral view, evenly pubescent; apical hairs longer than basal ones, beyond the apex, but not forming hair tuft; length of projecting part of hypopygial spine approximately equal to height.

Bisexual gall (Fig. 3): Single-chambered, irregularly globulous with a thin wall, smooth, tinged with yellow or red, on the surface of the catkins; solitary or several galls clustered per catkin; maximum diameter 2–3mm.

Unisexual gall (Fig. 3): Usually coalesced but often separate, roughly spherical, closely covered with spines, light green at the beginning of appearance and becoming fulvous, maximum diameter 10–20 mm; the larval chamber single with an air space between this and the outer woody gall wall, ovoid, attached by its base to the gall wall, 5 mm in maximum diameter.

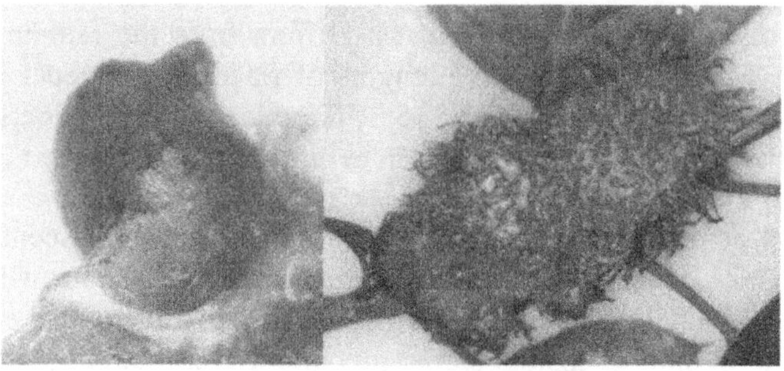

Fig. 3. Galls of *Neuroterus serratae*. *Left*, bisexual gall; *right*, cluster of unisexual galls.

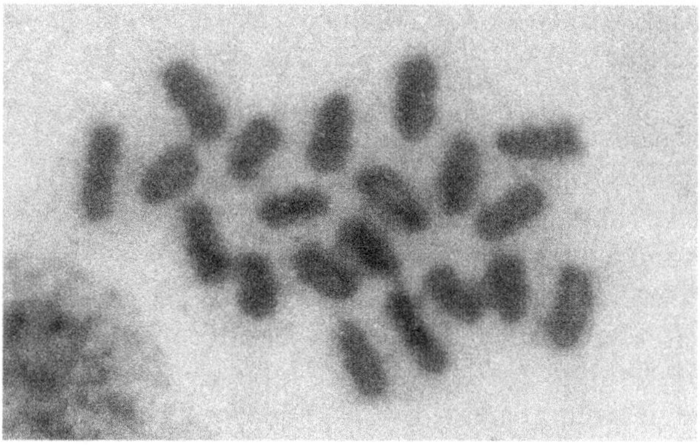

Fig. 4. Chromosomes from pupal ovary of unisexual generation of *Neuroterus serratae*.

Karyotype: Metaphase chromosomes of *N. serratae* are shown in Fig. 4. The karyotype was composed of ten pairs of acrocentrics in all the specimens examined.

Geographical distribution: Japan (Honshu, Shikoku and Kyushu), Korea.

Host plants: *Quercus acutissima* Carruth. and *Q. variabilis* Blume.

Remarks: According to the original description (Ashmead 1904), the type locality is Sapporo, Hokkaido, northern Japan, and the host plant is *Q. serrata*. However, the gall wasp and its gall have not been recorded from *Q. serrata* since the original description. Mukaigawa (1913) recorded *Q. acutissima* as a host plant of the gall wasp for the first time. Later, investigators of gall wasps confirmed *Q. acutissima* as a host plant (Masuda 1972; Sakagami 1952; Yukawa and Masuda 1996), and *Q. variabilis* was also recorded as a host plant (Abe 1992). *Q. acutissima* and *Q. variabilis* belong to the section Cerris, and *Q. serrata* belongs to the section Prinus. Moreover, *Q. serrata* is distributed in Hokkaido, whereas *Q. acutissima* and *Q. variabilis* are not (Kitamura and Murata 1984). Sakagami (1952) stated that the distribution of this gall wasp in Hokkaido is questionable.

Unisexual galls usually appear in August and pupation takes place in the following September, but some unisexual larvae prolong their larval duration a year (Yukawa and Masuda 1996). The collection and emergence date of unisexual females collected by Dr. Y. Murakami support the occurrence of prolonged diapause in *N. serratae*. Further study is needed to clarify the adaptive significance of the larval diapause.

25.3.2 Taxonomic Status of *Trichagalma*

The genus *Trichagalma* is monotypic. Melika and Abrahamson (2002) stated that the robust size and strongly arched and densely pubescent mesosoma differentiate *Trichagalma* from *Neuroterus*. However, these features are shared with the unisexual females of some members of other genera (e.g., *Andricus* and *Cynips*) among the tribe Cynipini. Moreover, as described by Melika et al. (1999), dense pubescence is found even in a member of *Neuroterus*, the unisexual female of *N. macropterus* (Hartig). The present results show that adults of the bisexual generation do not have the features diagnostic for *Trichagalma* mentioned by Melika and Abrahamson (2002).

Melika et al. (1999) listed four diagnostic features for *Neuroterus*: (1) absence of a transscutal fissure, (2) a body that is usually smooth and gracile, with delicate coriaceous or alutaceous sculpturing on the mesosoma, (3) notauli usually being absent, and (4) a long and narrow marginal cell of the forewing. These four features appear to be apomorphic among the Cynipini (Lilgeblad 2002). The unisexual female of *N. serratae* has all the above-mentioned diagnostic features except for the second one. Moreover, the present study revealed that females and males of the bisexual generation have all four features diagnostic for *Neuroterus*. No possible autapomorphies have been found for *Trichagalma* not only in the external morphology of the unisexual generation but also in that of the bisexual generation.

As reviewed by Gokhman and Quicke (1995), the haploid chromosome number of the three *Neuroterus* species examined is 10, as in most members of the Cynipini. The present study revealed that *N. serratae* also has $n = 10$, but is different from other *Neuroterus* species in having acrocentrics only.

Rokas et al. (2003) used statistical phylogenetic inference methods on sequence data for a fragment of the mitochondrial cytochrome *b* gene to reconstruct the relationships among 62 oak gall wasp species including *N. serratae* and 3 other *Neuroterus* species. Their results suggested that *N. serratae* and some *Neuroterus* species are in the same clade. Judging from the morphological, karyological and molecular data, *Trichagalma* is a synonym and is hereby dissolved.

25.4 Acknowledgements

I would like to thank Prof. Emeritus Y. Murakami, Prof. Emeritus J. Yukawa and Prof. O. Tadauchi for the gift or loan of many valuable specimens. Thanks are also due to the late Mr. H. Masuda for suggestion. I thank Dr. H. Hoshiba for advice on the air-drying method. Mr. M. Okunishi assisted kindly in the preparation of SEM pictures.

25.5 References

Abe Y (1992) A new host record of *Trichagalma serratae* (Ashmead) (Hymenoptera: Cynipidae). Akitu, New series 130:8

Ashmead WH (1903) Classification of the gall-wasps and the parasitic cynipoids, or the superfamily Cynipoidea. III. Psyche 10:140-155

Ashmead WH (1904) Description of new Hymenoptera from Japan. Journal of New York Entomological Society 12:65-84

Gokhman VE, Quicke DLJ (1995) The last twenty years of parasitic Hymenoptera karyology: an update and phylogenetic implications. Journal of Hymenoptera Research 4:41-63

Hartig T (1840) Ueber die Familie der Gallwespen. III. Zeitschrift fur Entomologie 2:176-209

Imai HT, Taylor RW, Crosland MWJ, Crozier RH (1988) Modes of spontaneous chromosomal mutation and karyotype evolution in ants with reference to the minimum interaction hypothesis. Japanese Journal of Genetics 63:159-185

Kitamura S, Murata G (1984) Colored illustrations of wood plants of Japan II (revised edition) (in Japanese). Hoiku-sha, Osaka

Levan A, Fredga K, Sandberg AA (1964) Nomenclature for centric position on chromosomes. Hereditas 52:201-220

Liljeblad J (2002) Phylogeny and evolution of gall wasps (Hymenoptera: Cynipidae). Ph D thesis Stockholm University, Stockholm

Masuda H (1972) Life of Japanese gall wasps (in Japanese). Insectarium 9:222-225

Mayr G (1907) Zwei Cynipiden. Marcellia 6:3-7

Melika G, Abrahamson WG (2002) Review of the world genera of oak cynipid wasps (Hymenoptera: Cynipidae: Cynipini). In: Melika G, Thuroczy C (eds) Parasitic wasps. Agroinform, Budapest, pp 150-190

Melika G, Stone GN, Csoka G (1999) Description of an oak gall-wasp, *Neuroterus ambrusi* sp. n. (Hymenoptera, Cynipidae) from Hungary. Acta Zoologica Academiae Scientiarum Hungaricae 45:335-343

Monzen K (1929) Studies on galls (in Japanese). Saito-hoonkai-jigyo-nenpo 5:295-368 + 20 pls

Mukaigawa Y (1913) Notes on the life histories of *Dryophanta mukaigawae* and *D. serratae* (in Japanese). Insect World 17:261-264, pl

Rokas A, Melika G, Abe Y, Nieves-Aldrey J-L, Cook J M, Stone GN (2003)Lifecycle closure, lineage sorting, and hybridization revealed in a phylogenetic analysis of European oak gallwasps (Hymenoptera: Cynipidae: Cynipini) using mitochondrial sequence data. Molecular Phylogenetics and Evolution 26:36-45

Sakagami SF (1952) Zur Cynipoidenfauna Japans und seiner Nachbarlander (Hymenoptera). Mushi 24:67-79

Usuba S (1995) An introduction to galls (in Japanese). Yasaka-shobo, Tokyo

Yukawa J, Masuda H (1996) Insect and mite galls of Japan in colors (in Japanese with English explanations for color plates). Zenkoku Nôson Kyôiku Kyôkai, Tokyo

26 Phylogenetic Position of the Genus *Wagnerinus* Korotyaev (Coleoptera: Curculionidae) Associated with Galls Induced by *Asphondylia baca* Monzen (Diptera: Cecidomyiidae)

Toshihide Kato, Hiraku Yoshitake, and Motomi Ito

Ito Laboratory, Department of General Systems Studies, Graduate School of Arts and Sciences, The University of Tokyo, 3-8-1 Komaba, Meguro-ku, Tokyo 153-8902, Japan

Summary. Based on molecular phylogenetic analysis using three regions of mitochondrial DNA, we investigated the phylogenetic position of the genus *Wagnerinus* (Curculionidae: Ceutorhynchinae), which includes a species that feeds obligatorily on galls induced by *Asphondylia baca* (Cecidomyiidae) on the axillary buds of *Weigela* species (Caprifoliaceae). Although *Wagnerinus* has been placed in either Ceutorhynchini or Scleropterini, the phylogenetic relationships inferred from the molecular data suggest that neither Ceutorhynchini nor Scleropterini are monophyletic, and *Wagnerinus* is separated from the other genera placed in the two tribes.

Key words. Cecidophagy, Phytophagous insect, Gall midge, Ceutorhynchinae, Caprifoliaceae

26.1 Introduction

Obligatory cecidophages are relatively rare among herbivorous insects, while many facultative cecidophages have been reported (Yukawa and Masuda 1996). Recently, Sugiura et al. (2004) reported that a Japanese weevil, *Wagnerinus costatus* (Hustache, 1916) (Coleoptera: Curculionidae), is associated with galls induced on the axillary buds of *Weigela hortensis* (Sieb. et Zucc.) K. Koch (Caprifoliaceae) by the gall midge *Asphondylia baca* Monzen, 1937 (Diptera: Cecidomyiidae). *W. costatus* females lay their eggs in the midge galls, and the hatched larvae feed exclusively on gall tissues. The inhabitation of the galls by *W. costatus* larvae has minimal effect on the survival of *A. baca* (Sugiura et al. 2004), unlike most obligatory cecidophages, which have fatal effects on the original gall inhabitants (e.g., Itô and Hattori 1983; Kopelke 1994; Sugiura et al. 2002; Yamazaki and Sugiura 2001).

At present, the genus *Wagnerinus* Korotyaev, 1980 is classified in the subfamily Ceutorhynchinae Gistel, 1848, which comprises four species from Northeast Asia (Colonnelli 2004), but the ecological traits of the other congeners are still uncertain and several undescribed species occur in Japan (H. Yoshitake, unpublished data). This has prevented progress in evolutionary studies of the cecidophagous habit of *Wagnerinus* weevils. Moreover, the systematic position of *Wagnerinus* within Ceutorhynchinae remains uncertain. Korotyaev (1980) established this genus in the subtribe Scleropterina (currently known as the tribe Scleropterini Schultze, 1902) of Ceutorhynchinae. Later, Colonnelli (1984) transferred *Wagnerinus* from Scleropterini to the tribe Ceutorhynchini Gistel, 1848 in the same subfamily. Although subsequent authors have followed this treatment (Colonnelli 2004; Morimoto 1989; Yoshitake et al. 2004), Korotyaev and Hong (2004) retained *Wagnerinus* in Scleropterini.

We sought to determine the phylogenetic position of *Wagnerinus* within Ceutorhynchinae as the first step in a comprehensive study of this genus. For this purpose, we investigated its phylogenetic position using a molecular phylogenetic analysis based on sequences of the mitochondrial 16S rRNA, tRNA-Val, and 12S rRNA genes.

26.2 Materials and Methods

26.2.1 Weevil Samples

The weevil samples used in this study are listed in Table 1. To investigate the phylogenetic position of *Wagnerinus*, we mainly selected representatives from the Ceutorhynchini and Scleropterini. In addition, two outgroup species, *Lobotrachelus minor* Hustache, 1921 and *Orobitis apicalis* Kôno, 1935 were selected from Conoderinae Schoenherr, 1833 and Orobitidinae Thomson, 1859, respectively, because these subfamilies are thought to be related to Ceutorhynchinae (Korotyaev et al. 2000). A single adult from each species was preserved from life in 99.5% ethanol or 99.5% acetone. All species were identified by H. Yoshitake, and voucher specimens have been deposited in the Kyushu University Museum, Fukuoka.

Table 1. Species list and DDBJ accession numbers

Subfamily, Tribe, Species	Accession No.*
Ceutorhynchinae Gistel, 1848	
Ceutorhynchini Gistel, 1848	
Cardipennis shaowuensis (Voss, 1958)	AB232957
Ceutorhynchoides styracis Yoshitake et Colonnelli, 2005	AB232958
Ceutorhynchus albosuturalis (Roelofs, 1875)	AB232956
Ceutorhynchus filiae Dalla Torre, 1922	AB232959
Ceutorhynchus ibukianus Hustache, 1916	AB232960
Coeliodes nakanoensis Hustache, 1916	AB232961
Coeliodinus etorofuensis (Kôno, 1935)	AB232962
Dieckmannius lewisi (Hustache, 1916)	AB232963
Hadroplontus ancora (Roelofs, 1875)	AB232964
Hainokisaruzo japonicus Yoshitake et Colonnelli, 2005	AB232965
Mogulones geographicus (Goeze, 1777)	AB232966
Nedyus quadrimaculatus (Linnaeus, 1758)	AB232967
Sirocalodes umbrinus (Hustache, 1916)	AB232968
Thamiocolus kraatzi (C. Brisout, 1869)	AB232969
Trichocoeliodes excavatus (Hustache, 1916)	AB232970
Wagnerinus costatus (Hustache, 1916)	AB232971
Wagnerinus harmandi (Hustache, 1916)	AB232972
Wagnerinus sp. 1-Kotamagawa	AB232973
Wagnerinus sp. 2-Yukomanbetsu	AB232974
Wagnerinus sp. 3-Kashikougen	AB232975
Zacladus geranii (Paykull, 1800)	AB232976
Egriini Pajni et Kohli, 1982	
Cyphosenus grouvellei Hustache, 1916	AB232977
Mecysmoderini Wagner, 1938	
Mecysmoderes nigrinus Hong et Woo, 1999	AB232978
Scleropterini Schultze, 1902	
Scleropterus serratus (Germar, 1824)	AB232979
Scleropteroides hypocrita (Hustache, 1916)	AB232980
Tapeinotus sellatus (Fabricius, 1794)	AB232981
Conoderinae Schoenherr, 1833	
Lobotrachelus minor (Hustache, 1921)	AB232982
Orobitidinae Thomson, 1859	
Orobitis apicalis (Kôno, 1935)	AB232983

*The collecting sites of all the samples have been deposited in the DDBJ with their sequence data under the accession numbers noted above.

26.2.2 DNA Extraction, Amplification, and Sequencing

Total genomic DNA was extracted from the entire body using a DNeasy Tissue Kit (Qiagen, Hilden, Germany), following the manufacturer's instructions. A DNA fragment consisting of the 16S rRNA, tRNA-Val, and 12S rRNA mitochondrial genes was amplified using the polymerase chain reaction (PCR) with the primers MtrA1adv and MtrK1 (Table 2). The template profile was as follows: 94.0°C for 5 min; 6 cycles at 94.0°C for 45 sec, 51.8°C for 45 sec, and 72.0°C for 90 sec; 29 cycles at 94.0°C for 45 sec, 49.8°C for 45 sec, and 72.0°C for 90 sec; and 72.0°C for 8 min. PCR was performed in a reaction volume of 20 μl using 10× Ex *Taq* Buffer (Takara Bio, Tokyo, Japan), 0.2 mM each dNTP, 0.5 μM each primer, 0.5 U/μl Ex *Taq* DNA polymerase, and 0.4 μl template DNA. The PCR products were purified using ExoSAP-IT (Amersham Pharmacia Biotech, Uppsala, Sweden) and served as template DNA for cycle sequencing reactions with CEQ Quick Start Mix (Beckman Coulter, Fullerton, CA, USA), following the manufacturer's instructions. The internal primers used are listed in Table 2. The cycle sequencing products were purified by ethanol precipitation and electrophoresed using the CEQ8000 Genetic Analysis System (Beckman Coulter). DNA sequences obtained in both directions were assembled and edited using ATGC version 4.0 (Genetyx, Tokyo, Japan).

All the DNA sequences determined herein have been deposited in the DDBJ Nucleotide Sequence Database under the accession numbers shown in Table 1.

26.2.3 Phylogenetic Analysis

The DNA sequences were aligned using ClustalX version 1.83 (Thompson et al. 1997) and the final alignment was adjusted manually on MacClade version 4.06 (Maddison and Maddison 2003). Ambiguously aligned re-

Table 2. List of primers used in this study

Primer	Sequence
MtrA1adv[a]	5'-AAA CTA GGA TTA GAT ACC CT-3'
MtrK1[b]	5'-CAT AAT AAG ATT CTA AAT C-3'
LR-N-13398[c]	5'-CAC CTG TTT ATC AAA AAC AT-3'
MtriD1r[d]	5'-TGG AAT AAG TCG TAA CAA AG-3'
MtriE1f[d]	5'-AAA ATA CCG CGG CTT TAA-3'
MtriE1r[d]	5'-TTA AAG CCG CGG TAT TTT-3'
MtriI1f[d]	5'-CCC TGA TAC ACA AGG TAC-3'
MtriI1r[d]	5'-GTA CCT TGT GTA TCA GGG-3'
MtriJ1f[d]	5'-TCT ATA GGG TCT TCT CGT C-3'

[a, b] PCR primers.

[c, d] Internal primers.

[a] Modified from Fukatsu et al. (2001).

[c] Xiong and Kocher (1991).

gions were excluded from the data set. The incongruence length difference test (ILD test; Farris et al. 1994) was conducted for the data set on PAUP* version 4.0b10 (Swofford 2002) to examine congruency among the three regions (16S rRNA, tRNA-Val, and 12S rRNA). A phylogenetic analysis was performed using PAUP* version 4.0b10 under the maximum parsimony criterion with the heuristic search algorithms (Swofford et al. 1996). The heuristic search parameters used for the parsimony analysis were 100 random stepwise addition replicates with tree bisection reconnection (TBR) branch swapping and saving multiple trees (MulTrees). All gaps in the data set were treated as missing data. Support for the tree topology was evaluated using a bootstrap analysis (Felsenstein 1985) with 1000 replications using heuristic algorithms with TBR branch swapping and 10 random stepwise addition sequences.

26.3 Results

A total of 1707 nucleotide characters for 28 species were determined and aligned. After excluding the ambiguous sites (75 characters: 57 for 16S rRNA, 10 for tRNA-Val, and 8 for 12S rRNA), the data set contained 1632 characters, of which 788 were variable and 571 were parsimoniously informative. Since the ILD test indicated no significant incongruence among the data partitions ($P = 0.35$), we combined these sequences into a single data set.

Parsimony analysis of the combined data set resulted in nine most parsimonious trees. The 50% majority rule consensus of these trees is shown in Fig. 1. In all trees, the ingroup taxa formed five major clades (A–E). Clade A (with 67% bootstrap support) comprising *Cyphosenus* Schultze, 1899 (Egriini) and *Mogulones* Reitter, 1916 (Ceutorhynchini) is the most basal lineage within the trees. *Wagnerinus* weevils were included in Clade B (with 100% bootstrap support), which is a sister clade to Clades C to E. Clade C (with 98% bootstrap support) is composed of *Zacladus* Reitter, 1913 (Ceutorhynchini), *Tapeinotus* Schoenherr, 1826 (Scleropterini), and *Mecysmoderes* Schoenherr, 1837 (Mecysmoderini) and is a sister clade to Clades D and E in six of the nine most parsimonious trees. Clade D (with 65% bootstrap support) consists of representatives of nine Ceutorhynchini genera. Within this clade, *Hainokisaruzo* Yoshitake et Colonnelli, 2005 and *Ceutorhynchoides* Colonnelli, 1979 formed the outermost branch (with 99% bootstrap support), but the relationships among the remaining genera were not clear. Clade E (with 63% bootstrap support) comprises three genera of Ceutorhynchini and two of Scleropterini.

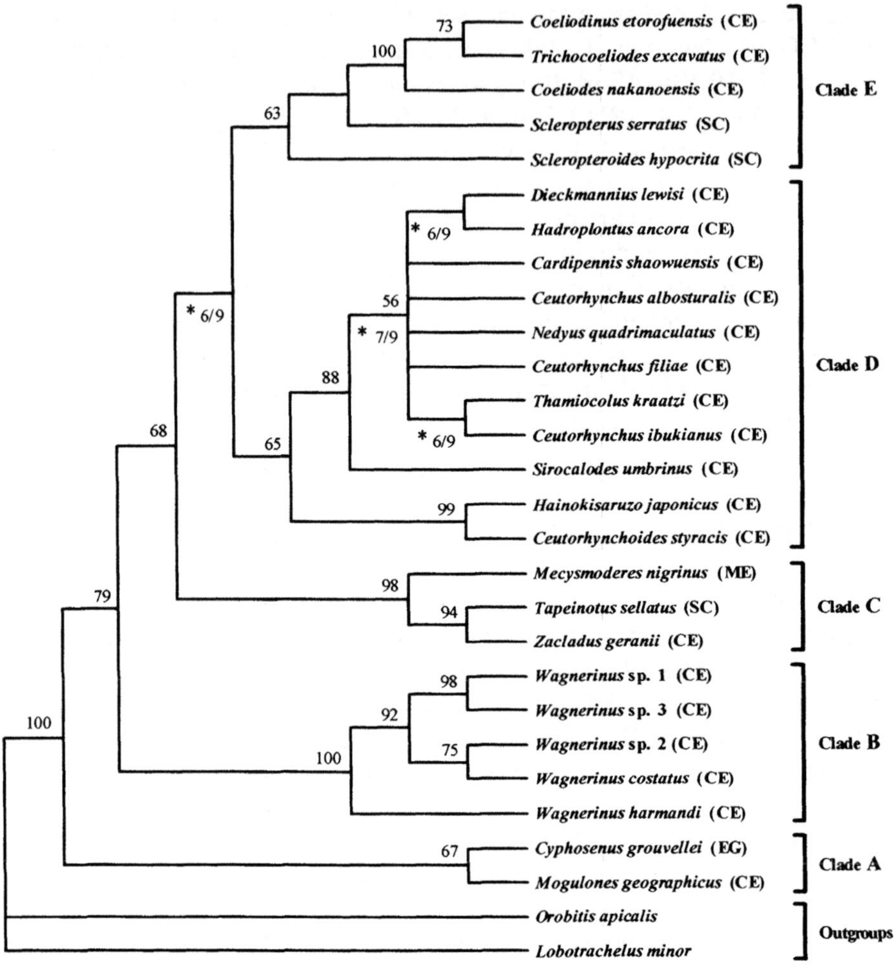

Fig. 1. The 50% majority rule consensus tree of the nine most parsimonious trees for Ceutorhynchinae weevils (*tree length* = 2985 steps, *consistency index* = 0.405, *retention index* = 0.442) based on the combined data set. Bootstrap values exceeding 50% are noted above the corresponding nodes. *Asterisks* indicate clades collapsed in the strict consensus tree, together with the number of most parsimonious trees supporting each clade (e.g., a clade with 6/9 means that it was supported by six of the nine most parsimonious trees). The taxonomic positions of the ingroup taxa are abbreviated as follows: *CE*, Ceutorhynchini; *EG*, Egriini; *SC*, Scleropterini; and *ME*, Mecysmoderini. The major clades in the tree are labeled Clades A–E according to the text (see Results).

26.4 Discussion

To date, systematic studies of Ceutorhynchinae have been at the stage of alpha taxonomy, with no hypotheses presented regarding phylogenetic relationships among higher taxa. This study is the first phylogenetic analysis of the relationships among the major genera of Ceutorhynchini and Scleropterini for elucidating the phylogenetic position of *Wagnerinus*.

Korotyaev (1980, 1981) and Korotyaev and Hong (2004) placed *Wagnerinus* in Scleropterini, while Morimoto (1989), Colonnelli (2004), and Yoshitake et al. (2004) placed it in Ceutorhynchini, as did Colonnelli (1984). However, our results do not support the monophyly of either Ceutorhynchini or Scleropterini and indicate that *Wagnerinus* is a unique genus representing a separate lineage from the other genera now included in the two tribes. The lack of detailed morphological studies of Ceutorhynchinae at the tribe level is thought to have caused this conflict between our result and previous taxonomic treatments. The morphological distinction between Ceutorhynchini and Scleropterini is still unclear due to the inconsistency in character states defining the two tribes (Colonnelli 1984, 2004). In addition, the diagnostic characteristics of *Wagnerinus* proposed by Korotyaev (1980, 1996), such as the slender rostrum, seven-segment antennal funicle, and sparse vestiture and minute granules on the elytral intervals, are insufficient to determine the systematic position of this genus because he compared these characteristics only with those of Scleropterini genera, which likely have no close relationship to *Wagnerinus*. Therefore, further studies at the tribe level based on morphological and molecular data are required to resolve this confusion.

Within Ceutorhynchinae, *Wagnerinus* is a unique taxon that is associated with Caprifoliaceae and includes a species exhibiting obligatory cecidophagy (Sugiura et al. 2004). Currently, this genus consists of four species from Northeast Asia (Colonnelli 2004) and except for *W. costatus*, little is known of their ecological traits. Our preliminary surveys suggest that more than 10 undescribed species of this genus occur mainly in Japan. A taxonomic revision combined with studies of fundamental ecological traits is also needed to understand the evolutionary processes involved in the cecidophagous habit of *Wagnerinus* weevils.

26.5 Acknowledgments

We thank J. Yukawa (Kyushu University) for reading an early draft of this paper, S. Aoki and Y. Kita (The University of Tokyo) for advice on the phylogenetic analysis, and T. Katsuda (Kyushu University) for his suggestions on the literature. Our thanks are also extended to M. Horikawa (Yokohama), I. Matoba (Kainan), Y. Notsu (Hiratsuka), P. Sprick (Hannover), and A. Yoshida (Inagi) for collecting the weevil samples.

26.6 References

Colonnelli E (1984) Notes sur quelques Ceutorhynchinae de l'Afrique tropicale (Coleoptera, Curculionidae). Annales Historico-naturales Musei Nationalis Hungarici 76:207-238

Colonnelli E (2004) Catalogue of Ceutorhynchinae of the world, with a key to genera (Insecta: Coleoptera: Curculionidae). Argania editio, Barcelona

Farris JS, Källersjö M, Kluge AG, Bult C (1994) Testing significance of incongruence. Cladistics 10:315-319

Felsenstein J (1985) Confidence limits on phylogenies: an approach using the bootstrap. Evolution 39:783-791

Fukatsu T, Shibao H, Nikoh N, Aoki S (2001) Genetically distinct populations in an Asian soldier-producing aphid, *Pseudoregma bambucicola* (Homoptera: Aphididae), identified by DNA fingerprinting and molecular phylogenetic analysis. Molecular Phylogenetics and Evolution 18:423-433

Itô Y, Hattori I (1983) Relationship between *Nola innocua* Butler (Lepidoptera: Nolidae), a kleptoparasite, and aphids which cause galls on *Distylium racemosum* trees. Applied Entomology and Zoology 18:361-370

Kopelke JP (1994) The parasite complex (parasitic inquilines and parasitoids) of *Pontania* galls (Insecta: Hymenoptera: Tenthredinidae). Senckenbergiana Biologica 73:83-133

Korotyaev BA (1980) Materials to the knowledge of Ceutorhynchinae (Coleoptera, Curculionidae) of Mongolia and the USSR (in Russian). Nasekomie Mongolii 7:107-282

Korotyaev BA (1981) New and little-known weevils of the subfamily Ceutorhynchinae (Coleoptera, Curculionidae) from the Palearctic, Indo-Malayan, and Australian regions (in Russian). Entomologicheskoe Obozrenie 60:126-159

Korotyaev BA (1996) A key to genera of the tribe Ceutorhynchini (in Russian). In: Ler PA (ed) Key to the insects of Russian Far East, vol. 3. Dal'nauka, Vladivostok, pp 455-468

Korotyaev BA, Hong K-J (2004) A revised list of the weevil subfamily Ceutorhynchinae (Coleoptera; Curculionidae) of the Korean fauna, with contribution to the knowledge of the fauna of neighbouring countries. Journal of Asia–Pacific Entomology 7:143-169

Korotyaev BA, Konstantinov AS, O'Brien CW (2000) A new genus of the Orobitidinae and discussion of its relationships (Coleoptera: Curculionidae). Proceedings of the Entomological Society of Washington 102:929-956

Maddison DR, Maddison WP (2003) MacClade 4.06. Sinauer Associates, Sunderland

Morimoto K (1989) Curculionoidea (in Japanese). In: Hirashima Y (ed) A check list of Japanese insects. Entomological Laboratory, Kyushu University, Fukuoka, pp 485-538

Sugiura S, Yamazaki K, Hishi T (2002) A cecidophagous weevil, *Curculio albovittatus* (Coleoptera: Curculionidae), in the gall of *Pontania* sp. (Hymenoptera: Tenthredinidae). Entomological Science 5:193-196

Sugiura S, Yamazaki K, Fukasawa Y (2004) Weevil parasitism of ambrosia galls. Annals of the Entomological Society of America 97:184-193

Swofford DL (2002) PAUP*. Phylogenetic Analysis Using Parsimony (*and Other Methods): version 4.0b10. Sinauer Associates, Sunderland

Swofford DL, Olsen GJ, Wadell PH, Hillis DM, (1996) Phylogenetic inference. In: Hillis DM, Mable BK, Moritz C (eds) Molecular Systematics, 2nd Edition. Sinauer Associates, Sunderland

Thompson JD, Gibbson TJ, Plewniak F, Jeanmougin J, Higgins DG (1997) The CLUSTAL-X windows interface: flexible strategies for multiple sequence alignment aided by quality analysis tools. Nucleic Acids Research 25:4876-4882

Xiong B, Kocher TD (1991) Comparison of mitochondrial DNA sequences of seven morphospecies of black flies (Diptera: Simuliidae). Genome 34:306-311

Yamazaki K, Sugiura S (2001) Bionomics of the gall-parasitic flea weevil *Rhynchaenus hustachei* (Coleoptera: Curculionidae). Entomological Science 4:239-242

Yoshitake H, Kojima H, Morimoto K (2004) Ceutorhynchinae. In: Kojima H, Morimoto K (2004) An online checklist and database of the Japanese weevils (Insecta: Coleoptera: Curculionoidea) (excepting Scolytidae and Platypodidae). Bulletin of the Kyushu University Museum 2:100-106

Yukawa J, Masuda H (1996) Insect and Mite Galls of Japan in Colors (in Japanese, with English explanations for the color plates). Zenkoku Nôson Kyôiku Kyôkai, Tokyo

Key Word Index

GPSR Compliance

The European Union's (EU) General Product Safety Regulation (GPSR) is a set of rules that requires consumer products to be safe and our obligations to ensure this.

If you have any concerns about our products, you can contact us on ProductSafety@springernature.com

In case Publisher is established outside the EU, the EU authorized representative is:

Springer Nature Customer Service Center GmbH
Europaplatz 3
69115 Heidelberg, Germany

Batch number: 10755284

Printed by Printforce, the Netherlands